Wide Bandgap Nanowires

Wide Bandgap Nanowires

Synthesis, Properties, and Applications

Tuan Anh Pham
Griffith University
Queensland, Australia

Toan Dinh
University of Southern Queensland
Queensland, Australia

Nam-Trung Nguyen
Griffith University
Queensland, Australia

Hoang-Phuong Phan
Griffith University
Queensland, Australia

WILEY

This edition first published 2022

Registered Office(s)
John Wiley & Sons, Inc., 111 River Street, Hoboken, NJ 07030, USA

Editorial Office
John Wiley & Sons, Inc., 111 River Street, Hoboken, NJ 07030, USA

For details of our global editorial offices, customer services, and more information about Wiley products visit us at www.wiley.com.

Wiley also publishes its books in a variety of electronic formats and by print-on-demand. Some content that appears in standard print versions of this book may not be available in other formats.

Library of Congress Cataloging-in-Publication Data

Names: Pham, Tuan Anh, author. | Dinh, Toan, author. | Nguyen, Nam-Trung, 1970- author. | Phan, Hoang-Phuong, author.
Title: Wide band gap nanowires : synthesis, properties, and applications / Tuan Anh Pham, Toan Dinh, Nam-Trung Nguyen, Hoang-Phuong Phan.
Description: Hoboken, NJ : Wiley, 2022. | Includes index.
Identifiers: LCCN 2022000571 (print) | LCCN 2022000572 (ebook) | ISBN 9781119774372 (cloth) | ISBN 9781119774457 (adobe pdf) | ISBN 9781119774389 (epub)
Subjects: LCSH: Nanowires. | Wide gap semiconductors.
Classification: LCC TK7874.85 .P43 2022 (print) | LCC TK7874.85 (ebook) | DDC 621.3815–dc23/eng/20220215
LC record available at https://lccn.loc.gov/2022000571
LC ebook record available at https://lccn.loc.gov/2022000572

Cover Design: Wiley
Cover Image: © Xuanyu Han/Getty Images

Set in 9.5/12.5pt STIXTwoText by Straive, Pondicherry, India

Contents

Preface

The last two decades have witnessed a remarkable progress in the field of semiconducting nanowires, which provide unique features such as high electron mobility, tunable mechanical properties, and high surface to volume areas. Starting with the bottom-up synthesis of silicon nanowires and nanorods in the early of 2000s, over one hundred types of nanoscale materials are today widely deployed in numerous nanoelectromechanical systems (NEMS) applications. Among these, wide bandgap nanowires such as III-nitride, silicon carbide, diamond, and oxide semiconductors have attracted a great deal of attention from the research community and the electronic industrial sector, due to their superior physical and chemical properties. The large breakdown voltage, electrical stability at elevated temperatures, and corrosive tolerance in this group of semiconductor nanowires enable a new class of NEMS sensing devices that function in extreme environments, where silicon-based counterparts fail to deliver a reliable operation.

This book systematically summarizes the state-of-the-art works on wide band gap semiconductor nanowires from material synthesis, to device development, to practical applications. The book starts with an introductory level and then develops further into deeper technical contents. This structure is highly suitable for a broad range of readership including undergraduate/postgraduate students as well as engineers and researchers working on the field of NEMS. The book aims to provide a fundamental guideline for NEMS designers, covering key parameters such as choice of materials, nanoarchitectures, fabrication and integration approaches, and functional mechanisms for particular electronics and sensing applications.

With the completion of this book, we would like to thank the generous supports from research funding agencies such as the Australian Research Council (ARC), the Innovative Manufacturing Cooperative Research Centre, Australia (IMCRC), as well as our industrial partners such as SPTS Technologies, Rio Tinto, and QuestSemi who have been greatly supporting our research on silicon carbide and gallium nitride (two representative materials of the wide band gap semiconductor group) over the last decade. We also acknowledge our long-standing collaborators including Prof. Dzung Viet Dao and Prof. Sima Dimitrijev at Griffith University, Prof. Takahiro Namazu at Kyoto University of Advanced Science, Prof. Toshihiro Itoh at The University of Tokyo, Prof. Debbie G. Senesky at Stanford University, Prof. Mina Rais-Zadeh at NASA, JPL, and Prof. John A. Rogers at Northwestern University for their dedication and effort in pushing the frontiers of this fast-growing research area.

Although we have edited and proofread this book carefully, mistake and typos are unavoidable. We welcome any feedbacks and suggestions from the readers.

Tuan Anh Pham, Toan Dinh, Nam-Trung Nguyen, and Hoang-Phuong Phan
Griffith University, Queensland, Australia
10 December 2021

1

Bottom-Up Growth Methods

1.1 Introduction

Wide bandgap semiconductor nanowires have recently attracted increasing attention from researchers and engineers due to their superior chemical inertness and mechanical properties together with interesting electronics and electrical behaviors [1–6]. In particular, the chemical inertness of wide bandgap semiconductors offers the reliable and long-term operation, while their large energy gap allows for high-temperature operation and unique functionalities that cannot be achieved with the conventional Si nanowires. Importantly, nanowires exhibit numerous advantages compared with their bulk counterparts at the nanoscale level, including tunable properties, large surface-to-volume ratios, excellent sensitivity, low power consumption, and the ability to integrate different functional elements into a single chip [7–10]. For example, piezoresistance of top-down and bottom-up fabricated Si nanowires was found to be much more significant compared with that of bulk Si [11, 12]. Moreover, the quantum confinement of nanowires may result in exotic properties in sensing and optoelectronic devices such as tunable band gap and fast operation speed [13–15]. Another vital benefit of nanowires is that, unlike conventional thin-film structures, lattice mismatch strain in nanowires can be relieved at the free surfaces by elastic deformation without dislocation defects [16–20]. These huge advantages of nanowires offer exciting opportunities for their integrations on diverse substrates for the fabrication of high-performance electronic and photonic devices, such as light-emitting diodes, sensors, photodetectors, and solar cells. In this context, it is realized that the resulting performance of such electronic devices strongly depends on the morphological quality and uniformity and the ability to control the doping types and concentration of synthesized wide bandgap semiconductor nanowires. However, these key parameters determining the electronic device performance are strongly associated with the choice of the synthesis and fabrication methods. Therefore, an in-depth understanding of the synthesis and fabrication methods of wide bandgap semiconductor nanowires is highly necessary to achieve the desired nanowires bearing a high degree of structural perfection with tunable electronic properties.

To date, various approaches have been employed to synthesize and fabricate a wide range of wide bandgap semiconductor nanowires. Among them, the bottom-up approach has been currently considered as one of the most effective methods for the nanowire's fabrications. The bottom-up method is usually defined as an additive process, in which atoms and molecules are the basic building for the construction of various sophisticated nanostructures. This method exhibits a huge flexibility to synthesize high-quality wide bandgap

Wide Bandgap Nanowires: Synthesis, Properties, and Applications, First Edition. Tuan Anh Pham, Toan Dinh, Nam-Trung Nguyen, and Hoang-Phuong Phan.

semiconductor nanowires bearing controllable and programmable properties. On the other hand, the most significant drawbacks of the bottom-up approach are the limitation of large-scale uniformity and integration of different elements into functional devices. Considering advantages and disadvantages of this approach, an in-depth understanding on the bottom-up growth of wide bandgap nanowires is highly necessary for researchers and engineers who are currently working toward the fabrication of semiconductor nanowires-based electronic devices. This chapter, therefore, will review the basic concepts as well as the recent progress on the bottom-up synthesis and fabrication of several widely used wide bandgap semiconductor nanowires. The chapter consists of three sections and its exposition is the following. After a brief introduction in Section 1.1, Section 1.2 contains an introduction to various growth mechanisms for the synthesis of wide bandgap nanowires using bottom-up approaches with an emphasis on advantages and limitations of each mechanism. Finally, Section 1.3 discusses the possibility to use different bottom-up synthesis techniques to effectively control over the atomic structures, crystalline orientation, chemical composition, defect densities, and doping concentration of wide bandgap semiconductor nanowires.

1.2 Bottom-Up Growth Mechanisms

1.2.1 Vapor–Liquid–Solid Growth Mechanism

The vapor–liquid–solid (VLS) mechanism has been the most widely exploited process among various effective processes developed for the bottom-up synthesis of semiconductor nanowires. This process was firstly introduced by Wagner and Ellis for large whisker growth in 1960s, where they employed Au particles as catalysts to grow crystalline silicon microwires from $SiCl_4$ and SiH_4 [21–25]. Over the past decades, considerable efforts have been devoted to uncovering many fundamental aspects of VLS growth mechanism, which could open-up new pathways to synthesize various wide bandgap semiconductor nanowires with an enhanced quality and uniformity [23, 26–28].

Because the growth process is extremely rapid under conventional growth conditions, probing the VLS growth of nanowires at different stages in real-time represents a huge technical barrier for researchers. In this regard, Wu et al. reported the first real-time observation of semiconductor Ge nanowire growth in an *in situ* high-temperature transmission electron microscope (TEM) [28]. Their results undoubtedly provided an insight into the VLS growth mechanism at nanometer scale. Figure 1.1 shows a series of *in situ* TEM images recorded during the growth process of Ge nanowires, providing direct evidence for the events at different VLS growth stages of Ge nanowires. In another pioneering work, Ross and coworkers systematically investigated the VLS growth of Au–Si system using a combination of ultra-high vacuum TEM and a capillary precursor delivery system. Their results provided much significant information on the growth kinetic of the VLS process, in which the growth proceeds in a layer-by-layer fashion [29]. In particular, the VLS mechanism relies on a three stage-process, including metal alloying, crystal nucleation, and axial growth stage as shown in Figure 1.2. In the alloying stage, a catalytic liquid alloy phase is formed on the substrate upon annealing. Subsequently, this alloy phase can rapidly absorb a vapor of precursors to its supersaturation level in the nucleation stage. Finally, nanowire growth can proceed from nucleated seeds at the solid/liquid interface in the growth stage.

Due to its great flexibility, the VLS process has been successfully employed to produce a wide range of wide bandgap semiconductor nanowires, such as silicon carbide (SiC), gallium

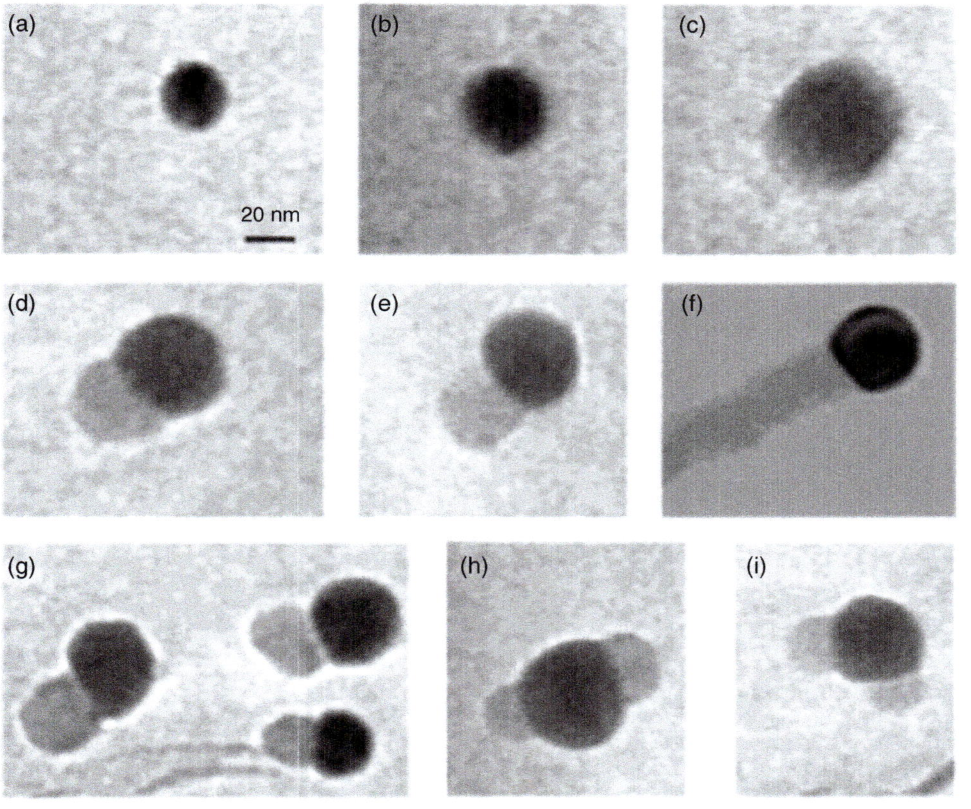

Figure 1.1 A series of *in situ* TEM images recorded during the process of Ge nanowire growth. (a) Au nanoclusters in solid state at 500 °C; (b) alloying initiates at 800 °C, at this stage Au exists in mostly solid state; (c) liquid Au/Ge alloy forms; (d) the nucleation of Ge nanocrystal on the alloy surface; (e) Ge nanocrystal elongates with further Ge condensation; (f) eventually a wire form; (g) several other examples of Ge nanowire nucleation. (h, i) TEM images show two nucleation events on single alloy droplet. *Source:* Reprinted from Wu and Yang [28].

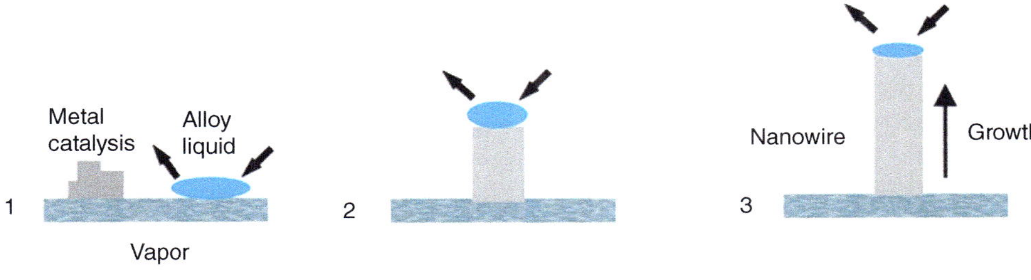

Figure 1.2 Schematic illustration of vapor–liquid–solid nanowire growth mechanism including three stages: (1) alloying, (2) nucleation, and (3) axial growth.

nitride (GaN), diamond nanowires, and others [30–34]. For example, SiC nanowires with 20 nm diameter and 20 μm length were grown directly on the surface of bulk SiC ceramic substrate at elevated temperature using aluminum as a catalyst (Figure 1.3a,b) [35]. In another example, diamond nanowires were grown on a Si substrate using the chemical vapor

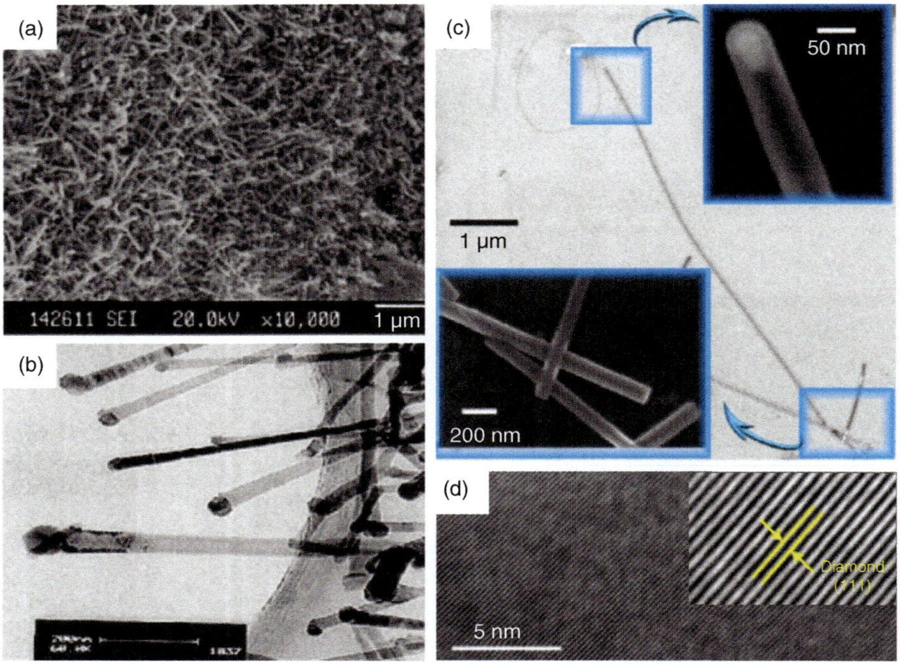

Figure 1.3 (a, b) SEM and TEM images of SiC nanowires growth by the VLS process, respectively. *Source:* Reprinted from Deng et al. [35]. (c) Electron microscopy of diamond nanowires encased within a carbon nanotube shell. The inset in (c) shows a catalyst embedded inside the tip of the nanowire. (d) The high-resolution TEM image of a single diamond nanowire. Inset in (d) shows a zoomed-in view of the crystalline structure of cubic diamond (111) surface. *Source:* Reprinted from Hsu et al. [36].

deposition (CVD) process based on the VLS mechanism with an iron catalyst. These diamond nanowires are straight, thin, and long over tens of micrometers with a high crystallinity and structural uniformity (Figure 1.3c,d) [36]. One of the main advantages of the bottom-up nanowire growth based on the VLS mechanism is that nanowires with different morphologies and structures can be synthesized in a controllable manner depending on the proper choices of the substrates, catalysts, and growth parameters (such as pressure, temperature, and time). For example, Wang et al. demonstrated the possibility to manipulate and control the morphology of VLS grown SiC nanowires from an Eiffel-tower shape to spindle shape by varying the pressure of the precursors. These new morphologies of SiC nanowires are completely different from cone or tip-shaped SiC nanowires commonly reported by other research groups as shown in Figure 1.4 [37]. In another work, Men et al. reported that under nickel (Ni)-based catalysis, SiC nanowires synthesized by VLS process were long and thin, and increased with increasing concentration. In contrast, under Fe-based catalysis, they were short and thick, and the influence of concentration on the morphology of nanowires is negligible [38]. On the other hand, Low et al. investigated the morphological and structural characteristics of GaN nanowires synthesized by Ni-catalyzed CVD method by varying the growth temperatures. The authors found that straight and smooth GaN nanowires with the best elemental composition were achieved at the optimal growth temperature of 950 °C [39]. In another interesting work reported by Krishnan et al., they found that the growth direction of SiC nanowires was dictated by the crystallographic

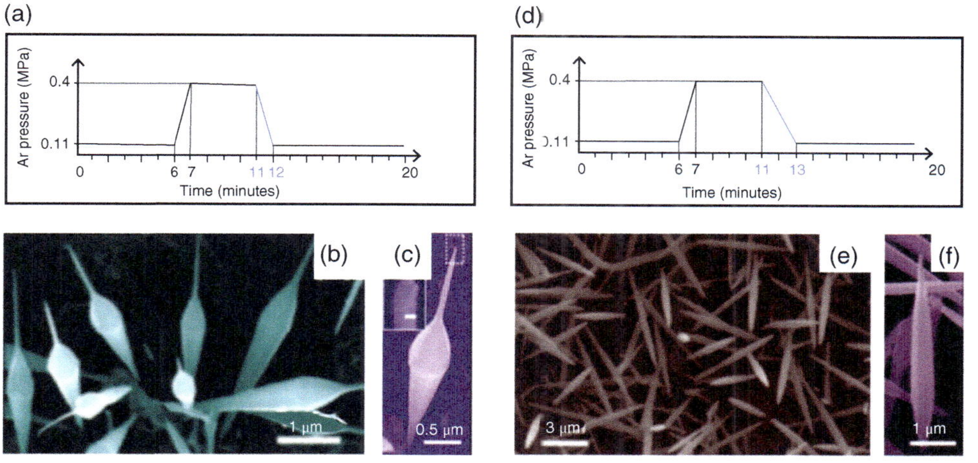

Figure 1.4 SiC nanowires with different morphologies ranging from an Eiffel tower to spindle shape depending on the pressure of the source species: (a) Schematic showing a three-step processing schedule. (b, c) SEM images of the SiC nanowires, showing an Eiffel-tower shape obtained using the processing schedule described in (a). (d) Processing schedule showing the difference in pressure as a function of time compared with the one shown in (a). (e) and (f) SEM image of SiC spindles obtained using the processing schedule described in (d). *Source:* Reprinted from Wang et al. [37].

orientation of the 4H-SiC substrates. This interesting result demonstrated that highly aligned SiC nanowires can be synthesized depending on the choice of substrates [40]. Another key merit of the bottom-up approach based on the VLS process is *in situ* doping by incorporating desired dopant precursors into the VLS synthesis procedure. This is very beneficial to the fabrication of high-performance electronic and photonic devices because such the bottom-up growth method eliminates the requirement for additional destructive processes, such as ion implantation, to introduce additional charge carriers to the as-grown nanowires. For instance, Gao et al. reported, for the first time, the synthesis of Al-doped SiC nanowires. According to their results, the doping levels of these nanowires were effectively controlled by tailoring the Al concentrations in the precursors, which caused redshifts of the photoluminescence bands [41]. In another example, Liu et al. successfully synthesized high-quality GaN nanowires doped with Si using the VLS mechanism, which showed a distinct blueshift from the bulk bandgap emission [42].

Despite these great benefits, the bottom-up nanowire synthesis based on the VLS process has some inherent drawbacks, which is basically induced by the presence of foreign metallic elements serving as catalysts in the synthesis process. These metallic particles usually result in the formation of unavoidable defects and thus lead to negative impacts on optical and electronic properties of the as-grown nanowires as unambiguously proven by numerous studies [43–47]. In this context, Karakostas and coworkers compared GaN nanowires grown by two different methods: catalyst-free or catalyst-induced methods utilizing Ni seeds. According to their findings, the catalyst-induced nanowires contain many more basal-plane stacking faults and thus, the photoluminescence of such nanowires is much weaker compared with that of the nanowires synthesized by the catalyst-free route as shown in Figure 1.5 [48]. A more critical problem is the incorporation of metallic particles within nanowires that typically causes detrimental impacts on device performance. For example, although gold (Au) is the most widely employed catalyst in the VLS process due to its

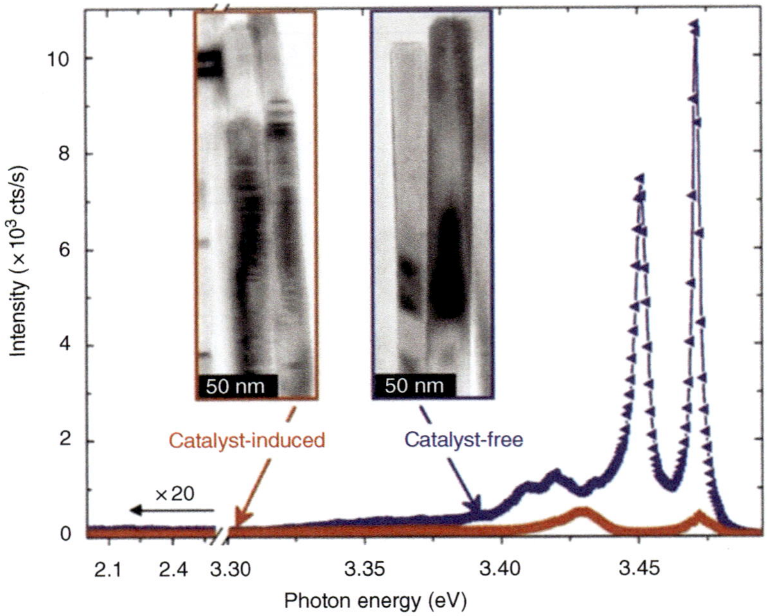

Figure 1.5 Low-temperature photoluminescence spectroscopy (PL) spectra of GaN nanowires grown by catalyst-induced and catalyst-free growth methods. Insets show the corresponding TEM images showing the stacking faults for the case of catalyst-induced growth method. *Source:* Reprinted from Cheze et al. [48].

superior catalyst properties, Au exhibits a fast diffusion from the surface into the underlying Si substrate and then acts as a deep-level trap, reducing the carrier lifetime, which is inherently incompatible with semiconductor technology [49, 50]. To overcome these daunting challenges, a considerable number of studies have been devoted to establishing new strategies for the catalyst-free synthesis of wide bandgap semiconductor nanowires [51–53]. For example, Noborisaka et al. reported the fabrication of GaAs nanowires on a GaAs(111) substrate using selective-area metal–organic vapor-phase-epitaxial growth. In this work, a circular mask opening with a diameter of 50–200 nm was defined on a thin SiO_2 partially coated on the GaAs substrate. As a result, GaAs nanowires with a typical diameter ranging from 50 to 200 nm and a height ranging from 2 to 9 μm were formed on the substrate without any catalysts depending on the opening size of the masked substrate [51]. In another interesting study, Paek et al. demonstrated the possibility to grow GaAs nanowires on a Si substrate using molecular beam epitaxy (MBE) techniques based on the VLS process with different Ga and As fluxes. The authors found that the diameter of GaAs nanowires can be tailored by controlling the Ga and As fluxes under ultra-high vacuum conditions [52]. Detailed information on the catalyst-free synthesis routes of various wide bandgap nanowires will be discussed in the Section 1.3.

1.2.2 Vapor–Solid–Solid Growth Mechanism

The vapor–solid–solid (VSS) mechanism is relatively similar to the VLS growth mechanism. The only difference between these mechanisms lies in their catalyst states. In the VSS process, the catalysts remain in a solid state, while that of the VLS process is in a liquid state. The VSS process was first introduced by Kamins et al. in 2000 when they synthesized silicon nanowires on Si substrates using $TiSi_2$ islands as catalysts [54]. According to their results, the $TiSi_2$ islands remain solid and anchored to the substrate during the growth, while the silicon nanowires grow out from the $TiSi_2$ islands remaining at their bases as shown in Figure 1.6. This differs from the VLS process, where the growth occurs at the tip of the nanowires when the solubility of Si in the liquid droplet reaches saturation. Compared with the VLS growth with liquid catalysts, the VSS growth possesses several advantages, which could benefit the bottom-up synthesis of wide bandgap semiconductor nanowires. First, the VSS growth can be carried out at reduced temperatures in comparison with the VLS growth due to the catalyst remains in a solid state. Second, the VSS growth can minimize the incorporation of unwanted impurities from the metal catalysts, which is based on the reduction of atom diffusion and solid solubility during the growth of nanowires at lower temperatures. Third, the crystalline structure of solid catalysts may offer a huge opportunity to obtain nanowires with a high degree of structural perfection, regardless of the chosen substrates. Finally, the VSS process may enable more abrupt interfaces in axial nanowire heterostructures by mitigating the so-called reservoir effect due to solid solubility, which is usually much lower than liquid solubility for the same materials [55–57]. For example, Wen et al. demonstrated that single interfaces in Si–SiGe heterostructures with defect-free and close to atomically abrupt can be obtained using solid Al–Au alloy catalyst particles instead of conventional liquid semiconductor–metal eutectic droplets [57]. On the other hand, the VSS growth still remains a shortcoming compared with the VLS growth is that the VSS growth rate is significantly slower than the VLS growth rate at similar growth conditions due to the weaker surface reactivity as well as lower diffusivity through the solid in the VSS process [58]. To overcome this technical barrier, Wen et al. added Al to the Au catalyst to raise the eutectic temperature so that the growth can be achieved at higher temperature with reasonable growth rate but still keep the alloy catalyst in the solid state [57, 59].

To date, the VSS growth mechanism has increasingly attracted researchers as an alternative to grow wide bandgap semiconductor nanowires. For example, Hou et al. reported the Au-assisted growth of GaN nanowires based on the VSS mechanism using the plasma-enhanced chemical vapor deposition (PECVD). The authors claimed that the detachment and migration of Au catalyst seeds in the growth can be controlled by the partial pressure of gallium and nitrogen radicals [60].

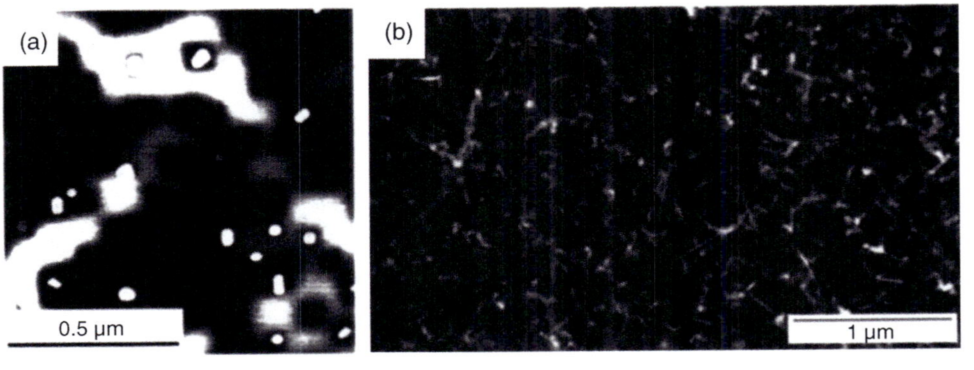

Figure 1.6 (a) Atomic-force micrographs of $TiSi_2$ islands formed by chemical vapor deposition at 640 °C on Si(001) and (b) scanning-electron micrograph of long, dense Si nanowires nucleated on Ti-containing islands grown on Si(001) at 670 °C. *Source:* Reprinted from Kamins et al. [54].

In another example, Wang and coworkers synthesized GaN nanowires via the VSS mechanism by Ni-assisted metal–organic chemical vapor deposition (MOCVD). According to this work, a sputtered nickel film on *c*-plane sapphire was used as a substrate, while trimethylgallium (TMG) and ammonia were employed as precursors. After introducing TMG into the reactor, the nickel film was transformed into Ni_3Ga nanoparticles via thermodynamically favored reaction between Ni and Ga at 900 °C. These nanoparticles keep capturing Ga species and become supersaturated. Consequently, the Ga species react with the decomposed ammonia at the surface of the Ni_3Ga particles to form GaN nanowires [61]. Despite these huge achievements, the VSS growth mechanism remains much less understood in both theoretical and experimental aspects. Therefore, considerable efforts have been devoted in the past years to obtaining a better understanding on the VSS mechanism. In this context, Cui et al. developed a general mass-transport-limited kinetic model to describe the VSS growth process of Si nanowires. In this model, the VSS mechanism proposed by the authors is illustrated in Figure 1.7 and can be described in following steps: (i) atoms diffuse from the vapor phase to adsorb at the surface of catalyst; (ii) the diffused atoms hit the sidewall areas of the nanowires at an incident angle of θ; (iii) the desorption of adatoms from the catalyst; (iv) adatoms diffuse along the sidewalls of the nanowires; (v) the desorption of adatoms from the sidewall; (vi) and (vii) the diffusion and desorption of adatoms from the substrate, respectively; and (viii) adatoms diffuse at the catalyst–nanowires interfaces. Thus, the authors considered three surface diffusion processes and a slow interface diffusion process. The former is associated with the atoms transport pathway, while the later relates to the rate-determining step for the VSS growth of nanowires [62].

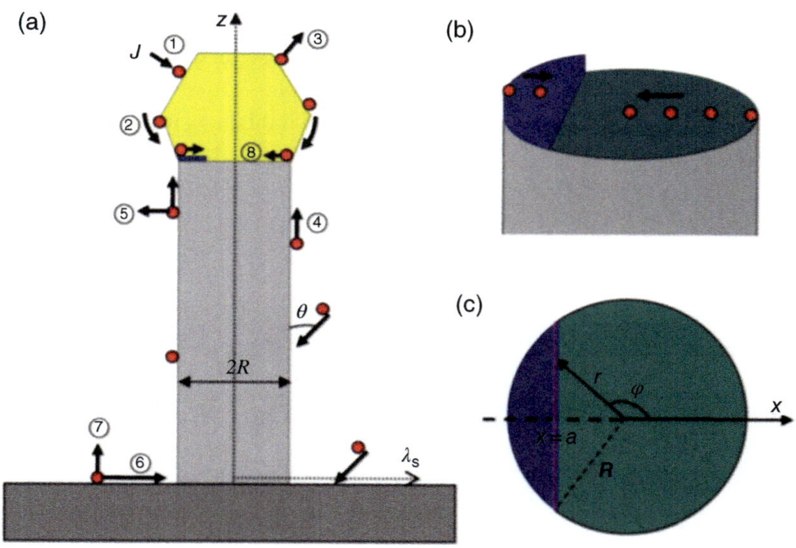

Figure 1.7 (a) Schematic illustration of the step-flow growth of nanowire via VSS mechanism: (1) impinging flux from the vapor, (2) adatoms diffusion flux on the surface of catalyst, (3) desorption from the catalyst, (4) adatoms diffusion flux on the sidewalls, (5) desorption from the sidewalls, (6) adatoms diffusion flux on the substrate, (7) desorption from the substrate, (8) atom diffusion flux in the interface. (b) Detailed atom diffusion and step movement at the catalyst–nanowire interface. (c) Coordinate system used in the calculation of step movement at the catalyst–nanowire interface. *Source:* Cui et al. [62] / American Chemical Society.

This model offers an exact solution to demonstrate the impact of various kinetic parameters on the VSS growth mechanism, which could be extended to various types of wide bandgap semiconductor nanowires. Together with theoretical studies, several experimental studies have also been conducted to get an insight into the VSS growth mechanism. For instance, Maliakkal et al. investigated the VSS mechanism via the nickel-assisted growth of GaN nanowires using MOCVD [63]. The authors proposed that unlike the traditional VLS mechanism where the material is collected directly from the vapor by the liquid alloy, there are three different modes by which reactants can reach the catalyst particles in the VSS process as shown in Figure 1.8a. According to the authors, the correlation between diameter and length of nanowires can provide information to determine the mode in which the reactants reach the catalyst. Interestingly, the authors observed an initial increase in length with

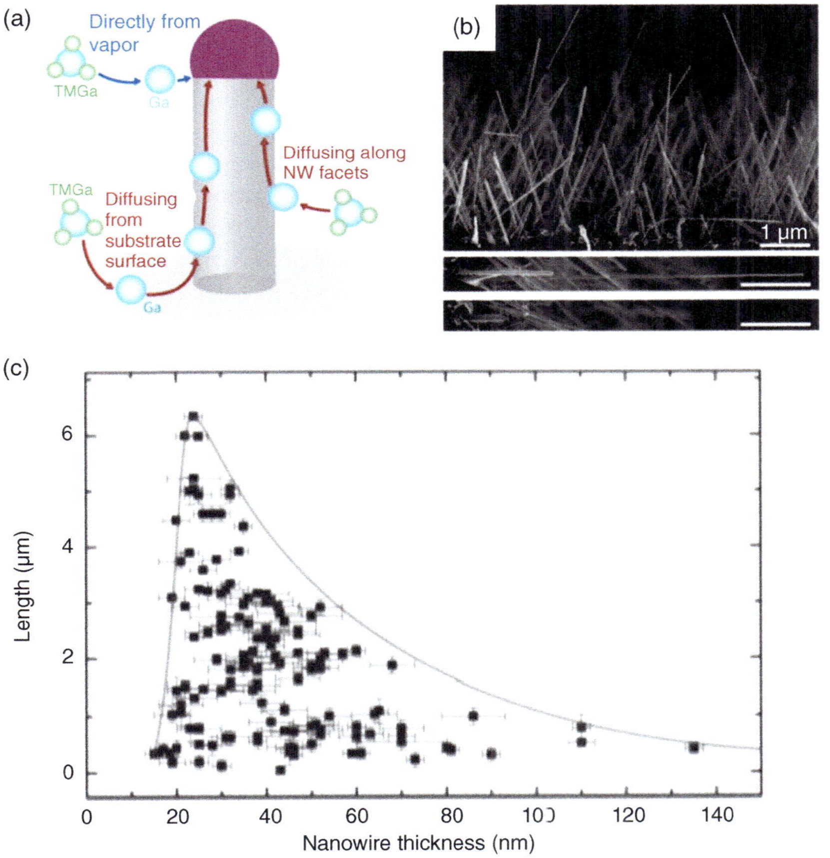

Figure 1.8 (a) Schematic showing the three different modes by which reactants reach the catalyst particle. The supplied precursor molecules can be collected (i) from the vapor phase directly into the catalyst, (ii) from adatoms on the NW side facets diffusing into the catalyst, or (iii) from adatoms on the substrate surface that diffuse along the substrate and NW side surfaces into the catalyst. (b) Side-view SEM image of the GaN NWs grown on *r*-plane sapphire. It is evident that the thick wires are short, and the thin wires are long. Enlarged versions of a couple of representative NWs are shown at the bottom. The scale bars correspond to 1 μm. (c) Plot showing the distribution of length and thickness for more than 130 NWs. *Source:* Reprinted from Maliakkal et al. [63].

increasing thickness of nanowires due to Gibbs–Thomson effect and subsequently, a decrease in length when the thickness of nanowires is larger than a certain value as the consequence of the diffusion-induced growth mechanism as shown in Figure 1.8b,c. In another example, Cheze et al. examined the nucleation of Ni-induced GaN nanowires driven by the VSS process through *in situ* monitoring employed quadruple mass spectroscopy and reflection-high energy electron diffraction [64]. According to their findings, three nucleation stages of the growth were evidenced. The first stage relates to Ga incorporation into the Ni collector. The second stage corresponds to the transformation of the crystal structures of the collector. The third stage is the nucleation of GaN nanowires under the Ni collectors. These findings are expected to provide significant information to control the growth of various wide bandgap semiconductor nanowires based on the VSS mechanism.

1.2.3 Vapor–Solid Growth Mechanism

As mentioned in the previous sections, the VLS and VSS mechanisms offer many huge advantages for the bottom-up synthesis of nanowires; however, the use of foreign metallic particles as catalysts remains an inherent challenge, which significantly limits the possibility to achieve high-performance wide bandgap semiconductor nanowires. Hence, the development of catalyst-free synthesis processes is highly needed for the bottom-up synthesis of nanowires in order to eliminate negative impacts of catalysts on the nanowire growth. In this regard, the vapor–solid (VS) mechanism is currently viewed as a reliable approach to synthesize a series of wide bandgap semiconductor nanowires without the presence of catalysts. Due to the absence of energetically favored catalyst sites to promote the growth, the VS process usually requires a higher synthesis temperature compared with that of the VLS and VSS mechanisms. On the other hand, the synthesis process based on the VS mechanism is relatively simple compared with other catalyst-based synthesis methods. In general, the VS process starts via the decomposition of precursor gases at the chosen substrates induced by high reaction temperatures or the evaporation of precursor powders upon thermal annealing at elevated temperatures [65]. During the VS growth process, three interfaces have to be taken into consideration, including vapor–substrate interface, vapor–solid interface, and solid–substrate interface. The vapor–substrate interface is the place where the nucleation occurs, which plays a crucial role for the VS growth process. The precursors adsorb on the substrate surface and, subsequently, condense into solid to form nucleus. At this interface, the nucleation step can be optimized via controlling various growth parameters, such as deposition temperatures, pressure, and substrate surface conditions. After the nucleus formed, the vapor–solid interface is also important, which could generate the one-dimensional epitaxial growth of nanowires. Otherwise, only nanoparticles can be formed on the substrate surface. The solid–substrate interface is generally associated with the influence of the chosen substrates on several important aspects, such as, the defect density, structure, orientation, and morphology of the as-grown nanowires via the VSS mechanism [66]. Another fundamental aspect influencing the VS growth of nanowires is the supersaturation ratio of the condensing species. In this context, Hsu et al. investigated the VS growth mechanism of Sn nanowires and found that the supersaturation ratio of the condensing species must be maintained below a critical value to ensure anisotropic growth to form nanowires. Otherwise, two-dimensional growth or even isotropic growth can occur. Moreover, homogeneous nucleation in the vapor phase can lead to the formation of powder forms, instead of nanowire forms, at high supersaturation ratios [67].

To date, various types of wide bandgap semiconductor nanowires have been synthesized using the VS mechanism [68–70]. For example, Chen and coworkers synthesized single crystalline 3C-SiC nanowires with lengths up to several tens of micrometers and diameters of 20–60 nm via a simple catalyst-free method based on the VSS mechanism using silicon powders and expandable graphite as raw materials. In this work, the authors proposed the VS mechanism to interpret the growth procedure of SiC nanowires as following: When temperature is up to 200 °C, the expandable graphite begins to expand and generates a porous and loose structure. Subsequently, oxygen in the furnace reacts with silicon and carbon to produce SiO and CO. In the next step, SiO reacts with carbon to form SiC embryos. Finally, the newly formed SiC nucleus deposits on the top of SiC nanowires form the long SiC nanowires [68]. Detailed steps of the VS growth process and the 3C-SiC nanowires obtained by this method are shown in Figure 1.9. In another work, Yazdi and coworkers synthesized highly oriented AlN nanowires with aspect ratio up to 600, diameter in the range of 40–50 nm, and 100 μm lengths via the VS growth mechanism. The as-grown nanowires are dislocation free single crystal and perfectly oriented along the *c*-direction of the wurtzite structure. The authors claimed that the AlN nanowires change in thickness after they have reached a critical length, which contributes to the understanding of the VS growth mechanism [69]. In another example reported by Zhou et al., they synthesized large-scale GaN nanowires by thermal evaporation of GaN powders under controlled conditions based on the VSS mechanism. The as-grown GaN nanowires are about 30 nm in diameter and several hundreds of microns in length, providing excellent dielectric properties. The main feature of the VS growth in this study is that the equilibrium phase diagrams can be used to predict a dislocation and growth conditions, thus enabling rational synthesis for large quantities of single crystalline GaN nanowires at relatively high purity and low cost [70].

Controlling and optimizing the uniformity, morphology, crystalline structures, and defect density of the as-grown nanowires via the VS process play an important role in the bottom-up fabrication of high-performance photonic and electronic devices. This challenging task can

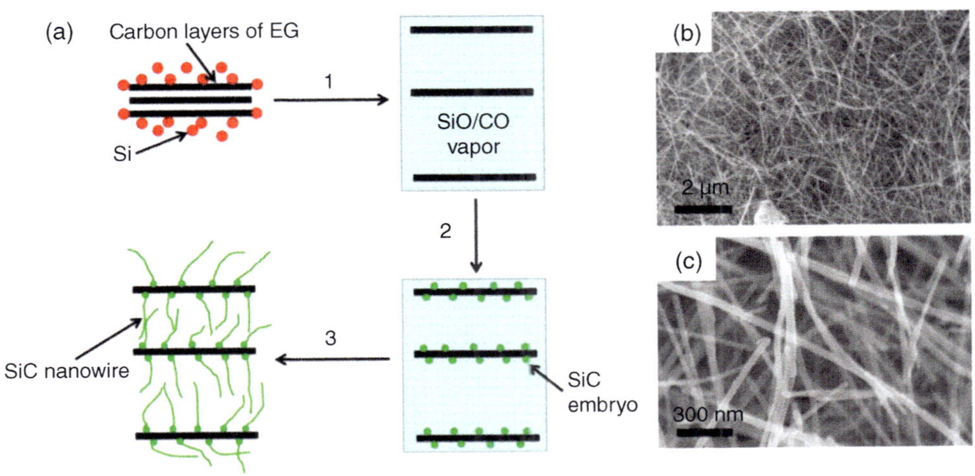

Figure 1.9 (a) A cartoon illustrates the VS growth process of 3C-SiC nanowires: (1) Expandable graphite expands, and SiO/CO vapor is formed; (2) SiC embryos are produced by heterogeneous nucleation; (3) long SiC nanowires grow along the (111) direction. (b, c) SEM images of SiC nanowires at different magnifications. *Source:* Reprinted from Chen et al. [68].

be achieved by tuning the evaporation of precursors, the vapor pressure, the substrate surface temperature as well as a proper selection of raw materials. For example, Wu et al. demonstrated that SiC nanowires synthesized by the VS mechanism can be tuned to cylinder, hexagonal prism, or bamboo shape by simply altering the reaction temperature from 1470, 1550 to 1630 °C, respectively [71]. More interestingly, Shen et al. reported the controlled synthesis of various high-quality GaS nanostructures, ranging from nanowires to nanobelts, hexagonal microplates, and heterostructures via a simple VS method by tailoring the substrate temperature and evaporation source as illustrated in Figure 1.10 [72]. According to their results, GaS nanostructures can be formed with different morphologies depending on the choice of the substrate temperature as shown in Figure 1.11. In particular, single crystal GaS nanowires with uniform diameters of around 20 nm are formed at a substrate temperature slightly less than 800 °C. When the substrate temperature reaches 800 °C, the products were composed of numerous nanobelts with typical lengths of several tens to hundred micrometers. When the substrate temperature is between 800 and 850 °C, many zigzag GaS nanobelts with widths of 1–2 μm were found. At the substrate temperature of 950 °C, GaS microbelts were formed instead of GaS nanobelts. Especially, in the case of lower substrate temperature, only hexagonal GaS microplates were formed at 650 °C [72]. These interesting results unambiguously demonstrated the possibility to control morphologies as well as structures of wide bandgap semiconductor nanowires via tailoring growth parameters in the VSS process.

1.2.4 Solution–Liquid–Solid Growth Mechanism

The solution–liquid–solid (SLS) mechanism is classified as one of the solution-based synthesis processes. This mechanism was firstly introduced in 1995 by Buhro and coworkers when

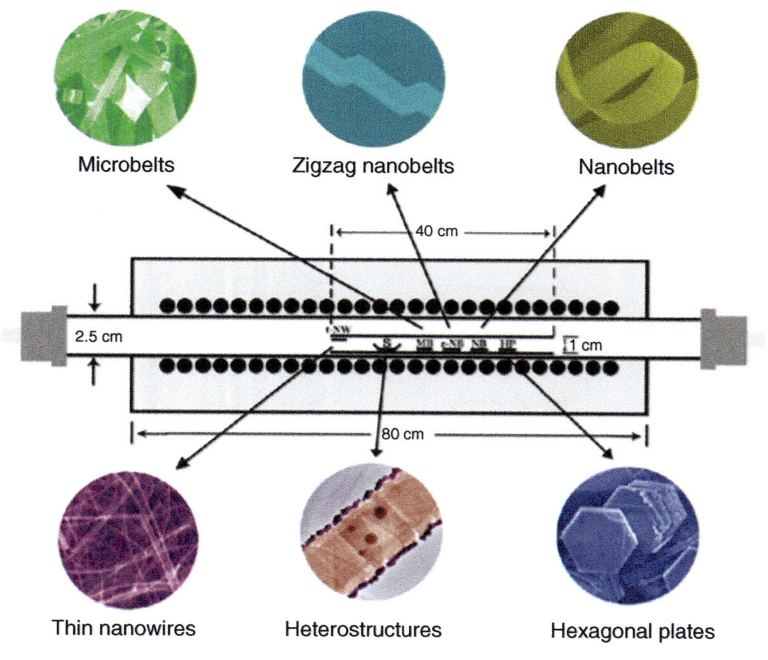

Figure 1.10 Vapor–solid method used for the controlled growth of one-dimensional GaS nanostructures, ranging from nanowires to various types of nanobelts, hexagonal plates, and heterostructures. *Source:* Reprinted from Shen et al. [72].

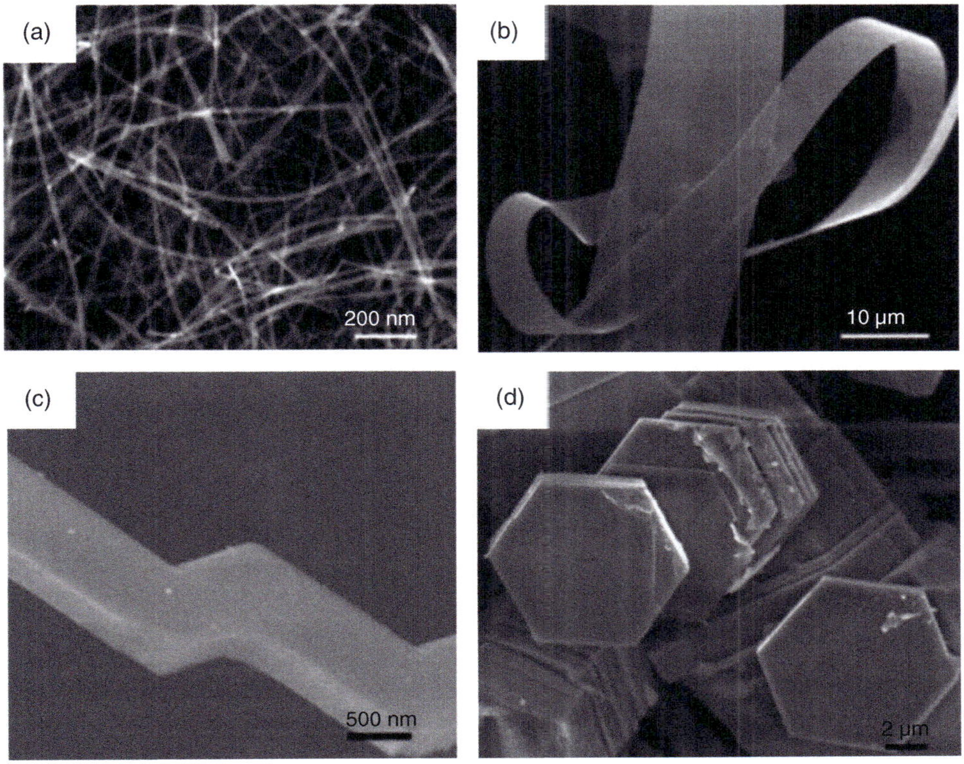

Figure 1.11 SEM images of (a) GaS nanowires; (b) GaS nanobelts; (c) GaS zigzag nanobelts; (d) GaS hexagonal microplates, obtained at different substrate temperatures. *Source:* Reprinted from Shen et al. [72].

they employed solution-phase reactions to synthesize a series of crystalline III–V semiconductor whiskers, such as InP, InAs, and GaAs, having widths of 10–150 nm and lengths of up to several micrometers [7, 73]. According to the authors, the SLS mechanism is analogous to the VLS mechanism discovered by Wagner and Ellis. These two mechanisms share common features, such as the role of catalysts in precursor decomposition and the presence of catalyst particles at the wire tips in the nanowires during the growth [26]. The key difference between the VLS and SLS mechanisms is that the VLS growth process is carried out under vapor phases at high temperature, while the SLS growth is conducted with a solution phase at relatively low temperatures, typically ranging from 200 to 350 °C. A comparison between the SLS and VLS mechanism is illustrated in the Figure 1.12 [26]. During the SLS growth, low-melting metallic nanoparticles catalyze the decomposition of molecular precursors at the solution–liquid interfaces and dissolve the semiconductor components. The solubility limitation of a semiconductor component in the liquid catalyst droplets leads to supersaturation, generating crystallization of the semiconductor phase from the catalyst droplets. As a result, the wire growth occurs at the liquid–droplet–solid crystal interface [74]. Compared with the VLS growth, the SLS growth offers several huge advantages for controlling and engineering the bottom-up synthesis to achieve desired nanowires bearing sophisticated structures and unique properties [75]. For example, this process enables the ability to synthesize nanowires having diameters down to a few nanometers with a narrow diameter distribution, which could open-up new avenues to schematically examine the size-dependent quantum

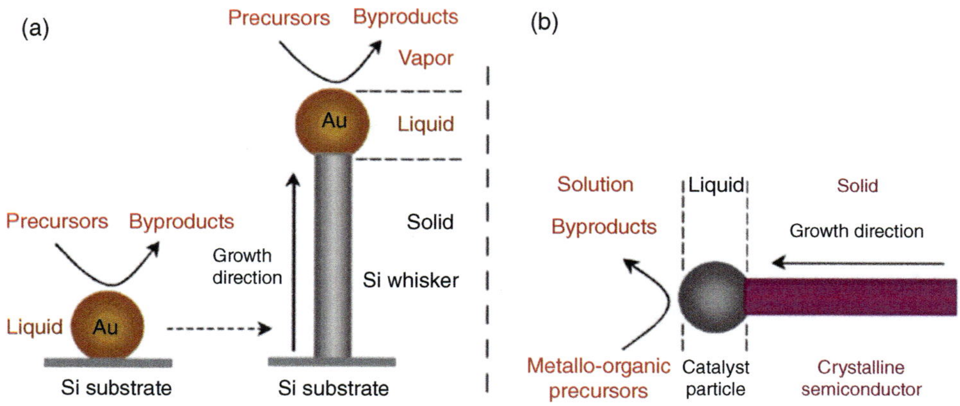

Figure 1.12 A comparison between two different growth mechanisms of nanowires: (a) VLS mechanism proposed by Wagner and Ellis under vapor phases and (b) SLS mechanism proposed by Buhro and coworkers for analogous growth from solution. *Source:* Wang et al. [26] / American Chemical Society.

confinement effect in nanowires. Such small diameters of nanowires cannot be easily accessible through the VLS growth process from vapor phases. However, the outcome of obtaining high-quality quantum nanowires strongly depends on the ability to employ near-monodisperse, low-melting catalyst particles with appropriate sizes to grow the nanowires in solution phases.

Notably, such catalyst particles were unavailable prior to 2001, when Buhro and coworkers made a huge progress in the syntheses of In, Bi, and Sn nanoparticles via a seeded-growth method [76]. Another advantage of the SLS growth over the VLS growth is possibility to form passivation layers on the surfaces of colloidal semiconductor nanowires to reduce trap states and enhance carrier transport, which is very beneficial to obtain high-performance nanowire-based optoelectronic devices. In this context, Janes and coworkers investigated the effect of molecular surface passivation on the transport properties of InAs nanowires grown by the SLS process. The authors found that the conduction mobility of the InAs nanowires increased dramatically from 740 to 1235 $cm^2/(V\,s)$ upon surface passivation with 1-octadecanethiol [77].

Over the past few years, many significant progresses have been made in improving the flexibility and efficiency of the SLS growth process. For example, a surfactant-free SLS growth process was recently demonstrated for the growth of colloidal GaP nanowires [78]. This process offers a new opportunity to overcome inherent drawbacks of the conventional SLS growth process when the absence of organic surfactants will avoid negative impacts on the resulting performance of the as-grown nanowires. Another important progress made in this context is the development of a new strategy that enabled the SLS growth of various semiconductor nanowires on solid substrates [79]. For example, Park and coworkers reported for the first time the SLS growth of nanowires on Si substrates. In this SLS process, Bi particles were deposited on Si substrates and, subsequently, were immersed in a solution containing the precursors at elevated temperature, which catalyzed the growth of nanowires. After the growth, the substrate and nanowires were withdrawn from the solution and washed carefully for further characterizations and fabrications [80]. In some cases, the SLS and VLS mechanism are combined in a synthesis process to grow nanowires on substrates. For instance, Yang et al. reported a new solution technique to grow 3C-SiC nanowires by simply heating Fe–Si melt on a graphite in argon atmosphere to 1600 °C for six hours. The 3C-SiC

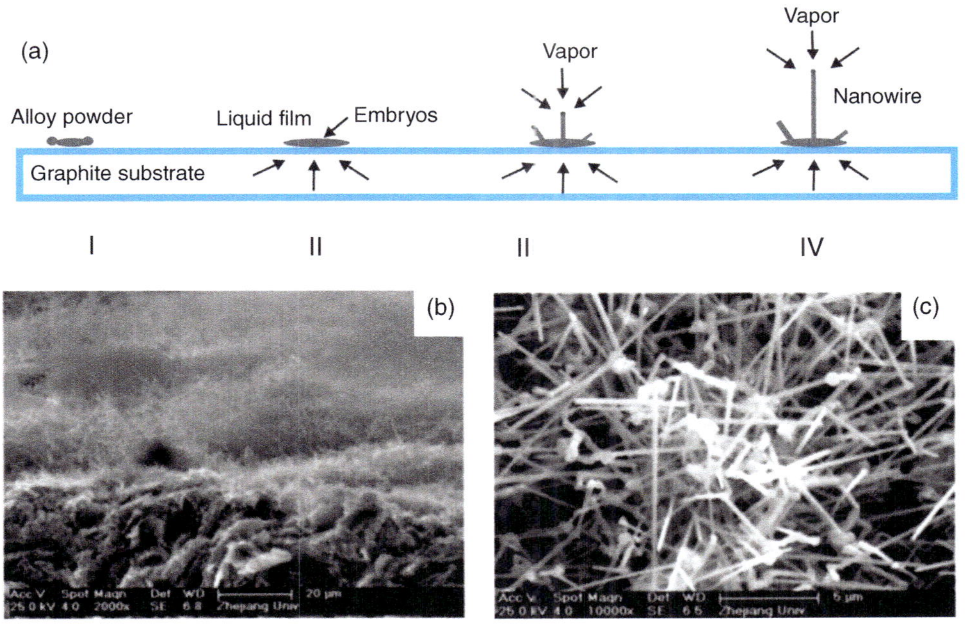

Figure 1.13 (a) The schematic steps of the growth mechanism of SiC nanowires: I and II is the SLS nucleation stage while III and IV is the VLS growth stage. (b, c) SEM images of 3C-SiC nanowires grown by a combination of the SLS and VLS mechanisms. *Source:* Reprinted from Yang et al. [81].

nanowires with diameters of 100 nm and lengths of several tens of micrometers formed on the surface of the melt. Interestingly, the authors proposed that the formation of 3C-SiC nanowires is a combination of the SLS reaction for nucleation and the VLS process for the nanowire growth [81]. Detailed steps for the combined SLS and VLS growth and the newly formed 3C-SiC nanowires obtained at 1600 °C are shown in Figure 1.13. Thus, it is expected that these improved synthesis processes could provide a new pathway for a reliable synthesis of various wide bandgap semiconductor nanowires, which is highly demanded for the development of high-performance electronic and optoelectronic devices.

1.3 Bottom-Up Growth Techniques

1.3.1 Chemical Vapor Deposition

CVD is a bottom-up synthesis technique, in which gas-phase or vapor-phase chemical precursors react on a hot substrate to form desired materials. CVD has been widely employed to synthesize diverse sophisticated structures, ranging from thin films to two- and one-dimensional nanostructures on various solid substrates, not only for scientific research but also for industrial applications. This technique enables the ability to achieve high-quality products with limited defects and excellent properties via a controllable modulation of a wide range of growth parameters. Besides, CVD allows a mass production of numerous nanomaterials through relatively reasonable-cost and facile procedures compared with other growth techniques [82, 83]. A typical CVD system consists of several key components, including quartz-tube furnace, thermocouple, heating elements, pumping systems, gas delivery system, mass flow controller, and pressure gauges as shown in Figure 1.14.

Over the past years, CVD has been considered as one of the most frequently used techniques to grow an array of wide bandgap semiconductor nanowires on solid substrates. In most of the cases, CVD growth of wide bandgap semiconductor nanowires is based on the VLS mechanism with the presence of catalysts [84–89]. For example, Chen et al. reported the CVD synthesis of GaN nanowires on a silicon substrate based on the VLS mechanism. In this study, gallium and ammonium were employed as precursors, while a metal complex of cobalt phthalocyanine was used as a catalyst. The resulting GaN nanowires exhibited wurtzite phase and were up to several micrometers in length with diameters from 10 to 50 nm. Figure 1.15 shows the TEM images of GaN nanowires revealing the presence of catalyst particles on the tip of the nanowire [84]. This strongly confirmed that the CVD growth of GaN in this study was indeed driven by the VLS mechanism. As mentioned previously, CVD represents huge flexibility for controlling and optimizing the morphology and structure of various wide bandgap semiconductor nanowires depending on the choice of catalysts as well as the modulation of substrate temperatures, pressure, growth location, reaction time, gas flow rate, and so on. In this context, Harris and coworkers conducted a series of experiments to systematically study the effects of growth conditions on the size and shape of GaN single crystal grown by CVD technique using Ga and NH_3 as precursors [85]. The authors found that GaN nanowires with uniform diameter, clear crystal structure, and high yield can be obtained at optimal growth parameters. Otherwise, other types of GaN products, such as microwires or crystal platelets, can be formed during the

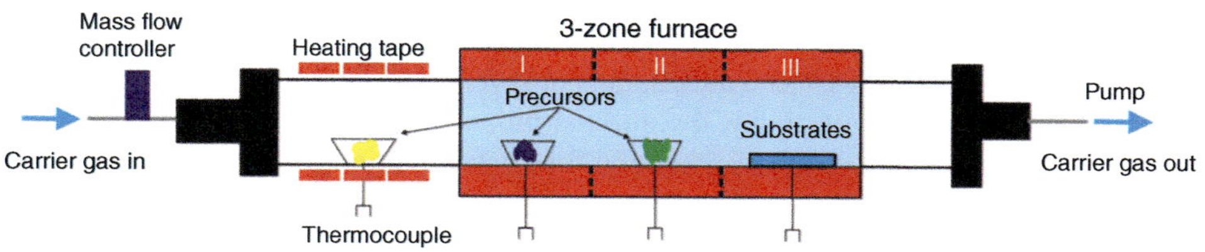

Figure 1.14 A typical experimental setup for the CVD growth of low-dimensional materials, showing several main components, such as heating tape, thermocouple, furnace, and pumping systems.

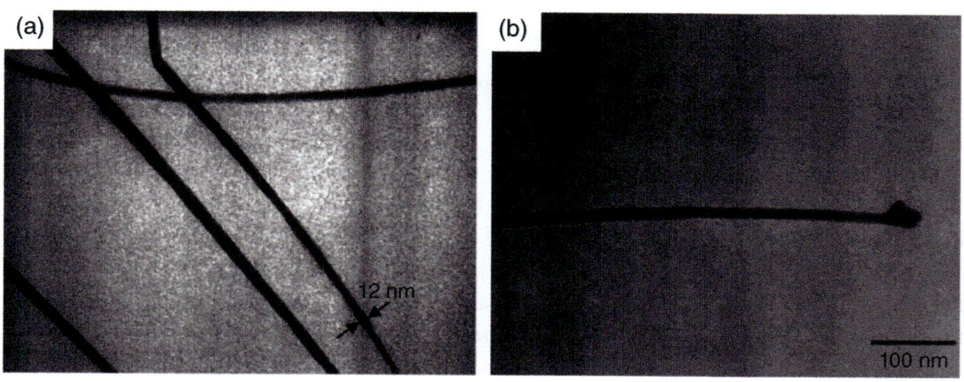

Figure 1.15 (a) TEM image of GaN nanowires obtained from the catalytic reaction when cobalt phthalocyanine was used as a catalyst and (b) TEM image showing the presence of catalyst particles on the tip of the nanowire. *Source:* Reprinted from Chen et al. [84].

growth. Besides the growth conditions, the catalytic materials also play a decisive role in the outcome of the CVD growth process. To date, a series of metal powders, metal alloys, and metal complexes, such as Au, In, Co, Ni, Fe, and Fe/Co, have been employed as catalysts for the CVD growth of wide bandgap semiconductor nanowires based on the VLS mechanism. For example, Park and coworkers synthesized GaN nanowires having single crystalline wurtzite structure with a few stacking faults via the reaction of gallium and gallium nitride mixture with ammonia catalyzed by Fe nanoparticles [86]. In another work, Yeh and coworkers reported a simple CVD method for the large-scale production of GaN nanowires using gallium and ammonia catalyzed by indium powder. The as-synthesized GaN nanowires showed a strong photoluminescence in the UV region [87]. Zheng and coworkers prepared GaN nanowires on a Si substrate through the Ni-catalyzed CVD method using Ga_2O_3/GaN mixtures as gallium sources. As a result, GaN nanowires with a diameter of 90 nm and the length of up to tens of micrometers were formed with uniform density [88]. Despite these great achievements, Liu and coworkers claimed that it is impossible to grow nanowires with predetermined diameters using In, Fe, Ni, or Co as catalysts due to the difficulties in producing such monodispersed nanoparticles. Thus, the authors employed monodispersed gold clusters as catalysts to grow GaN nanowires with controlled diameters around 10, 20, and 30 nm [89].

Although almost the CVD growth of nanowires is accomplished with the VLS mechanism in the presence of catalysts, few studies reported that the CVD growth can be performed via the VS mechanism without the need of catalyst [90–92]. For example, Lee and coworkers reported two distinctive temperature-dependent growth directions of GaN nanowires. According to this study, when the substrate temperature of 900–950 °C, the growth direction of GaN nanowires was perpendicular to the $\langle 10\bar{1}1 \rangle$ plane, while at 800–900 °C, the growth direction was perpendicular to the $\langle 0002 \rangle$ plane. Then, the authors proposed that these differences could be due to the influence of the VS growth mechanism instead of the VLS growth mechanism [90]. In another example, Lin synthesized large-scale centimeters-long single crystal β-SiC nanowires by catalyst-free CVD route under superatmospheric pressure conditions. According to the authors, the growth of the SiC nanowires was governed by the VS mechanism without the use of catalyst. In this study, the total gas pressure was an important factor that could be beneficial for increasing the yield and length of the as-grown nanowires [91]. Huang and coworkers reported a catalyst-free CVD synthesis of twinned SiC nanowires on a silicon wafer in a flowing Argon atmosphere at 1500 °C. The as-grown twinned SiC nanowires were up to several hundred microns long with relatively homogeneous diameters ranging from 20 to 100 nm. The catalyst-free CVD growth in this study was driven by the VS mechanism [92].

1.3.2 Metal–Organic Chemical Vapor Deposition

MOCVD is an enhanced variant of CVD, which has been viewed as one of the most reliable and powerful techniques to synthesize a wide range of wide bandgap semiconductor nanowires. The principle of MOCVD is almost similar to CVD, in which precursors are transported to the substrates by the carrier gas and subsequently absorbed on the heated substrate surfaces. These absorbed precursors are diffused over the surface and then react in heterogeneous phase to form the desired nanostructures. The key difference of MOCVD compared with CVD is the use of organo-metallic molecules as precursors for the growth process. Notably, the toxic and explosive nature of these organo-metallic precursors represents a main drawback of MOCVD, which raises several technical barriers to effectively establish MOCVD systems to grow desired nanowires at the laboratory level. Nevertheless, MOCVD offers huge advantages of growing high-quality semiconductor nanowires, including high growth rate, large-scale uniformity, abrupt interfaces, and effective control of dopant levels [93, 94].

Similar to the conventional CVD method, MOCVD can also be used to synthesize various wide bandgap semiconductor nanowires via either catalyst-assisted or catalyst-free processes. In terms of catalyst-assisted processes, Yang and coworkers synthesized high-quality GaN nanowires via metal-initiated MOCVD for the first time in 2003. In this pioneering study, the authors investigated the growth of GaN nanowires on a gold-coated *c*-plane sapphire substrate via the VLS mechanism using a series of metal catalysts, including such as gold, iron, or nickel. The as-grown GaN nanowires were single crystalline and oriented predominantly along the [210] and [110] direction with triangular cross-sections as shown in Figure 1.16. On the basis of these findings, the authors concluded that this

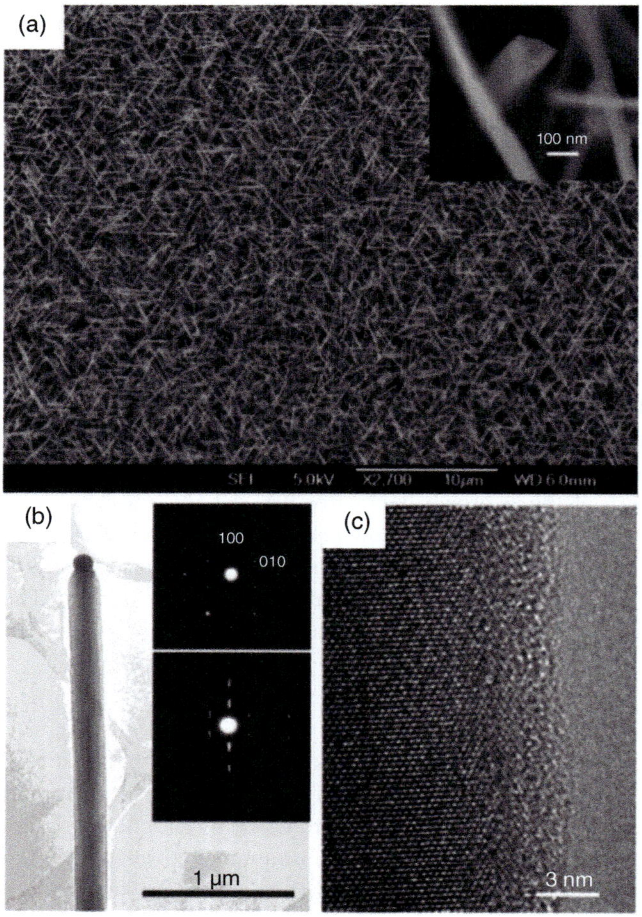

Figure 1.16 (a) SEM image of the GaN nanowires grown on a gold-coated *c*-plane sapphire substrate using MOCVD via the VLS mechanism. The inset shows a nanowire with its triangular cross section. (b) TEM image of a GaN nanowire with a gold metal alloy droplet on its tip. The insets are electron diffraction patterns taken along the [001] axis. (c) Lattice-resolved TEM image of the nanowire. *Source:* Reprinted from Kuykendall et al. [95].

MOCVD technique using organometallic precursors (TMG) offers several merits compared with early studies using Ga metal as the source material. First, MOCVD can be feasibly scaled up to wafer scale production. Second, this technique would enable facile doping of these GaN nanowires with standard precursors [95]. Since then, various studies have been devoted to synthesizing numerous wide bandgap semiconductor nanowires through the catalyst-assisted VLS growth using the MOCVD technique for diverse applications. For example, Yi and coworkers synthesized GaN nanowires with diameter ranging from 20 to 50 nm and lengths ranging from 4 to 6 μm using the Ni catalyst-assisted MOCVD technique. The as-grown GaN nanowire showed much better photocatalytic activities compared with that of GaN submicron dot arrays and thin films due to its high surface-to-volume ratio [96]. Notably, the initial formation of catalytic droplets strongly influences the size, morphology, and quality of the nanowires grown by the catalyst-assisted MOCVD technique. In this context, Lee and coworkers reported different growth behaviors of GaN nanowires grown with Au catalyst and (Au + Ga) solid solution nano-droplets on Si(111) substrates using MOCVD. According to their results, Au + Ga solid solution nano-droplets act as an efficient nucleation seed for the VLS growth and thus, leading to a higher density with uniform diameters of GaN nanowires in comparison to that of Au catalyst [97].

As previously mentioned, the presence of catalysts usually results in an unintentional incorporation of impurities, which generates a high density of structural defects of the as-grown nanowires. Moreover, some metallic catalysts, especially Au, should be strictly avoided in standard semiconductor technology because they could act as a deep-level trap and recombination centers, and thus, degrade the device performances. Therefore, although catalyst-assisted MOCVD growths offer great opportunity to achieve nanowires with controlled diameters as well as a high aspect ratio, considerable efforts have been devoted to establishing new catalyst-free MOCVD methods to synthesize wide bandgap semiconductor nanowires. In this regard, several reliable approaches, such as catalyst-free self-organized MOCVD growth, catalyst-free pulsed MOCVD growth, and catalyst-free selective area MOCVD growth, have been effectively employed as an alternative to synthesize a wide range of wide bandgap semiconductor nanowires without the use of foreign metallic catalysts [98–100]. For example, Heuken and coworkers reported the self-organized growth of catalyst-free GaN nanowires on Si(111) substrates by MOCVD. After the optimization process, GaN nanowires with high-density straight aligned perpendicular to the substrate were fully achieved. According to the authors, the use of a predose before AlN buffer deposition in this study plays an important role in the formation of high-performance GaN nanowires without the needs of catalysts [99]. In another pioneering study, Amano and coworkers developed a scalable process for the precise position-controlled selective growth of GaN nanowires arrays on *c*-GaN templates by MOCVD using a pulsed mode growth process [100]. The authors schematically investigated the influence of pulsed-mode growth parameters such as growth temperature, precursor injection, and interruption durations on the location, orientation, length, and diameter of each GaN nanowire. The effect of nucleation step on the morphology and shape of GaN nanowires was also examined via performing a so-called filling process under applied pulsed-mode growth cycles. During the filling process, TMG and NH_3 were introduced simultaneously into the reactor as precursors. The authors found that in the case of no filling process, a pyramidal structure emerged as the dominant shape after a few pulsed-mode growth cycles, but when the filling process was performed, the surface morphology on the top of nanowires showed a hexagonal structure as shown in Figure 1.17.

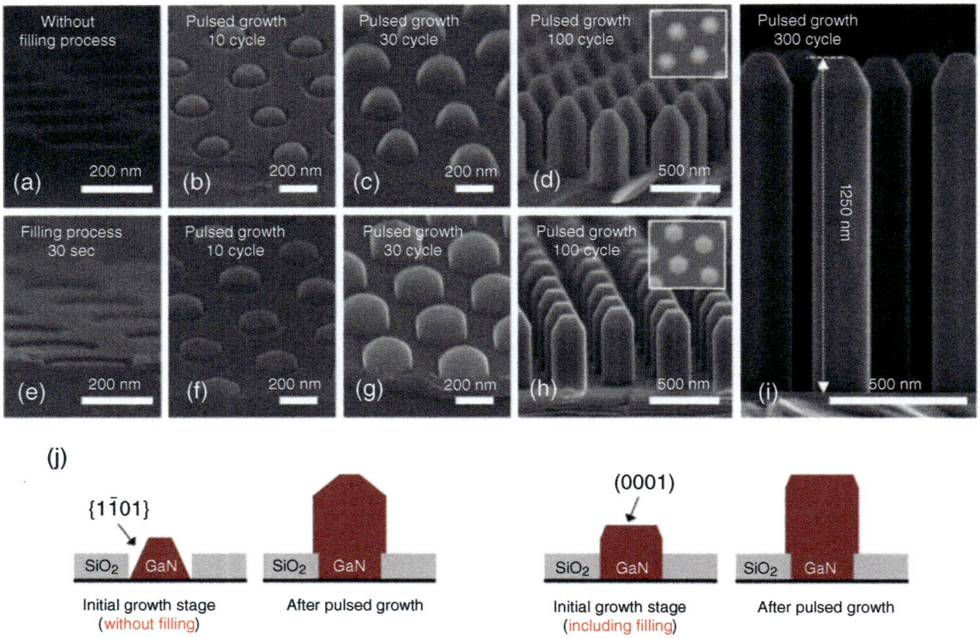

Figure 1.17 Cross-sectional SEM images of GaN nanowire array samples at each stage. (a–d) Results of pulsed-mode growth without a filling process. (e–h) Results of pulsed-mode growth including a 30 seconds filling process. (i) Result of GaN nanowire arrays after 300 pulsed-mode growth cycles. (j) Schematic diagram for pulsed-mode growth of GaN nanowires with and without a filling process. *Source:* Reprinted from Jung et al. [100].

1.3.3 Plasma-Enhanced Chemical Vapor Deposition

PECVD is another variant of CVD in which high-quality materials can be grown on diverse substrates at a lower temperature compared with conventional CVD by taking advantages of plasma. The plasma is usually ignited by applying a high frequency voltage to gases at low pressure. The principle of PECVD is almost the same compared with the low-pressure CVD, except for the lower substrate temperatures. PECVD is a widely employed technique to produce a wide range of semiconductor thin films or carbon-related materials for device fabrications, at both industrial and laboratory levels [101–103]. On the other hand, many research groups have intensively employed PECVD to synthesize numerous wide bandgap semiconductor nanowires due to the feasibility to achieve high-quality nanowires at relatively low temperatures. For example, Poursalehi and coworkers synthesized GaN nanowires on porous silicon by PECVD at 600 °C using gallium and nitrogen plasma as precursors. In this synthesis process, direct chemical reactions between gallium atoms and excited nitrogen atoms in the plasma environment via the self-induced mechanism resulted in the formation of GaN nanowires without using any catalyst [104]. In a similar work, Yang and coworkers synthesized GaN nanowires on graphite substrates by the direct reaction of gallium atoms with excited nitrogen plasma without the use of catalyst. The growth was obtained using PECVD under moderate vacuum conditions and temperatures below 900 °C [105].

Besides the elimination of catalysts in the growth process, the flexibility to control the growth and the elimination of catalysts are additional merits of PECVD compared with other

CVD methods. For instance, Yan and coworkers synthesized GaN nanowires on Si(100) substrates without the presence of catalysts at different hydrogen gas flow rates using PECVD. Interestingly, the authors demonstrated the ability to effectively control the size and morphology of the as-grown GaN nanowires via modulating the hydrogen atmosphere. According to their results, the diameters of the GaN nanowires ranged from 53 to 221 nm and their morphologies transformed from a hexagonal prism to a triangular pyramid via proper tuning the hydrogen atmosphere conditions as shown in Figure 1.18. The authors proposed that the changes in nanowire morphologies originated from the competitive equilibrium between surface and interfacial diffusion induced by hydrogen in the VLS growth mechanism [106]. In another work, Hong and coworker investigated the growth of GaN nanowires by controlling the surface diffusion of Ga species on sapphire using PECVD. The authors found that under nitrogen-rich growth conditions, the growth rate is strongly dependent on the surface diffusion of Ga species. However, adding of 5% hydrogen to nitrogen plasma significantly diminishes the surface diffusion effect [107]. More interestingly, PECVD enables the opportunity to develop a low-cost and green process to synthesize wide bandgap semiconductor nanowires using nontoxic precursors under mild growth conditions. In this regard, Zhao et al. introduced a simple and green route to synthesize GaN nanowires based on PECVD using gallium oxide and ionized nitrogen as Ga and N precursors. Notably, hydrogen gas and carbon were introduced into the reactor to create oxygen-free environment and reduce the vacuum level for the reaction. As a result, GaN nanowires with a maximum diameter about 140 nm and length up to 20 μm with excellent optical properties were achieved [108].

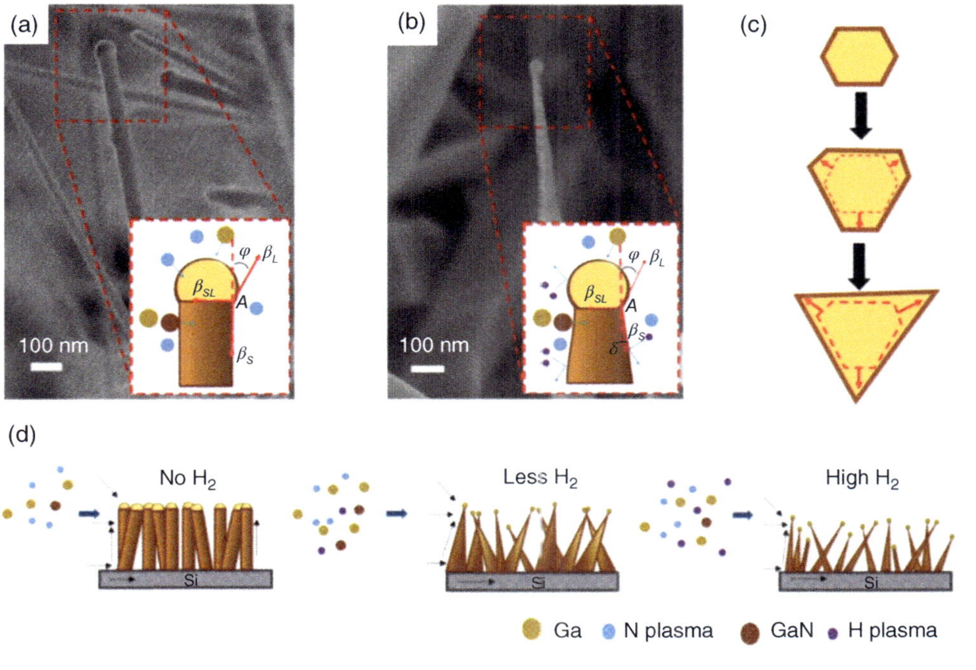

Figure 1.18 (a) SEM image of a hexagonal prism NW with inset showing the growth model. (b) SEM image of a triangular pyramid NW with inset showing the growth model. (c) Schematic diagram of the transformation from a hexagonal to triangular cross section. (d) Schematic diagram of the growth of NWs at different hydrogen flow rates. *Source:* Reprinted from Ji et al. [106].

1.3.4 Hydride Vapor Phase Epitaxy

Hydride vapor phase epitaxy (HVPE) is a type of CVD technique, which has been widely employed over the past decades to grow numerous wide bandgap semiconductor films. The main advantage of HVPE compared with other CVD growth techniques is the possibility to produce high-quality materials with relatively less defects at a high growth rate and reasonable cost. Moreover, HVPE can be carried out under atmospheric pressure, which is an additional benefit for mass-production of wide bandgap semiconductors at industrial levels [109–111]. The principle of HVPE is almost similar with respect to CVD as previously described. The key difference of HVPE compared with other CVD techniques is that HVPE usually employs hydride (such as PH_3 and NH_3) and chloride ($GaCl_3$ and GaCl) as source materials for the chemical reactions in the synthesis. To date, various wide bandgap semiconductor nanowires have been successfully synthesized by HVPE, especially group III–V and III-nitrides nanowires, such as AlN, GaN, and InN nanowires. For instance, Trassoudaine and coworkers synthesized GaN nanowires with exceptional lengths using near-equilibrium HVPE technique on *c*-plane sapphire substrates. In this study, GaN nanowires were grown at a significantly high growth rate of 130 μm/h. This was achieved by the high decomposition of GaCl precursors and a direct supply of Ga through the catalyst particle. High-resolution transmission electron microscopy and selective area diffraction demonstrated high-quality and defect-free structures of the as-grown GaN nanowires as shown in Figure 1.19 [112]. Notably, HVPE can be driven by either VLS or VS mechanisms with and without catalysts, respectively, depending on particular cases. For example, Hwang and coworkers reported the growth of single-crystalline AlN nanostructures in various forms, such as thin films, nanoneedles, nanocolumns, and nanowires, depending on the controlled gas-flow ratio.

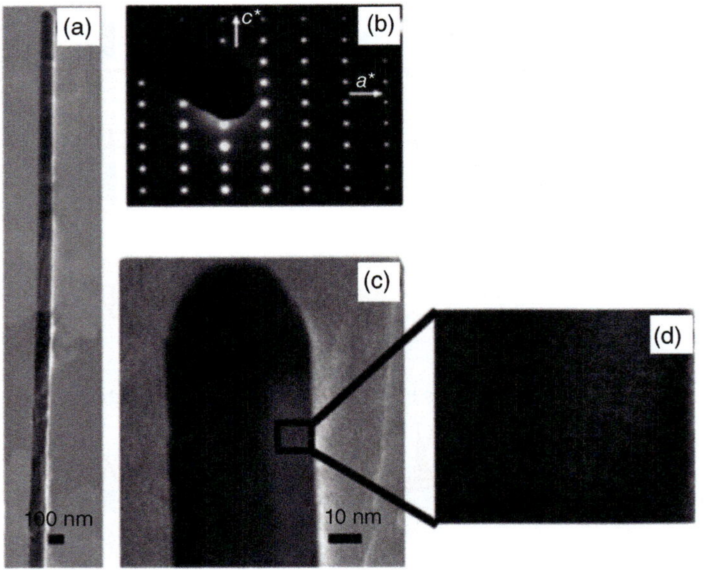

Figure 1.19 TEM images of GaN nanowires grown by HVPE: (a) low-magnification image on a long GaN nanowires; (b) selected area diffraction (SAED) pattern from the same nanowire; (c) high-resolution TEM image showing the top of the nanowire; (d) zoom of (c) nanowire sidewall. *Source:* Reprinted from Avit et al. [112].

According to their results, well-aligned AlN nanorod arrays with diameters below 20 nm were grown on a catalyst-free Si substrate using HVPE based on the VS mechanism [113]. In another work, Narayanamurti and coworkers synthesized GaN nanowires on sapphire by HVPE using nickel–gold as a catalyst. The authors found that the nanowire growth was limited to areas patterned with catalyst, and thus unambiguously confirming the crucial role of catalysts in the HVPE growth technique [114].

On the other hand, HVPE enables new pathways to engineer the nanowire growth directions, orientations, architectures, and structures. For example, horizontal nanowires are highly necessary to enhance the integration of nanowires into nanophotonic systems. However, the synthesis of such nanowires remains a huge challenging for other CVD growth techniques. Using HVPE with an Au/Ni thin film as the catalyst Li and coworkers successfully synthesized horizontal GaN nanowires on a sapphire by simply flipping the substrate. The growth rate for the horizontal nanowires was as fast as 400 μm/h, benefiting from the high decomposition of GaCl precursors using in the HVPE growth. Moreover, the authors were able to control the growth directions with various facet orientations [115]. Wu et al. reported the crystallographic orientation control of GaN nanowires using HVPE based on the VLS mechanism. In this study, the growth orientations were tailored from the polar c-axis to the non-polar m-axis by simply varying the supply of precursors. Thus, the authors suggested that the growth orientation of the nanowires can be effectively controlled by the flow rate of the precursor, regardless of the catalyst size and the substrate adopted [116]. More interestingly, HVPE offers great opportunity to achieve sophisticated heterostructures based on the combination of various wide bandgap semiconductor nanowires. For example, Rho and coworkers synthesized GaN/AlN core–shell nanowires with high crystal quality on sapphire substrates via a two-step HVPE method using Nickel as a catalyst and GaCl/AlCl as precursors [117]. Dubrovskii and coworkers reported an innovative method to produce GaN/AlN core–shell nanowires on sapphire using HVPE without any external aluminum source. In this study, aluminum was etched out from the sapphire substrate and then re-deposited onto the sidewalls of the GaN core. This approach is expected to resolve the main drawback in HVPE technique induced by the fast reaction of the AlCl precursor and the hot walls of the quartz reactor [118].

1.3.5 Molecular Beam Epitaxy

MBE has been well-known as one of the most advanced and reliable epitaxial growth methods for producing various high-quality thin films of semiconductors, superconductors, metals, and insulators, at both the laboratory and industrial production levels. A key advantage of MBE is that it allows crystal growth at temperatures lower than those used in conventional growth techniques. Moreover, this technique offers the possibility to precisely control the composition, doping profiles, interfaces, and thickness of the desired materials at a nanometer scale [119, 120]. The principle underlying MBE growth can be briefly described as following: Beams of atoms or molecules pointing to a target substrate surface are generated by heating up different source materials. This results in the adsorption of atoms and molecules onto a hot substrate surface at specific sites depending on the particular growth and substrate conditions. Following the surface diffusion and rotation, these atoms and molecules eventually incorporate into the growing films. It should be noted that the MBE processes take place under ultra-high vacuum conditions, which results in much higher material purity compared with other growth techniques [121]. A typical MBE system is presented in Figure 1.20 and consists of several main components as following. The first important part is a stainless-steel

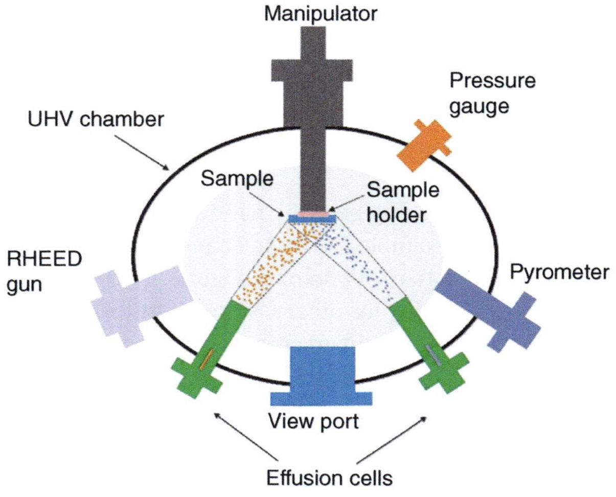

Figure 1.20 A typical setup of an MBE system consisting of several basic components: ultra-high vacuum (UHV) chamber, manipulator, reflection high-energy electron diffraction (RHEED) gun, effusion cells, pressure gauges, pyrometers, sample holders, and pumping system.

growth chamber where the growth can take place under ultra-high vacuum conditions. This chamber is connected to another so-called preparation chamber, where the substrates are degassed prior to growth. The pumping system is connected to the chamber to maintain an ultra-high vacuum environment for the growth. Effusion cells are another key component of an MBE system where the source materials are heated up to form a beam of atoms and molecules for the growth. These effusion cells should provide excellent flux stability and material purity for the growth. Pressure gauges and pyrometers are connected directly to the growth chamber to monitor the growth conditions, such as vacuum levels and substrate temperatures. Moreover, the MBE chamber is usually equipped with a reflection high energy electron diffraction gun to monitor and analyze the thin film quality during the growth.

Inspired by the success of MBE in the thin film production, many research groups have employed MBE to synthesize various wide bandgap semiconductor nanowires over the past years. In this regard, MBE enables a reliable and effective catalyst-free synthesis route and thus, avoiding negative impacts of catalyst particles on the performance of the as-grown nanowires. In general, the plasma-assisted MBE growth is the most frequently used technique for producing high-quality wide bandgap semiconductor nanowires without the use of catalysts. To date, a considerable number of research efforts have been devoted to a better understanding of the plasma-assisted MBE growth mechanism, both theoretical and experimental aspects [122–125]. For example, Lüth and coworkers synthesized GaN nanowires on Si(111) substrates without the use of external catalyst by plasma-assisted MBE. The authors proposed an explanation for the growth mechanism by making use of the diffusion-induced mechanism, in which different kinetic adatom processes, such as adsorption, desorption, and surface diffusion, were taken into consideration. According to their results, the top surface of nucleated wires acts like an efficient collector of adatoms. Thus, besides the direct impingement of atoms at the tip of the nanowire, the diffusion of adatoms toward the tip on side faces also contributes to the total growth of GaN nanowires [122].

To effectively control and optimize the MBE growth of nanowires, the influence of growth parameters, the role of the substrate, the influence of possible buffer layer, the detailed nucleation process as well as the importance of kinetics on the quality and morphology of the resulting nanowires are key factors, which should be seriously considered. In this regard, numerous studies have been carried out to examine the influence of such key factors on the plasma-assisted MBE growth. For instance, Daudin and coworkers investigated the influence of temperature on the growth of GaN nanowires using plasma-assisted MBE method. The authors found that density, diameter, and length of the as-grown nanowires were strongly dependent on temperature due to the temperature dependence of Ga adatom diffusion and GaN dissociation rate [123]. In another work, the same research team found that the addition of indium flux to the plasma-assisted MBE growth strongly enhances the Ga adatom diffusion, which results in reducing the growth temperature and preventing the formation of a two-dimensional layer between the as-grown nanowires [124]. Besides the impacts of adatom diffusion, the choice of substrates also plays a decisive role in the growth of wide bandgap semiconductor nanowires. In this context, Zytkiewicz and coworkers demonstrated that tuning of the substrate microstructure is an efficient tool to engineer the nucleation rate and, thus, the resulting outcome of the plasma-assisted MBE growth of GaN nanowires. The authors compared the nucleation and growth of GaN nanowires on different crystalline and amorphous substrates under the same growth conditions. They found that the deposition of an amorphous Al_2O_3 buffer layer on Si substrates significantly enhances the nucleation of GaN nanowires. On the contrary, no growth of nanowires was found on sapphire substrates and slower nucleation was observed on partially amorphous silicon nitride films. These differences were explained by a low density of defects on the monocrystalline substrate surface where the nanowires nucleate [125]. Moreover, several studies demonstrated a strong dependence of GaN nanowires morphologies on the characteristics of the AlN buffer layer, which offers a powerful tool to control over the shape, diameters, and density of GaN nanowires [126–129]. For instance, Berness and coworkers investigated the role of the low-temperature AlN buffer layer in GaN nanowire nucleation. The authors varied V/III flux ratios to produce a range of AlN buffer layers with different structural and polarity configurations. Their findings demonstrated that the characteristics of AlN buffer layers strongly determine the growth of low density, straight, and coalescence-free GaN nanowires on Si(111) substrates [129].

On the other hand, MBE offers a reliable and powerful tool to enhance the nanowire performances by effectively controlling the doping profiles of individual nanowires as well as producing new heterostructures based on the combination of dissimilar wide bandgap semiconductor nanowires [130–132]. This is currently considered as a stepping-stone for the integration of wide bandgap semiconductor nanowires into the next-generation electronic and photonic devices. For instance, Jiang and coworkers reported the synthesis of Mg-doped AlN nanowires on Si substrates by plasma-assisted MBE technique under nitrogen rich conditions. The as-grown Mg-doped AlN nanowires exhibited excellent properties with the free hole concentration was estimated to be on the order of $10^{16}\,cm^{-3}$ or higher [133]. Gonda and coworkers synthesized GaN/AlN heterostructures on sapphire substrates by reactive MBE. The authors found that the crystalline quality of the GaN was enhanced by producing GaN/AlN heterostructures, which resulted from the small mismatch of lattices and the small difference of thermal expansion coefficients between GaN and AlN. The heterostructures exhibited larger Hall mobilities and more intense cathodoluminescence at the peak of 360 nm in comparison with the GaN films grown on sapphire [134]. In another work, Mishra and coworkers employed plasma-assisted MBE to grow a set of two-dimensional electron gas AlN/GaN structures with AlN barrier thicknesses varied between 24 and 50 Å. Their findings showed that the increase in the electron sheet density was accompanied by the decrease in electron mobility relating to the tensile strain relaxation and interface roughness scattering [135].

1.3.6 Laser Ablation

Laser ablation is one of the most versatile laser-assisted growth techniques for producing high-quality epitaxial thin films and nanoparticles. This technique is also widely used to synthesize various wide bandgap semiconductor nanowires [136–138]. In order to grow nanowires using this technique, pulsed laser is employed to vaporize solid precursors and catalysts in a crucible at elevated temperatures. This results in liquid nanoclusters, which promote and direct the growth of crystalline nanowires via the VLS growth mechanism. A pioneering study reported by Lieber and coworker in 2000 demonstrated the feasibility to use the laser ablation for the bulk synthesis of single crystalline GaN nanowires. In this work, a GaN/Fe composite target was positioned with a quartz tube at the center of a furnace. Subsequently, a pulsed laser was employed to ablate the GaN/Fe target, resulting in the formation of Fe nanoclusters at a high temperature. This promoted the growth of GaN nanowires via the VLS mechanism as shown in Figure 1.21 [136]. The authors examined the morphology, structure, and composition of the as-grown GaN nanowires using transmission electron microscopy and convergent beam electron diffraction. The results showed that the single crystalline nanowires were straight with uniform diameters and a unique [100] growth direction as shown in Figure 1.22. Notably, the nanoparticle observed at the end of the nanowire was faceted due to the crystallization of the liquid nanocluster. The presence of Fe only in the terminal nanoparticle strongly confirmed the catalytic nature of Fe and, thus, proving that the laser ablation synthesis in this study was driven by the VLS mechanism. Using the same laser ablation technique, Lee and co-workers synthesized SiC nanowires with an average diameter of 80 nm and a length of tens of micrometers on a graphite substrate at 900 °C. The grown mechanism was also a typical VLS mechanism, in which Fe nanoparticles acted as catalysts at the tip of each SiC nanowires [137]. In another work, Hong and coworkers synthesized GaN nanowires with lengths ranging from 300 to 500 nm with diameters of around 50 nm on a gold-coated sapphire substrate by pulsed laser ablation of a target of GaN powder in low-pressure nitrogen gas. In this study, the laser ablation induced gallium and nitrogen plasma directly toward the substrate to initialize the VLS mechanism, in which a catalytic reaction of Au and GaN plasma quenched with nitrogen was employed to produce GaN nanowires homogeneously.

On the other hand, although the studies described earlier obtained some very impressive results, the presence of metal catalysts during the laser ablation growth process normally

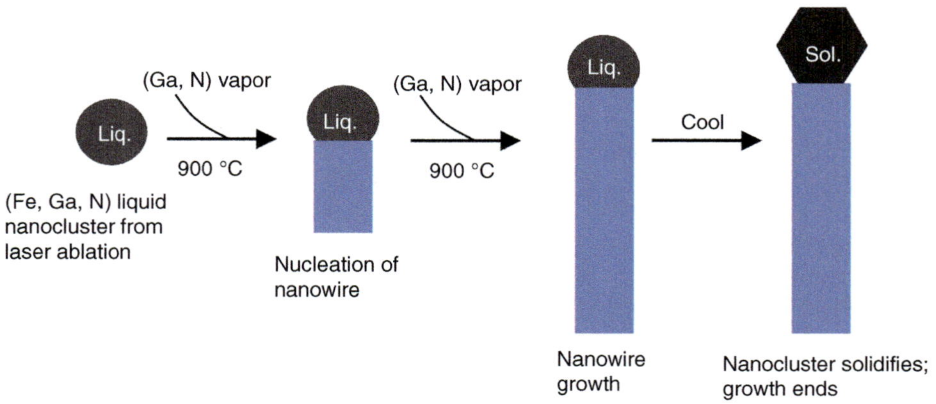

Figure 1.21 A cartoon illustrates the growth of GaN nanowires using laser ablation technique based on the VLS mechanism with the generation of the catalytic nanocluster. *Source:* Reprinted from Duan et al. [136].

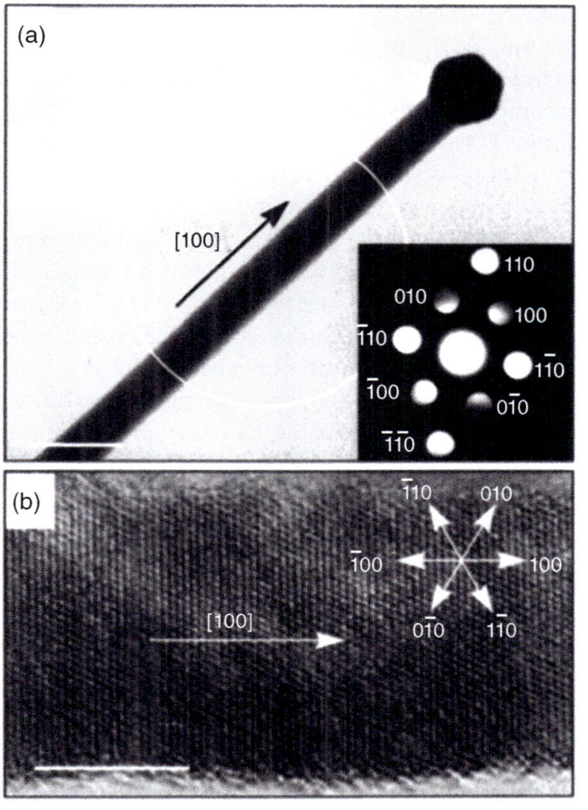

Figure 1.22 (a) Diffraction contrast TEM image of a GaN nanowire that terminates in a faceted nanoparticle of higher (darker) contrast. The inset in (a) shows convergent beam electron diffraction (CBED) pattern recorded along the ⟨001⟩ zone axis over the region indicated by the white circle. The white scale bar corresponds to 50 nm. (b) High-resolution transmission electron microscopy (HRTEM) image of another GaN nanowire with a diameter of c. 10 nm. The [100], [010], and [−110] directions are indicated with the [100] parallel to the wire axis. The white scale bar corresponds to 5 nm. *Source:* Reprinted from Duan et al. [136].

results in negative impacts on the resulting performance of the as-grown nanowires. In order to overcome this bottleneck, an improved laser ablation technique, so-called oxide-assisted method, was developed to synthesize wide bandgap semiconductor nanowires without the use of catalysts. In this context, Lee and coworkers synthesized binary compound semiconductor nanowires using the compound materials mixed with an oxide as the target. In this approach, the critical role of the oxide in the target is to react with an element induced by the laser ablation of the target to form a volatile oxide. This chemical reaction is responsible for the nucleation and growth of nanowires, which differentiates the conventional oxide-assisted growth process based on the metal-catalyst VLS mechanism. In particular, the authors synthesized GaN nanowires using the oxide-assisted method by laser ablating a target of GaN mixed with gallium oxide. Their results showed that the as-grown GaN nanowires were smooth and straight with a core–sheath structure of 80 nm in average diameter and tens of micrometers in length [139, 140]. The elimination of the catalysts in this approach represents a huge benefit for the purification and subsequent integrations of the as-grown nanowires into numerous high performance electronic and photonic devices for diverse applications.

1.3.7 Thermal Evaporation

Thermal evaporation is one of the simplest variants of the physical vapor deposition (PVD) technique, which is commonly used to grow thin films of various inorganic and organic materials on solid substrates under vacuum conditions. In this process, the precursors are heated to evaporate in vacuum conditions and subsequently adsorbed on target substrates to form desired products. There are many parameters influencing on the outcome of this technique, such as vacuum levels, heating facilities (resistive or electron beam heating), decomposition temperatures and purification of the precursors, and substrate surface conditions [141–144]. Figure 1.23 shows a schematic of a typical thermal evaporation system including several basic components, such as manipulator, heater, substrate holder, vacuum chamber, and pumping system. Together with the deposition of a thin layer of materials on substrates, the thermal evaporation technique is also widely employed to grow of various wide bandgap semiconductor nanowires. The main advantages of the thermal evaporation technique are the possibility to control the synthesis outcomes, low-cost of the equipment, high growth rate, and excellent purity of the final products [71]. Notably, the thermal evaporation process can be driven by both the VS and VLS mechanisms depending on the choice of precursors and growth conditions. For instance, Yin and coworkers reported the synthesis of SiC nanowires on a Si substrate by the thermal evaporation technique without the assistance of catalysts. Their results showed that the SiC nanowires grew along the [111] direction and had diameters of 50–100 nm with lengths of several hundreds of microns. The SiC nanowire growth process in this study is driven by the VS mechanism [145]. On the other hand, Yong and coworker synthesized cubic SiC nanowires on a Si substrate by the thermal evaporation of WO_3 and graphite powders at high temperature using NiO catalyst. The authors were able to control the densities of the as-grown SiC nanowires by varying the NiO catalyst concentration. The synthesis process in this study is driven by the VLS mechanism with the assistance of NiO catalyst [146]. The thermal evaporation technique also offers advantages to control the doping profiles of the as-grown nanowires, which is expected to provide new platforms for the fabrication of novel photonic devices bearing unprecedented performances [147–149]. For example,

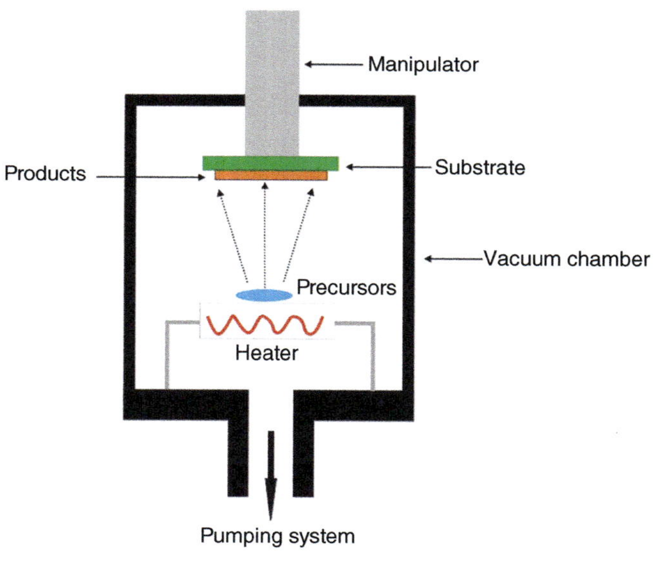

Figure 1.23 Schematic of a typical thermal evaporation system, showing main components, such as heater, pumping system, vacuum chamber, manipulator, and substrate holder.

Naseem and coworkers recently synthesized well-shaped Cr-doped GaN nanowires grown by the thermal evaporation method [148]. The authors found a strong electron–phonon coupling induced by Cr ion. The presence of ferromagnetic domains plays an important role in spin ordering due to mixed spin–spin, exciton–spin, and spin–carrier interaction. This finding could open-up new pathways toward the realization of new spin-polarized electronic devices.

1.3.8 Carbothermal Reduction

Beside advanced synthesis techniques widely employed to synthesize wide bandgap semiconductor nanowires, carbothermal reduction is also a currently used technique to achieve such nanowires due to its low-cost of equipment, simplicity of operation, and ability to scale-up at industrial levels. In this process, the chosen precursors are normally placed in a furnace and subsequently undergone carbothermic reactions at relatively high temperatures using carbon sources as the reducing agent. In order to control and maximize the outcome of the carbothermal reduction process, various parameters have to be considered in the synthesis, such as the mixing ratio of precursors, temperature set point, holding time, heating rate, cooling rate, and gas flow rate [150–152]. In this context, Choi and coworkers systematically examined the influences of various growth parameters on the synthesis of SiC nanowires by the carbothermal reduction technique using a mixture of low-purity SiO_2 and carbon [151]. The authors found that SiC nanowires with diameters of 60 nm and lengths of several hundred micrometers started to grow at 1400 °C for two hours in an argon flow rate of 2 l/min driven the VLS mechanism. The growth rate of SiC nanowires increased with increasing holding time and flow gas rate. Interestingly, no metal catalyst was found at the tip of the grown nanowires above 1500 °C due to the decomposition of iron metal at the tip at these high temperatures. This means that the carbothermal reduction process enables the possibility to remove the metal impurities without additional processes despite the growth driven by the VLS mechanism. This result is viewed as a huge advantage of this technique compared with other advanced techniques based on the VLS mechanism. Interestingly, the carbothermal reduction technique can be combined with another fabrication technique to produce high-quality wide bandgap semiconductor nanowires with unique characteristics in some cases, which cannot be achieved if it stands alone. For example, Qin and coworkers combined electrospinning and carbothermal reduction techniques to fabricate ultra-long and flexible SiC nanowires membrane [153]. In this study, the authors used silica sol and phenolic resin as the starting materials. Firstly, silica was mixed with phenolic resin in a molar ratio of silicon to carbon of 1 : 3 using ultrasonic dispersing and magnetic stirring. Subsequently, the mixed solution was dissolved in polyvinyl alcohol, followed by vigorously stirring for 12 hours to form a homogenous solution for the electrospinning. The electrospun precursor nanowires were collected on aluminum foil and cured in air for two hours at 210 °C. Finally, the cured nanowires were annealed at temperatures ranging from 1400 to 1600 °C in air for 60 minutes to achieve the flexible SiC nanowire membranes as shown in Figure 1.24. SEM and TEM images clearly showed that the as-grown SiC nanowires have a smooth surface along with an excellent flexibility without any breakage. This finding is very promising for extending the use of wide bandgap semiconductor nanowires in the development of new flexible electronic and photonic devices.

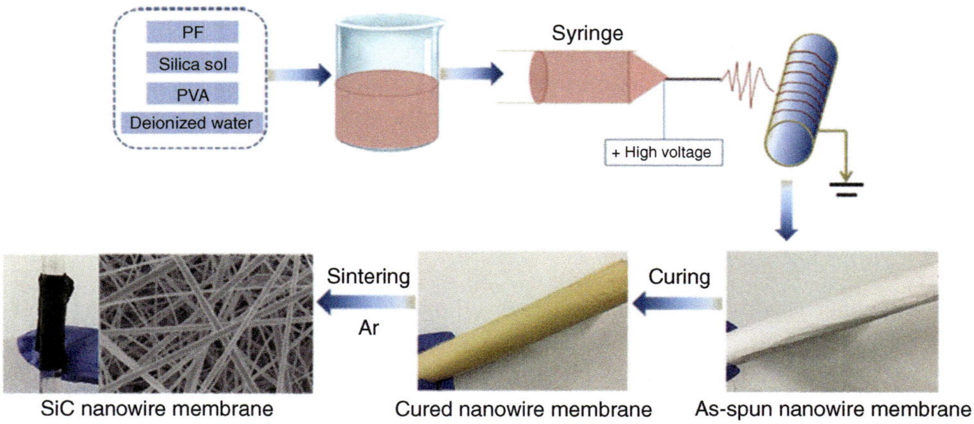

Figure 1.24 The schematic illustration of the combined electrospinning and carbothermal reduction to produce SiC nanowire membrane. *Source:* Reprinted from Wei et al. [153].

References

1 Pham, T.A., Qamar, A., Dinh, T. et al. (2020). Nanoarchitectonics for wide bandgap semiconductor nanowires: toward the next generation of nanoelectromechanical systems for environmental monitoring. *Science Advances* 7: 2001294.

2 Lu, W. and Lieber, C.M. (2006). Semiconductor nanowires. *Journal of Physics D: Applied Physics* 39: R387.

3 Dobrokhotov, V., Mcllroy, D.N., Grant Norton, M. et al. (2006). Principles and mechanisms of gas sensing by GaN nanowires functionalized with gold nanoparticles. *Journal of Applied Physics* 99: 104302.

4 Abdullah, Q.N., Yam, F.K., Hassan, Z., and Bououdina, M. (2014). Hydrogen gas sensing performance of GaN nanowires-based sensor at low operating temperature. *Sensors and Actuators B: Chemical* 204: 497–506.

5 Yang, N., Uetsuka, H., Osawa, E., and Nebel, C.E. (2008). Vertically aligned diamond nanowires for DNA sensing. *Angewandte Chemie, International Edition* 47 (28): 5183–5185.

6 Peng, X., Chu, J., Wang, L. et al. (2017). Boron-doped diamond nanowires for CO gas sensing application. *Sensors and Actuators B: Chemical* 241: 383–389.

7 Patolsky, F. and Lieber, C.M. (2005). Nanowire nanosensors. *Materials Today* 8 (4): 20–28.

8 He, B., Morrow, T.J., and Keating, C.D. (2008). Nanowire sensors for multiplexed detection of biomolecules. *Current Opinion in Chemical Biology* 12 (5): 522–528.

9 Hsu, J.F., Huang, B.R., Huang, C.S., and Chen, H.L. (2005). Silicon nanowires as pH sensor. *Japanese Journal of Applied Physics* 44: 2626.

10 Zhou, X.T., Hu, J.Q., Li, C.P. et al. (2003). Silicon nanowires as chemical sensors. *Chemical Physics Letters* 369: 220–224.

11 Toriyama, T., Tanimono, Y., and Sugiyama, S. (2002). Single crystal silicon nano-wire piezoresistors for mechanical sensors. *Journal of Microelectromechanical Systems* 11: 605.

12 He, R. and Yang, P. (2006). Giant piezoresistance effect in silicon nanowires. *Nature Nanotechnology* 1 (1): 42–46.

13 Hu, L., Yan, J., Liao, M. et al. (2011). Ultrahigh external quantum efficiency from thin SnO_2 nanowire ultraviolet photodetectors. *Small* 7 (8): 1012–1017.

14 LaPierre, R.R., Robson, M., Azizur-Rahman, K.M. and Kuyanov, P. (2017). A review of III–V nanowire infrared photodetectors and sensors. *Journal of Physics D: Applied Physics* 50: 123001.

15 Mathur, S., Barth, S., Shen, J. et al. (2005). Size-dependent photoconductance in SnO_2 nanowires. *Small* 1 (7): 713–717.

16 Ye, H., Lu, P., Yu, Z. et al. (2009). Critical thickness and radius for axial heterostructures nanowires using finite-element method. *Nano Letters* 9 (5): 1921–1925.

17 Kavanagh, K.L. (2010). Misfit dislocations in nanowire heterostructures. *Semiconductor Science and Technology* 25: 024006.

18 Biermanns, A., Breuer, S., Trampert, A. et al. (2012). Strain accommodation in Ga-assisted GaAs nanowires grown on silicon(111). *Nanotechnology* 23: 305703.

19 Dioxon, C.J. and Curtines, O.W. (2010). *Nanotechnology: Nanofabrication, Patterning and Self-Assembly*. New York: Nova Science Publishers.

20 Cui, Z. (2008). *Nanofabrication: Principles, Capabilities and Limits*. Springer.

21 Wagner, R.S. and Ellis, W.C. (1964). Vapor–liquid–solid mechanism of single crystal growth. *Applied Physics Letters* 4 (5): 89–90.

22 Milewski, J.V., Gac, F.D., Petrovic, J.J., and Skaggs, S.R. (1985). Growth of beta-silicon carbide whiskers by the VLS process. *Journal of Materials Science* 20: 1160.

23 Wacaser, B.A., Dick, K.A., Johansson, J. et al. (2009). Preferential interface nucleation: an expansion of the VLS growth mechanism for nanowires. *Advanced Materials* 21 (2): 153–165.

24 Kolasinski, K.W. (2006). Catalytic growth of nanowires: vapor–liquid–solid, vapor–solid–solid, solution–liquid–solid and solid–liquid–solid growth. *Current Opinion in Solid State & Materials Science* 10 (4): 182–191.

25 Hannon, J.B., Kodambaka, S., Ross, F.M., and Tromp, R.M. (2006). The influence of the surface migration of gold on the growth of silicon nanowires. *Nature* 440: 69–71.

26 Wang, F., Dong, A., Sun, J. et al. (2006). Solution–liquid–solid growth of semiconductor nanowires. *Inorganic Chemistry* 45 (19): 7511–7521.

27 Wang, Y., Schmidt, V., Senz, S., and Gösele, U. (2006). Epitaxial growth of silicon nanowires using an aluminium catalyst. *Nature Nanotechnology* 1: 186–189.

28 Wu, Y. and Yang, P. (2001). Direct observation of vapor–liquid–solid nanowire growth. *Journal of the American Chemical Society* 123 (13): 3165–3166.

29 Kim, B.J., Tersoff, J., Wen, C.Y. et al. (2009). Determination of size effects during the phase transition of a nanoscale Au–Si eutectic. *Physical Review Letters* 103: 155701.

30 Feng, W., Ma, J., and Yang, W. (2012). Precise control on the growth of SiC nanowires. *CrystEngComm* 14: 1210–1212.

31 Attolini, G., Rossi, F., Negri, M. et al. (2014). Growth of SiC NWs by vapor phase technique using Fe as catalyst. *Materials Letters* 124: 169–172.

32 Hsu, C.H. and Xu, J. (2012). Diamond nanowire – a challenge from extremes. *Nanoscale* 4: 5293–5299.

33 Cheng, G.S., Chen, S.H., Zhu, X.G. et al. (2000). Highly ordered nanostructures of single crystalline GaN nanowires in anodic alumina membranes. *Materials Science and Engineering A* 286 (1): 165–168.

34 Gottschalch, V., Wagner, G., Bauer, J. et al. (2008). VLS growth of GaN NWs on various substrates. *Journal of Crystal Growth* 310: 5132–5128.

35 Deng, S.Z., Wu, S.Z., Zhou, J. et al. (2002). Synthesis of silicon carbide nanowires in a catalyst-assisted process. *Chemical Physics Letters* 356 (5): 511–514.

36 Hsu, C.H., Cloutier, S.G., Palefsky, S., and Xu, J. (2010). Synthesis of diamond nanowires using atmospheric-pressure chemical vapor deposition. *Nano Letters* 10 (9): 3272–3276.

37 Wang, H., Xie, Z., Yang, W. et al. (2008). Morphology control in the vapor–liquid–solid growth of SiC nanowires. *Crystal Growth & Design* 8 (11): 3893–3896.

38 Men, J., Liu, Y., Luo, R. et al. (2016). Growth of SiC nanowires by low pressure chemical vapor infiltration using different catalysts. *Journal of the European Ceramic Society* 36 (15): 3615–3625.

39 Low, L.L., Yam, F.K., Beh, K.P., and Hassan, Z. (2011). The influence of growth temperatures on the characteristics of GaN nanowires. *Applied Surface Science* 258 (1): 542–546.

40 Krishnan, B., Thirumalai, R.V.K.G., Koshaka, Y. et al. (2011). Substrate-dependent orientation and polytype control in SiC nanowires grown on 4H-SiC substrates. *Crystal Growth & Design* 11 (2): 538–541.

41 Gao, F., Yang, W., Wang, H. et al. (2008). Controlled Al-doped single-crystalline 6H-SiC nanowires. *Crystal Growth & Design* 8 (5): 1461–1464.

42 Liu, J., Meng, X.M., Jiang, Y. et al. (2003). Gallium nitride nanowires doped with silicon. *Applied Physics Letters* 83: 4241.

43 Hemesath, E.R., Schreiber, D.K., Gulsoy, E.B. et al. (2012). Catalyst incorporation at defects during nanowire growth. *Nano Letters* 12 (1): 167–171.

44 Kehagias, T., Komninou, P., Dimitrakopulos, G.P. et al. (2008). Atomic-scale configuration of catalyst particles on GaN nanowires. *Physica Status Solidi C* 5: 3716–3719.

45 Shin, N., Chi, M., Howe, J.Y., and Filler, M.A. (2013). Rational defect introduction in silicon nanowires. *Nano Letters* 13 (5): 1928–1933.

46 Zhang, C., Liu, X., Li, J. et al. (2020). Investigation of catalyst-assisted growth of nonpolar GaN nanowires via a modified HVPE process. *Nanoscale* 12: 4293–4399.

47 Shen, G., Bando, Y., Ye, C. et al. (2006). Synthesis characterization and field-emission properties of bamboo-like β-SiC nanowires. *Nanotechnology* 17 (14): 3468.

48 Cheze, C., Geelhaar, L., Brandt, O. et al. (2010). Direct comparison of catalyst-free and catalyst-induced GaN nanowires. *Nano Research* 3: 528–536.

49 Zhou, K., Zhao, Z., Pan, L., and Wang, Z. (2019). Silicon nanowire pH sensors fabricated with CMOS compatible sidewall mask technology. *Sensors and Actuators B: Chemical* 279: 111–121.

50 Kim, D.M. and Jeong, Y.H. (2014). *Nanowire Field Effect Transistors: Principles and Applications*. New York: Springer.

51 Noborisaka, J., Motohisa, J., and Fukui, T. (2015). Catalyst-free growth, of GaAs nanowires by selective-area metalorganic vapor-phase epitaxy. *Applied Physics Letters* 86: 213102.

52 Paek, J.H., Nishiwaki, T., Yamaguchi, M., and Sawaki, N. (2009). Catalyst free MBE-VLS growth of GaAs nanowires on (111)Si substrate. *Physica Status Solidi C* 6 (6): 1436–1440.

53 Mattila, M., Hakkarainen, T., and Lipsanen, H. (2006). Catalyst-free growth of In(As)P nanowires on silicon. *Applied Physics Letters* 89: 063119.

54 Kamins, T.I., Williams, R.S., Chen, Y. et al. (2000). Chemical vapor deposition of Si nanowires nucleated by $TiSi_2$ islands on Si. *Applied Physics Letters* 76: 562.

55 Lensch-Falk, J.L., Hemesath, E.R., Perea, D.E., and Lauhon, L.J. (2009). Alternative catalyst for VSS growth of silicon and germanium nanowires. *Journal of Materials Chemistry* 19: 849–857.

56 Dick, K.A., Bolinsson, J., Borg, B.M., and Johansson, J. (2012). Controlling the abruptness of axial heterojunctions in III–V nanowires: beyond the reservoir effect. *Nano Letters* 12 (6): 3200–3206.

57 Wen, C.W., Reuter, M.C., Bruley, J. et al. (2009). Formation of compositionally abrupt axial heterojunctions in silicon-germanium nanowires. *Science* 326 (5957): 1247–1250.

58 Kodambaka, S., Tersoff, J., Reuter, M.C., and Ross, F.M. (2007). Germanium nanowire growth below the eutectic temperature. *Science* 316 (5825): 729–732.

59 Chen, L., Lu, W., and Lieber, C.M. (2014). Chapter 1: Semiconductor nanowire growth and integration. In: *Semiconductor Nanowires: From Next-Generation Electronics to Sustainable Energy*, 1–53. https://doi.org/10.1039/9781782625209-00001.

60 Hou, W.C., Chen, L.Y., Tang, W.C., and Hong, F.C N. (2011). Control of seed detachment in Au-assisted GaN nanowires growths. *Crystal Growth & Design* 11 (4): 990–994.

61 Weng, X., Burke, R.A., and Redwing, J.M. (2009). The nature of catalyst particles and growth mechanisms of GaN nanowires grown by Ni-assisted metal-organic chemical vapor deposition. *Nanotechnology* 20: 085610.

62 Cui, H., Lu, Y.Y., Wang, G.W. et al. (2015). Step-flow kinetics model for the vapor–solid–solid Si nanowires growth. *Nano Letters* 15 (5): 3640–3645.

63 Maliakkal, C.B., Hatui, N., Bapat, R.D. et al. (2016). The mechanism of Ni-assisted GaN nanowires growth. *Nano Letters* 16 (12): 7632–7638.

64 Cheze, C., Geelhaar, L., Trampert, A. et al. (2010). Collector phase transitions during vapor–solid–solid nucleation of GaN nanowires. *Nano Letters* 10 (9): 3426–3431.

65 Fan, H.J., Werner, P., and Zacharias, M. (2006). Semiconductor nanowires: from self-organization to patterned growth. *Small* 2 (6): 700–717.

66 Suo, G., Jiang, S., Zhang, J. et al. (2004). Synthetic strategies and application of GaN nanowires. *Advances in Condensed Matter Physics* 2014: 456163.

67 Hsu, Y.J. and Lu, S.Y. (2005). Vapor–solid growth of Sn nanowires: growth mechanism and superconductivity. *The Journal of Physical Chemistry B* 109 (10): 4398–4403.

68 Chen, J., Shi, Q., Xin, L. et al. (2011). A simple catalyst-free route for large-scale synthesis of SiC nanowires. *Journal of Alloys and Compounds* 509 (24): 6844–6847.

69 Yazda, G.R., Persson, P.O., Gogova, D. et al. (2009). Aligned AlN nanowires by self-organized vapor–solid growth. *Nanotechnology* 20: 495304.

70 Zhou, S.M., Feng, Y.S., and Zhang, L.D. (2003). A physical evaporation synthetic route to large-scale GaN nanowires and their dielectric properties. *Chemical Physics Letters* 369: 610–614.

71 Wu, R., Li, B., Gao, M. et al. (2008). Tuning the morphologies of SiC nanowires via the control of growth temperature, and their photoluminescence properties. *Nanotechnology* 19: 335602.

72 Shen, G., Chen, D., Chen, P.C., and Zhou, C. (2009). Vapor–solid growth of one-dimensional layer-structured gallium sulfide nanostructures. *ACS Nano* 3 (5): 1115–1120.

73 Trentler, T.J., Hickman, K.M., Goel, S.C. et al. (1995). Solution–liquid–solid growth of crystalline III–V semiconductors: an analogy to vapor–liquid–solid growth. *Science* 270 (5243): 1791–1794.

74 Wang, F., Dong, A., and Buhro, W.E. (2016). Solution–liquid–solid synthesis, properties, and applications of one-dimensional colloidal semiconductor nanorods and nanowires. *Chemical Reviews* 116 (8): 10888–10933.

75 Dasgupta, N.P., Sun, J., Liu, C. et al. (2014). 25th anniversary article: semiconductor nanowires – synthesis, characterization, and applications. *Advanced Materials* 26 (14): 2137–2184.

76 Yu, H., Gibbons, P.C., Kelton, K.F., and Buhro, W.E. (2001). Heterogeneous seeded growth: a potentially general synthesis of monodisperse metallic nanoparticles. *Journal of the American Chemical Society* 123 (37): 9198–9199.

77 Hang, Q., Wang, F., Carpenter, P.D. et al. (2008). Role of molecular surface passivation in electrical transport properties of InAs nanowires. *Nano Letters* 8 (1): 49–55.

78 Sun, J., Liu, C., and Wang, P. (2011). Surfactant-free, large-scale, solution–liquid–solid growth of gallium phosphide nanowires and their use for visible light driven hydrogen production from water reduction. *Journal of the American Chemical Society* 133 (48): 19306–19309.

79 Wang, F., Wayman, V.L., Loomis, R.A., and Buhro, W.E. (2011). Solution–liquid–solid growth of semiconductor quantum-wire films. *ACS Nano* 5 (6): 5188–5194.

80 Quyang, L., Maher, K.N., Yu, C.L. et al. (2007). Catalyst-assisted solution–liquid–solid synthesis of CdS/CdSe nanorod heterostructures. *Journal of the American Chemical Society* 129 (1): 133–138.

81 Yang, G., Wu, R., Chen, J. et al. (2007). Growth of SiC nanowires/nanorods using a Fe–Si solution method. *Nanotechnology* 18: 155601.

82 Sun, L., Yuan, G., Gao, L. et al. (2021). Chemical vapour deposition. *Nature Reviews Methods Primers* 1: 1–5.

83 Pierson, H.O. (1999). *Handbook of Chemical Vapor Deposition: Principles, Technology, and Applications*. New Jersey: Noyes Publications.

84 Chen, C.C., Yeh, C.C., Chen, C.H. et al. (2001). Catalytic growth and characterization of gallium nitride nanowires. *Journal of the American Chemical Society* 123 (12): 2791–2798.

85 ElAhl, A.M.S., He, M., Zhou, P., and Harris, G.L. (2003). Systematic study of effects of growth conditions on the (nano-, meso-, micro) size and (one-, two-, three-dimensional) shape of GaN single crystals grown by a direct reaction of Ga with ammonia. *Journal of Applied Physics* 94: 7749.

86 Seo, H.W., Bae, S.Y., and Park, J. (2002). Strained gallium nitride nanowires. *The Journal of Chemical Physics* 116: 9492.

87 Chen, C.C. and Yeh, C.C. (2000). Large-scale catalytic synthesis of crystalline gallium nitride nanowires. *Advanced Materials* 12 (10): 738–741.

88 Cheng, H., Li, J., Wu, D. et al. (2015). Effects of precursor distances on the growth of GaN nanowires. *Journal of Nanomaterials* 2015 (2): 1–6. https://doi.org/10.1155/2015/343541.

89 Han, S., Jin, W., Tang, T. et al. (2003). Controlled growth of gallium nitride single-crystal nanowires using a chemical vapor deposition method. *Journal of Materials Research* 18 (2): 245–249.

90 Peng, H.Y., Wang, N., Zhou, X.T. et al. (2002). Control of growth orientation of GaN nanowires. *Chemical Physics Letters* 359 (3): 241–245.

91 Lin, L. (2011). Synthesis and optical property of large-scale centimetres-long silicon carbide nanowires by catalyst-free CVD route under superatmospheric pressure conditions. *Nanoscale* 3: 1582–1591.

92 Huang, Z., Liu, H., Chen, K. et al. (2014). Synthesis and formation mechanism of twinned SiC nanowires made by a catalyst-free thermal chemical vapor deposition method. *RSC Advances* 4: 18360–18364.

93 Dapkus, P.D. (1982). Metalorganic chemical vapor deposition. *Annual Review of Materials Science* 12: 243–269.

94 Williams, J.O. (1989). Metal organic chemical vapor deposition (MOCVD) perspectives and prospects. *Angewandte Chemie, International Edition* 28 (8): 1110–1120.

95 Kuykendall, T., Pauzauskie, P., Lee, S. et al. (2003). Metalorganic chemical vapor deposition route to GaN nanowires with triangular cross sections. *Nano Letters* 3 (8): 1063–1066.

96 Jung, H.S., Hong, J., Li, Y. et al. (2008). Photocatalysis using GaN nanowires. *ACS Nano* 2 (4): 637–642.

97 Navamathavan, R., Ra, Y.H., Song, K.Y. et al. (2011). Different growth behaviors of GaN nanowires grown with Au catalyst and Au + Ga solid solution nano-droplets on Si(111) substrate sby using MOCVD. *Current Applied Physics* 11 (1): 77–81.

98 Alloing, B. and Perez, J.Z. (2016). Metalorganic chemical vapor deposition of GaN nanowires: from catalyst-assisted to catalyst-free growth, and from self-assembled to selective-area growth. *Materials Science in Semiconductor Processing* 55 (15): 51–58.

99 Foltynski, B., Giesen, C., and Heuken, M. (2015). Self-organized growth of catalyst-free GaN nano- and micro-rods on Si(111) substrates by MOCVD. *Physica Status Solidi B: Basic Solid State Physics* 252 (5): 1132–1137.

100 Jung, B.O., Bae, S.Y., Kato, Y. et al. (2014). Morphology development of GaN nanowires using a pulsed-mode MOCVD growth technique. *CrystEngComm* 16: 2273–2282.

101 Meyyappan, M., Delzeit, L., Cassel, A., and Hash, D. (2003). Carbon nanotube growth by PECVD: a review. *Plasma Sources Science and Technology* 12 (2): 205.

102 Bo, Z., Yang, Y., Chen, J. et al. (2013). Plasma-enhanced chemical vapor deposition synthesis of vertically oriented graphene nanosheets. *Nanoscale* 5: 5180–5204.

103 Vasudev, M.C., Anderson, K.D., Bunning, T.J. et al. (2013). Exploration of plasma-enhanced chemical vapor deposition as a method for thin-film fabrication with biological applications. *ACS Applied Materials & Interfaces* 5: 3983–3994.

104 Gohlampour, M., Zadeh, A.A., Shekari, L., and Poursalehi, R. (2015). A catalyst free method to grow GaN nanowires on porous Si at low temperature. *Ceramics International* 41 (10): 13855–13860.

105 Feng, X.Y., Wang, R.Z., Liang, Q. et al. (2019). Direct growth of GaN nanowires by Ga and N_2 without catalysis. *Crystal Growth & Design* 19 (5): 2687–2694.

106 Ji, Y.H., Wang, R.Z., Feng, X.Y. et al. (2017). Modulation effects of hydrogen on structure and photoluminescence of GaN nanowires prepared by plasma-enhanced chemical vapor deposition. *The Journal of Physical Chemistry* 121 (44): 24804–24808.

107 Hou, W.C. and Hong, F.C.N. (2009). Controlled surface diffusion in plasma-enhanced chemical vapor deposition of GaN nanowires. *Nanotechnology* 20 (5): 055606.

108 Zhao, J.W., Zhang, Y.F., Li, Y.H. et al. (2015). A low cost, green method to synthesize GaN nanowires. *Scientific Reports* 5: 17692.

109 Hu, J., Wei, H., Yang, S. et al. (2019). Advances in new generation diluted magnetic semiconductors with independent spin and charge doping. *Journal of Semiconductors* 40: 101181.

110 Molnar, R.J., Gotz, W., Romano, L.T., and Johnson, N.M. (1997). Growth of gallium nitride by hydride vapor-phase epitaxy. *Journal of Crystal Growth* 178 (1): 147–156.

111 Imnishi, M., Yoshida, T., Kitamura, T. et al. (2017). Homoepitaxial hydride vapor phase epitaxy growth on GaN wafers manufactured by the Na-flux method. *Crystal Growth & Design* 17 (7): 3806–3811.

112 Avit, G., Lekhal, K., Andre, Y. et al. (2014). Ultralong and defect-free GaN nanowires grown by the HVPE process. *Nano Letters* 14 (2): 559–562.

113 Byeun, Y.K., Telle, R., Jung, S.H. et al. (2010). The growth of one-dimensional single-crystalline AlN nanostructures by HVPE and their field emission properties. *Chemical Vapor Deposition* 16 (1): 72–79.

114 Seryogin, G., Shalish, I., Moberlychan, W., and Narayanamurti, V. (2005). Catalytic hydride vapour phase epitaxy growth of GaN nanowires. *Nanotechnology* 16: 2342.

115 Wu, S., Wang, L., Liu, Z. et al. (2018). Ultrafast growth of horizontal GaN nanowires by HVPE through flipping the substrate. *Nanoscale* 10: 5888–5896.

116 Wu, S., Wang, L., Yi, X. et al. (2018). Crystallographic orientation control and optical properties of GaN nanowires. *RSC Advances* 8: 2181–2187.

117 Lee, T., Kim, J.H., Choi, Y.J. et al. (2019). Polarized Raman studies of single GaN nanowire and GaN/AlN hetero-nanowire structures. *Thin Solid Films* 671 (1): 147–151.

118 Trassoudaine, A., Roche, E., Bougerol, C. et al. (2016). Spontaneous formation of GaN/AlN core–shell nanowires on sapphire by hydride vapor phase epitaxy. *Journal of Crystal Growth* 454 (15): 1–5.

119 Cho, A.Y. and Arthur, J.R. (1975). Molecular beam epitaxy. *Progress in Solid State Chemistry* 10 (3): 157–191.

120 Joyce, B.A. (1985). Molecular beam epitaxy. *Reports on Progress in Physics* 48: 1637.

121 Arthur, J.A. (2002). Molecular beam epitaxy. *Surface Science* 500 (1): 189–217.

122 Debnath, R.K., Meijers, R., Richter, T. et al. (2007). Mechanism of molecular beam epitaxy growth of GaN nanowires on Si(111). *Applied Physics Letters* 90: 123117.

123 Mata, R., Hestroffer, K., Budagosky, J. et al. (2011). Nucleation of GaN nanowires grown by plasma-assisted molecular beam epitaxy: the effect of temperature. *Journal of Crystal Growth* 334 (1): 177–180.

124 Landre, O., Songmuang, R., Renard, J. et al. (2008). Plasma-assisted molecular beam epitaxy growth of GaN nanowires using indium-enhanced diffusion. *Applied Physics Letters* 93: 183109.

125 Sobanska, M., Klosek, K., Borysiuk, J. et al. (2014). Enhanced catalyst-free nucleation of GaN nanowires on amorphous Al_2O_3 by plasma-assisted molecular beam epitaxy. *Journal of Applied Physics* 115: 043517.

126 Landre, O., Bougerol, C., Renevier, H., and Daudin, B. (2009). Nucleation mechanism of GaN nanowires grown on (111) Si by molecular beam epitaxy. *Nanotechnology* 20: 415602.

127 Bertness, K.A., Roshko, A., Mansfield, L.M. et al. (2007). Nucleation conditions for catalyst-free GaN nanowires. *Journal of Crystal Growth* 300 (1): 94–99.

128 Sekiguchi, H., Nadazato, T., Kikuchi, A., and Kishino, K. (2007). Structural and optical properties of GaN nanocolumns grown on (0001) sapphire substrates by rf-plasma-assisted molecular beam epitaxy. *Journal of Crystal Growth* 300 (1): 259–262.

129 Brubaker, M.D., Levin, I., Davydov, A.V. et al. (2011). Effect of AlN buffer layer properties on the morphology and polarity of GaN nanowires grown by molecular beam epitaxy. *Journal of Applied Physics* 110: 053506.

130 Stoica, T. and Calarco, R. (2011). Doping III-nitride nanowires grown by molecular beam epitaxy. *IEEE Journal of Selected Topics in Quantum Electronics* 17: 859.

131 Limbach, F., Caterino, R., Gotschke, T. et al. (2012). The influence of Mg doping on the nucleation of self-induced GaN nanowires. *AIP Advances* 2: 012157.

132 Furtmayr, F., Vielemeyer, M., Stutzmann, M. et al. (2008). Optical properties of Si- and Mg-doped gallium nitride nanowires grown by plasma-assisted molecular beam epitaxy. *Journal of Applied Physics* 104: 074309.

133 Connie, A.T., Zhao, S., Sadaf, S.M. et al. (2015). Optical and electrical properties of Mg-doped AlN nanowires grown by molecular beam epitaxy. *Applied Physics Letters* 106: 213105.

134 Yoshida, S., Misawa, S., and Gonda, S. (1983). Epitaxial growth of GaN/AlN heterostructures. *Journal of Vacuum Science and Technology B* 1: 250.

135 Smorchakova, I.P., Keller, S., Heikman, S. et al. (2000). Two-dimensional electron-gas AlN/GaN heterostructures with extremely thin AlN barriers. *Applied Physics Letters* 77: 3998.

136 Duan, X. and Lieber, C.M. (2000). Laser-assisted catalytic growth of single crystal GaN nanowires. *Journal of the American Chemical Society* 122 (1): 188–189.

137 Shi, W., Zheng, Y., Peng, H. et al. (2004). Laser ablation synthesis and optical characterization of silicon carbide nanowires. *Journal of the American Ceramic Society* 83 (12): 3228–3230.

138 Ng, D.K.T., Tan, L.S., and Hong, M.H. (2006). Synthesis of GaN nanowires on gold-coated substrates by pulsed laser ablation. *Current Applied Physics* 6 (3): 403–406.

139 Shi, W.S., Zheng, Y.F., Wang, N. et al. (2001). Microstructures of gallium nitride nanowires synthesized by oxide-assisted method. *Chemical Physics Letters* 345 (5): 377–380.

140 Shi, W.S., Zheng, Y.F., Wang, N. et al. (2001). A general synthetic route to III–V compound semiconductor nanowires. *Advanced Materials* 13 (8): 591–594.

141 Pham, T.A., Song, F., and Stöhr, M. (2014). Supramolecular self-assembly of metal-free naphthalocyanine on Au(111). *Physical Chemistry Chemical Physics* 16: 8881–8885.

142 Pham, T.A., Tran, B.V., Nguyen, M.T., and Stöhr, M. (2017). Chiral-selective formation of 1D polymers based on Ullmann-type coupling: the role of the metallic substrate. *Small* 13 (13): 1603675.

143 Pham, T.A., Song, F., Alberti, M.N. et al. (2015). Heat-induced formation of one-dimensional coordination polymers on Au(111): an STM study. *Chemical Communications* 51: 14473–14476.

144 Pham, T.A., Song, F., Nguyen, M.T., and Stöhr, M. (2014). Self-assembly of pyrene derivatives on Au(111): substituent effects on intermolecular interactions. *Chemical Communications* 50: 14089.

145 Tao, K., Gong, Y., and Lin, J. (2018). Low-temperature synthesis of NiS/MoS_2/C nanowires/nanoflakes as electrocatalyst for hydrogen evolution reaction in alkaline medium via sulfurizing metal-organic frameworks. *Electrochimica Acta* 274: 74–83.

146 Senthil, K. and Yong, K. (2008). Enhanced field emission from density-controlled SiC nanowires. *Materials Chemistry and Physics* 112 (1): 88–93.

147 Das, S.N., Patra, S., Kar, J.P. et al. (2013). Growth and characterization of Mg-doped GaN nanowire synthesized by the thermal evaporation method. *Materials Letters* 106 (1): 352–355.

148 Farooq, M.U., Zou, B., Ghaffar, A. et al. (2019). The contribution of Cr(III)-doping on the modulation of magnetic and luminescence properties of GaN nanowires. *Superlattices and Microstructures* 132: 106159.

149 Xu, C., Chung, S., Kim, M. et al. (2005). Doping of Si into GaN nanowires and optical properties of resulting composites. *Journal of Nanoscience and Nanotechnology* 5: 530–535.

150 Shen, Z., Chen, J., Li, B. et al. (2019). A novel two-stage synthesis for 3C-SiC nanowires by carbothermic reduction and their photoluminescence properties. *Journal of Materials Science* 54: 12450–12462.

151 Lee, J.S., Byeun, Y.K., Lee, S.H., and Choi, S.C. (2008). In situ growth of SiC nanowires by carbothermal reduction using a mixture of low-purity SiO_2 and carbon. *Journal of Alloys and Compounds* 456 (1): 257–263.

152 Chen, Z., Cao, C.B., and Zhu, H.S. (2007). Oriented bicrystalline GaN nanowire arrays suitable for field emission applications. *Chemical Vapor Deposition* 13: 527–532.

153 Wei, J., Li, X., Wang, Y. et al. (2020). Photoluminescence property of inexpensive flexible SiC nanowires membrane by electrospinning and carbothermal reduction. *Journal of the American Ceramic Society* 103 (11): 6187–6197.

2

Top-Down Fabrication Processes

2.1 Introduction

Wide bandgap semiconductor nanowires have been emerging as one of the most important building blocks for the development of complex multifunctional electronic devices with diverse applications, covering various engineering fields from energy, CMOS electronics, sensing to biomedical technologies [1–5]. With the continuous and rapid development of science and technology over the past decades, various advanced fabrication methods and techniques have been effectively employed to synthesize and fabricate high-quality wide bandgap semiconductor nanowires. These methods can be divided into two major categories: bottom-up and top-down approaches depending on the processing steps involved in creating the final nanostructures. In general, the bottom-up approach is an additive process in which atoms and molecules are the basic building blocks for the formation of complex and sophisticated nanostructures. On the other hand, the top-down approach is a subtractive process, where excess material is physically or chemically removed to produce the desired objects with a controllable shape and size. Compared with the bottom-up approach, the top-down approach offers several huge advantages for mass-producing of wide bandgap semiconductor nanowires [6–10]. For example, unlike the bottom-up approach where the grown nanowires might be distributed randomly after fabricating, the top-down approach enables precise positioning to produce highly aligned nanowire arrays across the wafer and thus offers the large-scale integration for high-performance nano-electronic circuits with standard industry processes. Nevertheless, the resolution limitation and high operation cost are the main challenges for the use of the top-down technology in the fabrication of wide bandgap semiconductor nanowires. However, with the huge progress on advanced manufacturing techniques in recent years, it is expected that these drawbacks will be not main challenges to achieve high-quality wide bandgap semiconductor nanowires at a relatively cost-effective rate anymore.

To date, various top-down techniques have been employed to fabricate a wide range of wide bandgap semiconductor nanowires bearing complex nanostructures. The most common top-down approach to the nanowire fabrication involves lithographic patterning techniques using a short wavelength light produced from UV or X-ray sources [11–13]. It should be noted that the conventional photolithographic techniques only can make patterns with a resolution of about 1 μm. For applications requiring a better resolution of nanostructures,

Wide Bandgap Nanowires: Synthesis, Properties, and Applications, First Edition. Tuan Anh Pham, Toan Dinh, Nam-Trung Nguyen, and Hoang-Phuong Phan.

researchers and engineers need employ more advanced photolithographic techniques. In this regard, scanning beam techniques such as electron beam lithography (EBL) offers a great opportunity to achieve patterns with a resolution of about 20 nm. This technique is basically based on the sweeping a finely focused electron beam across the wafer, which allows high consistency and high controllability on patterns size and shape [14, 15]. However, the major shortcoming of EBL comes from the high cost of equipment and limited access to the fabrication facilities. Besides, focused ion beams (FIB) can also be used for direct-write patterning at a high level of resolution. This technique is almost similar compared with EBL in its advantages and limitations. The main difference between EBL and FIB is that FIB patterning employs high-mass ions as the energy carriers, instead of electrons in EBL and therefore, FIB patterning is an inherently destructive process [16–18]. On the other hand, the lithographic techniques can be used not only alone, but also in combination with other top-down fabrication techniques such as dry etching or chemical wet etching methods for scalable, size and shape-controlled fabrication of an array of wide bandgap semiconductor nanowires [19–21]. Compared with the wet etching methods, the dry etching methods offer huge beneficial in the fabrication of nanowires, such as the ability of automation, the less of material consumption, and the flexibility to tune the etching outcome via using different etch gases. Among various dry etching methods, reactive ion etching (RIE) is widely used in the nanowire fabrication processes.

In this chapter, we will present a brief review of several top-down techniques, which have been widely employed to fabricate high-quality semiconductor nanowires. After the introduction in the Section 2.1, the Section 2.2 discusses recent progresses on advanced photolithographic as well as dry etching techniques. The focus is placed on a brief introduction of the working principle of these top-down techniques. Moreover, several outstanding results of the top-down nanowire fabrication obtained in the recent years will be highlighted. Finally, Section 2.3 deals with the possibility to combine different top-down techniques, which is expected to enhance the efficiency of the top-down approach for producing high-quality crystalline nanowires of wide bandgap semiconductor materials.

2.2 Top-Down Fabrication Techniques

2.2.1 Focused Ion Beam

Among various popular top-down nanofabrication techniques in semiconductor technology, FIB is one of the most powerful tools for surface modifications and prototyping functional structures of semiconductor materials at the micro- and nanoscale [22, 23]. The principle of FIB is almost identical to scanning electron microscopy but uses a beam of ions in place of electrons. It applies highly FIB to modify the target surface via the sputtering process. By properly controlling the energy and intensity of the ion beams, FIB offers a great opportunity to perform very precise nanomachining for producing sophisticated nanostructures [24, 25]. The basic components of a standard FIB system include the ion source, lens system, suppressor, extractor, and MCP for imaging are shown in Figure 2.1. Inspired by these advantages, FIB has been widely employed to fabricate wide bandgap semiconductor nanowires out of pre-patterned microstructures for the development of sensing and optoelectronic devices [26–28]. For example, Phan et al. investigated the piezoresistive effect of 3C-SiC nanowires fabricated using the FIB technique for mechanical sensing applications [29]. This

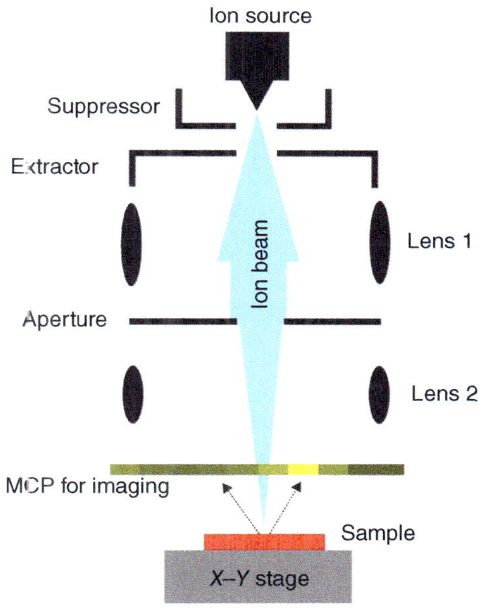

Figure 2.1 Schematic illustration of a typical FIB system, showing important components of the system: Ion source, suppressor, extractor, lens, X–Y stage.

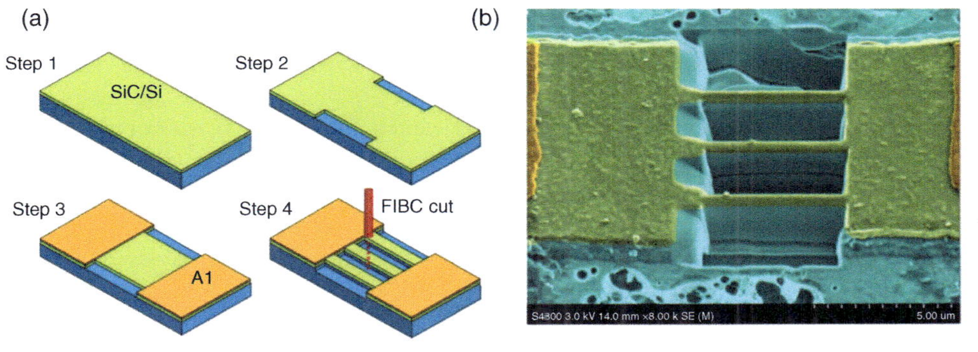

Figure 2.2 (a) Schematic illustration of the fabrication of SiC NWs using focused ion beam and (b) the corresponding SEM image of the as-fabricated SiC NWs in which the thickness, width, and length of the NWs were 300 nm, 300 nm, and 5 μm, respectively. *Source:* Reprinted from Phan et al. [29].

finding demonstrated a new pathway to mass production of SiC nanowires on Si wafers and their integrations with CMOS devices. Detailed information of the processing steps for the top-down fabrication of SiC nanowires using FIB and the corresponding SEM image of resulting SiC nanowires are shown in Figure 2.2. In detail, a SiC thin film with a thickness of 300 nm was grown on a Si wafer using the low-pressure CVD method. Afterwards, a thin layer of aluminum was deposited and patterned to form electrode pads. Finally, the SiC nanowires were formed using FIB, in which gallium ion beam was used to bombard the target material.

2.2.2 Electron Beam Lithography

Along with the FIB, EBL is also a well-known technique for designing and creating wide bandgap semiconductor nanowires, which are impossible to achieve with traditional photolithography methods due to the diffraction limits of light [30–32]. A typical EBL system

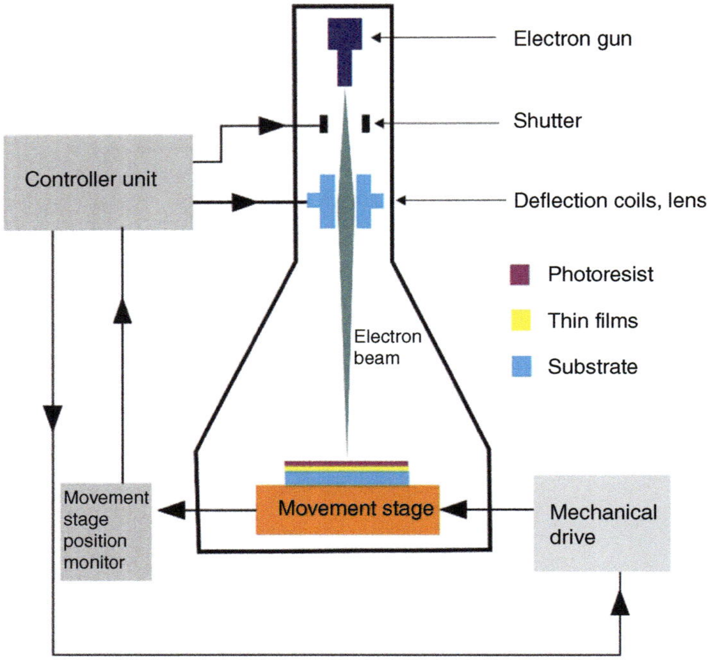

Figure 2.3 A schematic view of a typical electron beam lithography setup.

consists of a column of electron optics for producing and tuning the beam, control electronics, and a sample stage. Figure 2.3 shows main components of a typical EBL system, including electron gun, shutter, deflection coils, lens, movement stage, and controller unit. The EBL technique allows the creation of nanoscale patterns by directly modifying the solubility of a resist material that is sensitive to electron beams. The resolution of EBL depends on the spot size of the focused beam and the scattering of primary and secondary electrons in the resist film and the substrate. The main advantage of this technique is the ultra-high resolution of sub-10 nm that can be achieved via the well-controlled process [33].

EBL has been used as a standalone tool, or in combination with other top-down techniques for manufacturing of wide bandgap semiconductor nanowires integrated into sensing devices. In one of the early studies, Juhasz et al. reported a precisely controlled fabrication of SiC nanowires with their widths approaching the 10-nm regime using a combined process of EBL and electrochemical size reduction [34]. More interestingly, Roukes and coworkers developed very thin SiC nanowires using two different top-down nanofabrication processes based on EBL for low voltage nanoelectromechanical switches, as shown in Figure 2.4 [35]. These processes enable the thinnest nanowires as well as air gaps in their devices with thickness and widths of 16–25 nm and a minimum gap of 9 nm. The extremely tiny air gap and widths allowed for an excellent contact-mode in nanomechanical switches, offering very low turn-on voltages down to the level of approximately 1 V along with exceptionally short switching times in the sub-microsecond range.

2.2.3 Reactive Ion Etching

RIE is a dry etching technique employing chemically reactive plasma to etch specific deposited thin films away from the mother substrate surface. RIE is a synergistic process that

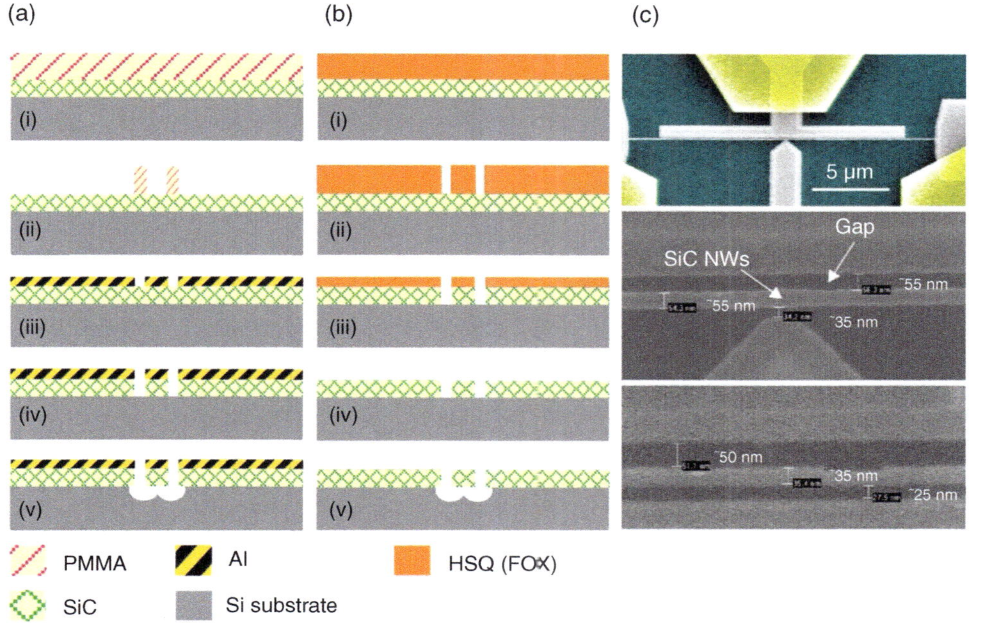

Figure 2.4 (a, b) Illustrations of the lift-off and negative-mask surface nanomachining processes for making the SiC nanowires using EBL and (c) a 20 μm long SiC nanowires with a 15 μm wide lateral gate and a point-contact gate, respectively. *Source:* Reprinted from Feng et al. [35].

involves both physical and chemical parts: energetic ion bombardment and chemically reactive species, respectively [36, 37]. This etching technique has been widely used for various semiconductors in device manufacturing. A key merit of RIE compared with other types of etching is that RIE offers an anisotropic etching capability, which reduce lateral etch rate and thus, enabling vertical sidewalls of desired structures. Besides, RIE exhibits several disadvantages such as a slow etch rate and a high cost for operation and maintenance [38–40]. In RIE process, the plasma is generated under vacuum environment by an electromagnetic field at a certain frequency (normally at 13.56 MHz) to the cathode, whereas the anode is grounded. The electric field creates the plasma upon ionizing the gas molecules. These ions are accelerated toward and then react at the surface of thin films being etched, eliminating gaseous materials. On the other hand, if the energy of these ions is high enough, they can knock atoms out from the surface of thin films being etched without any chemical reactions. Because the chemical etching part in RIE is isotropic and the physical etching part is anisotropic, optimizing these etching parts offer a huge possibility to achieve the sidewalls of the structures having diverse shapes at a high level of resolution. Figure 2.5 shows a schematic view of a typical RIE setup consisting of vacuum chambers, cathode, anode, pump, and sample holders.

In recent years, several research groups have employed RIE to produce wide bandgap semiconductor nanowires. For instance, Hardtdegen and coworkers reported the fabrication of GaN nanowires using RIE of maskless GaN layers [41]. According to their results, densely packed nanowires with a length of more than 1 μm and a diameter of about 60 nm were achieved by RIE of a 2.5 μm thick GaN layer. The authors proposed that the formation of GaN nanowires was originated from a defect selective etching process. In another example, Loncar and coworkers reported a top-down fabrication method for realizing diamond nanowires in both bulk single crystal and polycrystalline diamond using an inductively

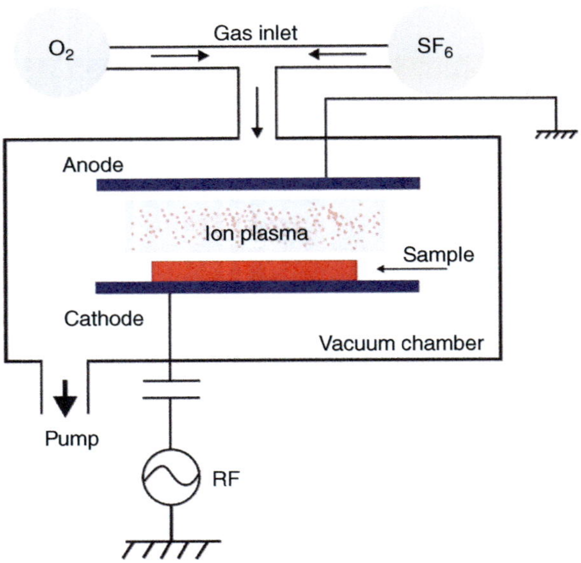

Figure 2.5 A schematic view of a typical reactive ion etching setup.

coupled plasma RIE technique [42]. The authors were able to obtain diamond nanowires with near-vertical sidewalls in both the polycrystalline and single crystal diamond. According to their results, the heights and diameters of the polycrystalline nanowires were approximately 1 μm and 120–340 nm, respectively, obtained at an etching rate of 200 nm/min. In the case of single crystal diamond nanowires, the heights and diameters were 1–2.4 μm and 120–490 nm, respectively, achieved with etching rates between 190 and 240 nm/min. In some cases, RIE can be combined with wet chemical etching to produce high-quality wide bandgap semiconductor nanowires. For example, Peiner and coworkers fabricated vertically aligned GaN nanowires for the development of novel optoelectronic devices using RIE and wet etching techniques [43]. In this work, after RIE etching, the GaN nanowires were subsequently treated in different wet chemical etching solutions to form hexagonal and smooth sidewalls. Their experimental results revealed that the orientations of nonpolar GaN nanowire sidewalls are dependent on the etchant concentration. Importantly, the authors proposed a model concerning surface bonding configurations on crystallography facets to explain the anisotropic wet etching mechanism, which is responsible for the formation of the GaN nanowires in this study.

2.3 Combined Lithography Techniques

Despite the FIB, EBL and RIE techniques exhibit many huge advantages for the top-down fabrication of wide bandgap semiconductor nanowires, and these techniques cannot maximize the flexibility and efficiency in producing high-performance wide bandgap semiconductor nanowires if they work individually. Therefore, the combination of different conventional lithography techniques is expected to create and establish reliable and efficient protocols that advance the top-down fabrication of wide bandgap semiconductor nanowires [44–46]. Examples of this approach include the work of Yang et al., which formed vertically aligned diamond nanowires with precisely controllable geometrical properties using

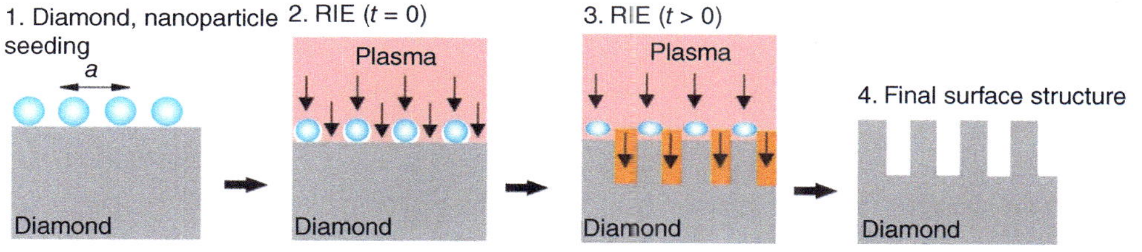

Figure 2.6 Schematic illustration of the fabrication of diamond nanowires using a combination of EBL and reactive ion etching techniques with nano-diamond particles as a hard mask. *Source:* [45], Yang et al., (2008) /American Chemical Society.

nano-diamond particles as a hard mask for subsequent anisotropic RIE [45]. This etching mask in the form of diamond nanoparticles was deposited on a single crystalline diamond film and then, RIE was applied to remove unprotected areas, as shown in Figure 2.6. As a result, vertically aligned diamond nanowires with optimized geometrical dimensions as small as 10 nm in diameter and a median distance of 11 nm were achieved. Using a similar combination of EBL and etching techniques, Liao et al. successfully fabricated suspended single-crystal diamond nanowires for high-performance nanoelectromechanical switches that can operate under harsh environments [46]. In this work, EBL was employed to define the cantilevers structures from a homoepitaxial boron-doped p-type diamond film. Subsequently, an aluminum layer with a thickness of 150 nm was deposited on the patterned diamond epilayer as a mask for RIE. After that, the sample was treated in a boiling HNO_3/H_2SO_4 acid solution to form high-quality suspended single-crystal diamond nanowires for nanoelectromechanical switches. In another work, Sumant and coworkers reported the fabrication of horizontally aligned ultra-nanocrystalline diamond nanowires using a combined top-down approach of EBL and RIE techniques with a photo resist layer as an etch mask. Using this top-down approach, the authors were able to fabricate diamond nanowires with 50 μm long and 40 nm width bearing excellent optical and electronic properties [47].

References

1 Zhou, K., Zhao, Z., Pan, L., and Wang, Z. (2019). Silicon nanowire pH sensors fabricated with CMOS compatible sidewall mask technology. *Sensors and Actuators B: Chemical* 279: 111–121.

2 Noborisaka, J., Motohisa, J., and Fukui, T. (2015). Catalyst-free growth of GaAs nanowires by selective-area metalorganic vapor-phase epitaxy. *Applied Physics Letters* 86: 213102.

3 Duan, X. and Lieber, C.M. (2000). General synthesis of compound semiconductor nanowires. *Advanced Materials* 12 (4): 298–302.

4 Yang, P. (2005). The chemistry and physic of semiconductor nanowires. *MRS Bulletin* 30: 85–91.

5 Mattila, M., Hakkarainen, T., and Lipsanen, H. (2006). Catalyst-free growth of In(As)P nanowires on silicon. *Applied Physics Letters* 89: 063119.

6 Biswas, A., Bayer, I.S., Biris, A.S. et al. (2012). Advances in top-down and bottom-up surface nanofabrication: techniques, applications, and future prospect. *Advances in Colloid and Interface Science* 170: 2–27.

7 Fu, X., Cai, J., Zhang, X. et al. (2018). Top-down fabrication of shape-controlled monodisperse nanoparticles for biomedical applications. *Advanced Drug Delivery Reviews* 132: 169–187.

8 Shtein, Z. and Shoseyov, O. (2017). When bottom-up meets top-down. *Proceedings of the National Academy of Sciences of the United States of America* 114 (3): 428–429.

9 Yu, H.D., Regulacio, M.D., Ye, E., and Han, M.Y. (2013). Chemical routes to top-down nanofabrication. *Chemical Society Reviews* 42: 6006–6018.

10 Woflsteller, A., Geyer, N., Nguyen-Duc, T.K. et al. (2010). Comparison of the top-down and bottom-up approach to fabricate nanowire-based silicon/germanium heterostructures. *Thin Solid Films* 518 (9): 2555–2561.

11 Gates, B.D., Xu, Q., Stewart, M. et al. (2005). New approaches to nanofabrication: molding, printing and other techniques. *Chemical Reviews* 105 (4): 1171–1196.

12 Grigorescu, A.E. and Hagen, C.W. (2009). Resists for sub-20 nm electron beam lithography with a focus on HSQ: state of the art. *Nanotechnology* 20: 292001.

13 Baek, D., Lee, S.H., Jun, B.H., and Lee, S.H. (2021). Lithography technology for micro and nanofabrication. *Advances in Experimental Medicine and Biology* 1309: 217–233.

14 Chen, K.S., Lin, I.K., and Ko, F.H. (2005). Fabrication of 3D polymer microstructures using electron beam lithography and nanoimprinting technologies. *Journal of Micromechanics and Microengineering* 15: 1894–1903.

15 Subramania, G. and Lin, S.Y. (2004). Fabrication of three-dimensional photonic crystal with alignment based on electron beam lithography. *Applied Physics Letters* 85: 5037–5039.

16 Melngailis, J. (1987). Focused ion beam technology and applications. *Journal of Vacuum Science and Technology B* 5: 469.

17 Volkert, C.A. and Minor, A.M. (2007). Focused ion beam microscopy and micromachining. *MRS Bulletin* 32: 389–399.

18 Reyntjens, S. and Puers, R. (2007). A review of focused ion beam applications in microsystem technology. *Journal of Micromechanics and Microengineering* 11: 287–300.

19 Yih, P.H., Saxena, V., and Steckl, A.J. (1997). A review of SiC reactive ion etching in fluorinated plasmas. *Physica Status Solidi B: Basic Solid State Physics* 202: 605.

20 Revell, P.J. and Goldspink, G.F. (1984). A review of reactive ion beam etching for production. *Vacuum* 34 (3, 4): 455–462.

21 Karouta, F. (2014). A practical approach to reactive ion etching. *Journal of Physics D: Applied Physics* 47: 233051.

22 Li, H.W., Kang, D.J., Blamire, M.G., and Huck, W.T.S. (2003). Focused ion beam fabrication of silicon print masters. *Nanotechnology* 14: 220.

23 Watt, F., Bettiol, A.A., Van Kan, J.A. et al. (2005). Ion beam lithography and nanofabrication: a review. *International Journal of Nanoscience* 4 (3): 269–286.

24 Tseng, A.A. (2005). Recent developments in nanofabrication using focused ion beams. *Small* 1 (10): 924–939.

25 Langord, R.M., Nellen, P.M., Gierak, J., and Fu, Y. (2007). Focused ion beam micro- and nanoengineering. *MRS Bulletin* 32: 417–423.

26 Lee, K.M., Choi, T.Y., Lee, S.K., and Poulikakos, D. (2010). Focused ion beam-assisted manipulation of single and double β-SiC nanowires and their thermal conductivity measurements by the four-point-probe 3-ω method. *Nanotechnology* 21: 125301.

27 Phan, H.P., Dinh, T., Kozeki, T. et al. (2016). Piezoresistive effect in p-type 3C-SiC at high temperatures characterized using Joule heating. *Scientific Reports* 28 (6): 284499.

28 Phan, H.P., Dowling, K.M., Nguyen, T.K. et al. (2018). Highly sensitive pressure sensors employing 3C-SiC nanowires fabricated on a free standing structure. *Materials and Design* 156 (15): 16–21.

29 Phan, H.P., Dinh, T., Kozeki, T. et al. (2016). The piezoresistive effect in top-down fabricated p-type 3C-SiC nanowires. *IEEE Electron Device Letters* 37 (8): 1029–1032.

30 Vieu, C., Carcenac, F., Pepin, A. et al. (2000). Electron beam lithography: resolution limits and applications. *Applied Surface Science* 164 (1): 111–117.

31 Chen, Y. (2015). Nanofabrication by electron beam lithography and its applications: a review. *Microelectronic Engineering* 135 (5): 57–72.

32 Kang, M.S., Lee, J.H., Bahng, W. et al. (2014). Top-down fabrication of 4H-SiC nano-channel field effect transistors. *Journal of Nanoscience and Nanotechnology* 14 (10): 7821–7823.

33 Liu, K., Avouris, P., Bucchignano, J. et al. (2002). Simple fabrication scheme for sub-10 nm electrode gaps using electron beam lithography. *Applied Physics Letters* 80: 865.

34 Juhasz, R., Elfström, N., and Linnros, J. (2005). Controlled fabrication of silicon nanowires by electron beam lithography and electrochemical size reduction. *Nano Letters* 5 (2): 275–280.

35 Feng, X.L., Matheny, M.H., Zorman, C.A. et al. (2010). Low voltage nanoelectrochemical switches based on silicon carbide nanowires. *Nano Letters* 10 (8): 2891–2896.

36 Cardinaud, C., Peignon, M.C., and Tessier, P.Y. (2006). Plasma etching: principles, mechanisms, application to micro- and nano-technologies. *Applied Surface Science* 164 (1): 72–83.

37 Franssila, S. and Sainiemi, L. (2008). *Encyclopedia of Microfluidics and Nanofluidics*. Boston, MA: Springer.

38 Ganjian, M., Modaresifar, K., Zhang, H. et al. (2019). Reactive ion etching for fabrication of biofunctional titanium nanostructures. *Scientific Reports* 9: 18815.

39 Bondur, J.A. (1976). Dry process technology (reactive ion etching). *Journal of Vacuum Science and Technology* 13: 1023.

40 Coldren, L.A., Miller, B.I., Iga, K., and Rentschler, J.A. (1981). Monolithic two-section GalnAsP/InP active optical resonator devices formed by reactive ion etching. *Applied Physics Letters* 38: 315.

41 Haab, A., Mikulics, M., Sutter, E. et al. (2014). Evolution and characteristics of GaN nanowires produced via maskless reactive ion etching. *Nanotechnology* 25: 255301.

42 Hausmann, B.J.M., Khan, M., Zhang, Y. et al. (2010). Fabrication of diamond nanowires for quantum information processing applications. *Diamond and Related Materials* 19 (5): 621–629.

43 Yu, F., Yao, S., Romer, F. et al. (2017). GaN nanowire arrays with nonpolar sidewalls for vertically integrated field-effect transistors. *Nanotechnology* 28 (9): 095206.

44 Wakui, K., Yonezu, Y., Aoki, T. et al. (2017). Simple method for fabrication of diamond nanowires by inductively coupled plasma ion etching. *Japanese Journal of Applied Physics* 56: 058005.

45 Yang, N., Uetsuka, H., Osawa, E., and Nebel, C.E. (2008). Vertically aligned nanowires from boron-doped diamond. *Nano Letters* 8 (11): 3572–3576.

46 Liao, M., Hishita, S., Watanabe, E. et al. (2010). Suspended single-crystal diamond nanowires for high-performance nanoelectromechanical switches. *Advanced Materials* 22 (47): 5393–5397.

47 Wang, X., Ocola, L.E., Divan, R.S., and Sumant, A.V. (2012). Nanopatterning of ultrananocrystalline diamond nanowires. *Nanotechnology* 23: 075301.

3

Hybrid Fabrication Techniques and Nanowire Heterostructures

3.1 Introduction

Wide bandgap semiconductor nanowires have been rapidly emerging over the last few decades as one of the most important building blocks for the developments of the next generation electronic and optoelectronic devices. To date, various bottom-up and top-down approaches have been effectively employed to synthesize an array of high-performance wide bandgap semiconductor nanowires [1–5]. Although these approaches have their own merits and inevitable disadvantages, they are very interesting from both scientific and practical point of view. In general, the bottom-up nanowires growth methods usually offer great flexibility to achieve sophisticated nanostructures with a high level of complex compositions as well as structural perfection, which may not be accessible by the top-down methods. Among various bottom-up approaches, the VLS mechanism-based growth techniques have been widely used to synthesize wide bandgap semiconductor nanowires because they enable great flexibility to control and modulate compositions and structures of the resulting nanowires and thus providing a huge opportunity to widen their applications. However, the use of foreign metallic droplets and particles as catalysts to grow wide bandgap semiconductor nanowires via the VLS mechanism might cause several negative impacts on the outcome of the growth process, such as the formation of stacking faults and point defects due to the possible integration of catalyst atoms into the as-grown nanowires. In this context, various reliable strategies have been introduced by researchers and engineers in recent years to minimize the unwanted impacts of metal catalysts on the resulting performance of wide bandgap semiconductor nanowires, such as reducing the synthesis temperatures, using self-catalyzing liquid droplets supplied by vapor-phase precursors instead of foreign metal particles or choosing solid catalysts rather than liquid catalysts for the growth. More interestingly, several catalyst-free synthesis routes based on the VSS or VS growth mechanisms have also been developed to eliminate the presence of metal catalysts in the growth process and thus, enabling the fabrication of high-quality wide bandgap semiconductor nanowires. For improving the capability of the bottom-up approaches, a more thorough understanding of the growth mechanism, further optimizing the growth parameters and continued improvements of the advanced nanomachining techniques should be constantly given a high priority to precisely control the crystal structure, dimensions, phase purity, interfaces, defects, compositions, and orientations of nanowires. In contrast, the top-down processes allow obtaining highly

Wide Bandgap Nanowires: Synthesis, Properties, and Applications, First Edition. Tuan Anh Pham, Toan Dinh, Nam-Trung Nguyen, and Hoang-Phuong Phan.

ordered arrays of semiconductor nanowires with perfect uniformity and reproducibility for the wafer-scale fabrication of electronic and optoelectronic devices. However, the key disadvantages of the top-down approaches can be viewed in the lack of composition control over the nanowire growth as well as the high-cost of operation and instrumentation [6–10]. From a practical standpoint, therefore, it is expected that a well-thought mix of both bottom-up and top-down approaches will be the future route to develop innovative nanowire-based electronic devices bearing unprecedented performances which may not be feasible by either technique alone.

With the continuous development of nanoscience and nanotechnology, nanowires-based electronic devices have become increasingly important for every aspect of our daily lives. Therefore, the current research efforts in term of the growth techniques have been devoted not only to optimizing the growth conditions for producing high-quality nanowires, but also to designing and constructing sophisticated architectures of wide bandgap semiconductor nanowire. This aims to broaden the nanowire applications in the fast-emerging fields, such as biomedical technology, nanoelectronics, robotics, and flexible sensors and nanogenerators. In this context, the growth of nanowire heterostructures consisting of different nanowires has attracted an increasing interest from researchers and engineers in recent years [11, 12]. This is because such hybrid structures of nanowires could be a powerful tool to push the limits of wide bandgap semiconductor nanowires via a band structure engineering, and thus taking their properties and performance to the next level that cannot be achieved by using individual nanowires. Despite the recent progress made in the growth of nanowire heterostructures, there are still many significant challenges that need to be fully addressed. For example, effective protocols to control the axial and radical growth of nanowire heterostructures have not been established due to the lack of a full understanding on the growth mechanisms of such heterostructures. Thus, more investigations on this context, both in experimental and in theoretical aspects, are highly needed. Another huge challenge for the development of nanowire heterostructures is the ability to achieve atomically clean and sharp interfaces as the interfaces determine the performance of the resulting electronic devices. To overcome this technical barrier, more experimental studies to optimize the growth conditions of nanowire heterostructures, such as flux rate of precursors, growth temperature, and substrate, are urgently needed. On the other hand, the integration of wide bandgap semiconductor nanowires with unconventional substrates of polymers, graphene, or other 2D materials is currently considered as a stepping-stone to establish a basic platform for the incremental developments of the future nanowire-based electronic devices as well as their integration into flexible electronics. In this regard, the transfer printing techniques have been widely employed to integrate various nanowires onto flexible substrates due to the lack of thermal stability of the polymer substrates.

Based on these observations, this chapter aims to summarize recent achievements on the fabrication of high-quality semiconductor nanowires and nanowire heterostructures as well as the integration of these nanowires on unconventional substrates. The first section focuses on new hybrid fabrication methods, which are based on the combination of the bottom-up and top-down approaches. The second section deals with the modifications and improvements on the transfer printing techniques, which enable the transfer of various nanowires in diverse forms and shapes onto polymeric substrates. The third section briefly discusses the fabrication of high-quality nanowires heterostructures as well as the integrations of these nanowires on unconventional low-dimensional substrates, such as graphene and related 2D materials.

3.2 Bottom-Up Meets Top-Down Approaches

Considering the significant advantages and limitations in each top-down and bottom-up approach, hybrid fabrication technologies that leverage the strength of the two approaches have been extensively investigated to create extremely low dimensional nanowires with precisely controllable properties. Hybrid fabrication processes offer an unprecedented possibility in moving semiconductor nanowires from the laboratory to the factory for large-scale fabrication of high-performance electronic and photonic devices. In this context, the bottom-up growth methods based on CVD and MBE techniques are widely employed in combination with conventional lithography and etching techniques to achieve precise control over the growth process of various wide bandgap semiconductor nanowires [13–17]. As an interesting example, the Stutzmann group reported the position-controlled growth of GaN nanowires on diamond substrates by a combination of MBE and EBL techniques [14]. In this study, the authors structured thin Ti masks using EBL, which enables the growth of perfectly homogeneous GaN nanowire arrays with miniaturized diameters and distances on diamond substrates. Figure 3.1 schematically shows the processing steps for the fabrication of the nanomask for the nanowire selective area growth, which was developed by Stutzmann et al. On the other hand, EBL and FIB can also be effectively employed to engineer the location of catalyst sites during the VLS growth and thus, allowing the precise control of the nanowire growth at desired sites on substrate surfaces [18, 19]. For instance, Liu and coworkers demonstrated the possibility to grow high-quality single crystal GaN nanowires at preferred sites on sapphire substrates using a combined technique of CVD and EBL [18]. Particularly, gold clusters as catalysts were deposited on the substrate at desired sites via e-beam patterning. Then, GaN nanowires were grown on the substrate decorated with gold clusters by CVD. This hybrid technique enables the selective growth of GaN nanowires in a near-vertical orientation on sapphire substrate due to the epitaxial nature of the GaN/sapphire interface. In another example, Fischer and coworkers employed FIB to achieve Pt nanopatterning for the position-selective growth of GaN nanowires on Pt catalyst dots. The authors proved that the locations and density of the as-grown GaN nanowires are dependent on the position, size, and thickness of the Pt catalyst dots deposited on Si substrates [19]. In another approach, pre-defined structures can be created on the chosen substrate surfaces by EBL or FIB to effectively control the growth of desired nanowires. For example, Tabassum et al. synthesized ultrahigh self-aligned SiC nanowires with tailored properties at predetermined locations on Si wafers without the requirement for a lithographic pattern transfer technique [17]. This synthesis route used EBL to form ribbon arrays of hydrogen silsesquioxane (HSQ) as pre-defined patterns on Si substrates. The SiC nanowires were subsequently grown on such patterns using a well-controlled thermal CVD process and ion etching techniques. Detailed processing steps for the

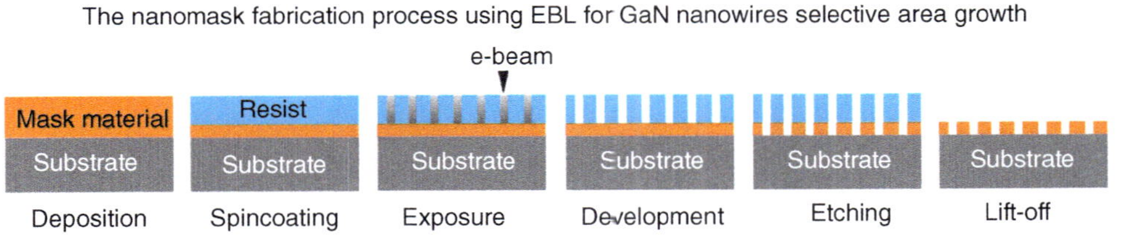

Figure 3.1 Illustration of the nanomask fabrication process using EBL for GaN nanowires selective area growth. *Source:* Schuster et al. [14]/American Chemical Society.

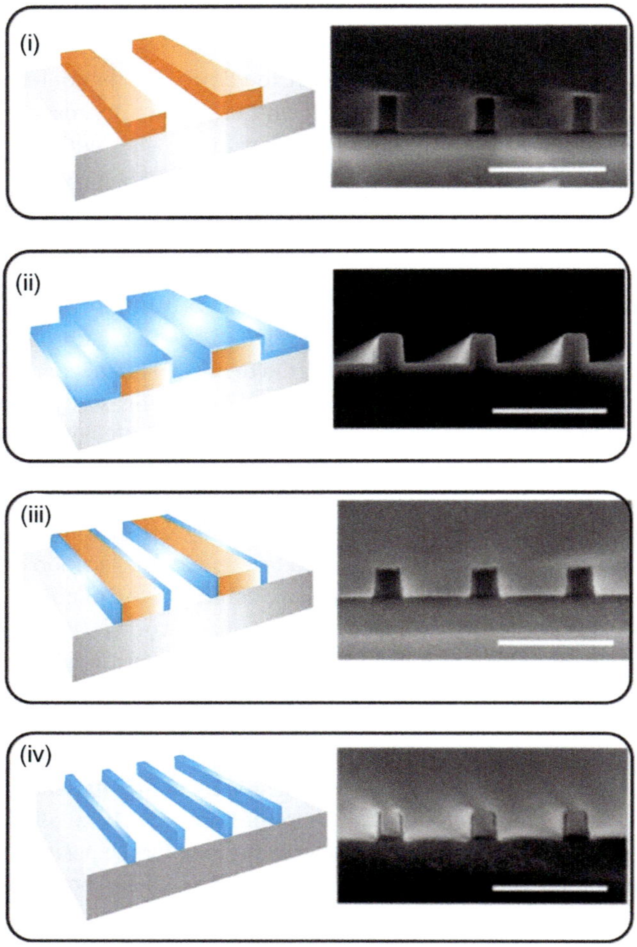

Figure 3.2 Illustration of the pre-defined substrate fabrication process using EBL and RIE for SiC nanowires self-aligned growth and corresponding SEM images for each processing step. [20], Li, et. al., (2005)/John Wiley & Sons, Inc.

fabrication of ultrathin self-aligned SiC nanowires are shown in Figure 3.2. In this work, Si wafer was spin-coated with HSQ followed by lithographic processes to achieve a ribbon array in the first step. Subsequently, an ultrathin conformal SiC layer was deposited using the well-known CVD method. In the third step, the ultrathin conformal SiC layer was etched to expose the HSQ ribbon array using RIE. Finally, the HSQ ribbon array was removed by wet etch in buffered hydrofluoric acid to yield SiC nanowire arrays.

3.3 Integration of Nanowires onto Unconventional Substrates

3.3.1 Transferring Nanowires onto Flexible Substrates

Recent advancements in printing technology and flexible electronics have also triggered the development of wide bandgap semiconductor nanowires integrated onto a soft platform. Flexible wide bandgap nanowire electronics offer many interesting properties for future multifunctional sensing devices, such as biocompatibility, light weight, transparency, and flexibility. Unfortunately, the direct growth of desired nanowires on plastic substrates is almost impossible due to the intolerance of the soft substrate to the elevated temperatures required for the growth process. The limitation of the bottom-up approach for flexible nanowire electronics has propelled studies on scalable top-down methods for transferring semiconductor nanowires onto plastic substrates at a high degree of structural perfection for their usage in biocompatible applications, such as in implantable and wearable sensing devices [21–25]. Among these methods, dry-transfer methods represent an effective top-down pathway to print a variety of highly ordered arrays of semiconductor nanowires onto plastic substrates. In a pioneering study, the Rogers group developed a dry-transfer printing process employing a PDMS-stamp to drop top-down fabricated GaAs nanowire arrays onto plastic substrates with good crystallographic orientation [23]. The whole process is schematically shown in Figure 3.3 and involves several key steps including: (i) anisotropic chemical etching of a GaAs wafer using SiO_2 strips as mask producing GaAs nanowire arrays; (ii) contact between PDMS and SiO_2 leading to the formation of siloxane bonds via a condensation reaction; (iii) peeling the PDMS stamps from the GaAs substrates lifts off the wires; (iv) polishing the remaining GaAs wafer which enables a new cycle of wire fabrication and lift-off; (v) contact between a PDMS stamps (with GaAs wires) and liquid film of PU cast on the polyethylene terephthalate (PET) sheet which causes the PU to flow and conforms to the wires; (vi) curing PU with UV light which bonds the wires to the PU and the PU to the PET, and then peeling off the PDMS stamp which leaves the wires embedded in PU; and (vii) buffered oxide etching, which removes the SiO_2 mask. Since this pioneering work, a considerable number of studies have been devoted to optimizing and enhancing the capability of the dry-transfer methods for the effective integration of nanowires on diverse flexible substrate surfaces. In this context, Zheng and coworkers have successfully developed a series of transfer-printing methods for fabricating Si nanowires on a wide range of flexible substrates, including PDMS, Petri dishes, Kapton tapes, thermal release tapes, and other adhesive

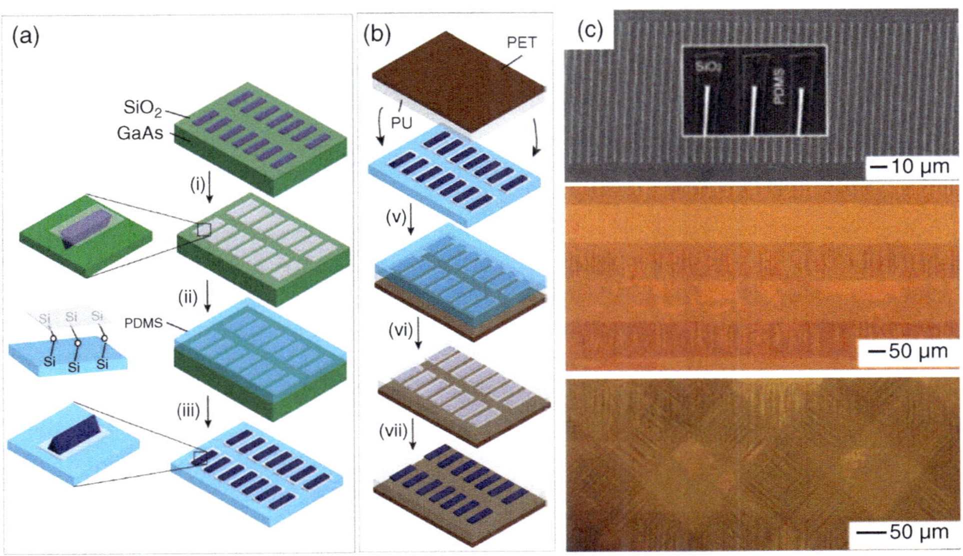

Figure 3.3 (a, b) Schematic illustration of the process of generating and transfer printing GaAs wire arrays onto plastic substrates. (c) SEM images of GaAs wire arrays bonded to PDMS via the SiO_2 stripes (top) and optical micrographs of GaAs wire arrays on PU/PET sheets with different numbers of layers of wires: single layer (middle) and triple layers (bottom). *Source:* (a–c) Reprinted from Sun and Rogers [23].

types [24]. In this excellent work, the authors categorized their transfer printing techniques into three different methods, namely, single transfer printing, double transfer printing, and multiple transfer printing. It should be noted that the three transfer printing methods basically rely on the differences in adhesion to transfer desired nanowires from weakly adhesive donor substrates to more strongly adhesive receiver substrates when these two substrates are brought into a physical contact. The differences between these methods come from the procedures of transferring steps. In the first method, Si nanowires can be transferred a Si wafer to receiving flexible substrates by a single peel-off step. In the second method, the nanowires were fabricated on adhesive substrate by transferring and printing nanowires and electrodes in sequence. Finally, the third method involves multiple transfers of electrodes by using thermal release tapes. According to their results, the flexible devices fabricated from the transferred nanowires exhibited several remarkable features, such as Ohmic contacts between the nanowires and metal electrodes, controllable density of nanowires per device as well as robustness against bending and heating. In another example, Zhou and coworkers reported an improved dry-transfer printing method compared with that developed by the Rogers group. In this study, the authors employed an anodized aluminum oxide membrane in the transfer process which allows annealing of nanowires at moderately high temperature to obtain nanowire films with extremely high conductivity and good transparency [26].

Despite the recent achievements made in the dry transfer printing method, the printing process functions efficiently only when the released microstructures attach to the mother wafer in a flat, planar configuration, such as cantilevers, membranes, or doubly clamped beams via defined anchors to maximize good conformal contact between the stamp and the microstructure. Thus, new transfer printing approaches that can be used to transfer complex micro and nanostructures onto flexible substrates is extremely needed. In this regard, Pham et al. recently developed a versatile transfer printing approach that works very well to transfer various SiC microstructures with diverse sizes and shapes on polymeric substrates without the need for defined anchor structures [27]. In this work, the authors employed a thin layer of aluminum on the backsides of a free-standing SiC nanomembrane on a Si wafer prior to an inductively coupled plasma etching process. The merit of this study is that the aluminum film can serve as a supporting layer that prevents the formation of wrinkles or buckling in the SiC microstructures induced by the residual stress gradients. Their results show that sophisticated SiC microstructures with a high degree of structural perfection without cracks and tears can be easily integrated into stretchable substrates by a standard transfer printing process as shown in Figure 3.4. Based on their results, the enhanced transfer printing approach is expected to apply to the transfer of various complex nanowire structures onto flexible substrates that could be very beneficial for the use of wide bandgap semiconductor nanowires in the hotspot of science and technology, such as bio-robotics, human–machine interfaces (HMT) or Internet of thing (IoT).

Using the dry transfer printing methods, various wide bandgap semiconductor nanowires have been successfully transferred onto diverse flexible substrates [28–30]. For example, Lei and coworkers fabricated flexible field emission devices by transferring SiC nanowires grown by the thermal evaporation method onto flexible substrates using a screen-printing method. In particular, the as-grown SiC nanowires were mixed with terpineol mixture and ethyl cellulose into a paste. Subsequently, the paste was printed onto glass substrate via conformal contact. After annealing and baking the printed paste at 390 °C to remove organic binders in a nitrogen environment, the SiC nanowires were transferred onto flexible substrates with conductive and adhesive tape. According to their findings, the flexible SiC nanowire emitters showed significant field

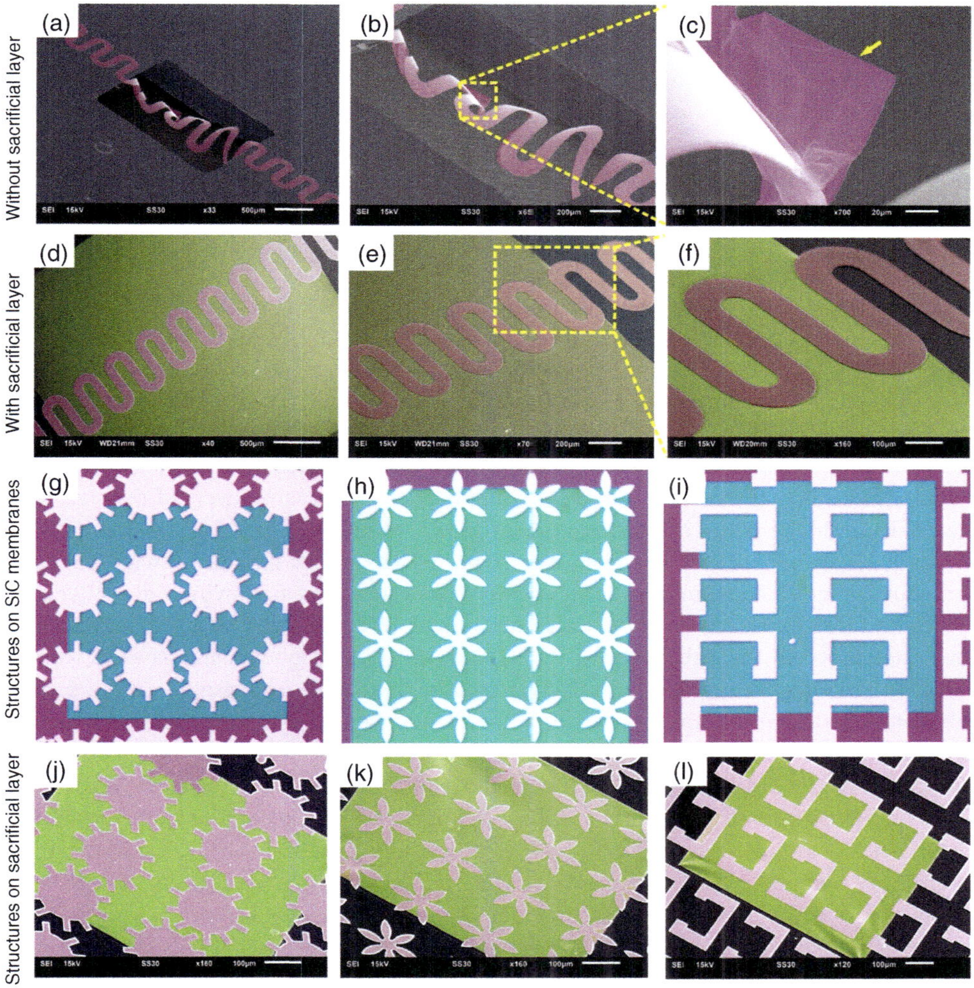

Figure 3.4 Thin SiC structures. (a–c) and (d–f) SEM images of the SiC micro-spring structures obtained without and with the sacrificial layer. (g–i) and (j–l) Optical microscopy images and SEM images of various microstructures laying on the aluminum sacrificial layer, respectively: gear (left), flower (middle), and U (right)-shaped structures. The yellow arrow in (c) marks the presence of residual SiC after the ICP etching. *Source:* Reprinted from Pham et al. [27].

emission characteristics, such as low turn-on field, threshold field, and high uniformity of the emission current [28]. In another example, Song and coworkers reported the integration of GaN nanowire-based light emitting diodes onto flexible platforms for potential applications in wearable electronics and bendable displays. The flexible devices were fabricated by transferring GaN nanowires from sapphire substrates onto flexible PET using a "paste-and-cut" integration process [29]. In this study, the GaN nanowires grown by the MBE technique were bonded to a silicon wafer using Crystalbond mounting adhesive. Following the bonding process, a laser beam was raster scanned through the transparent sapphire to decompose the GaN nanowires at the GaN/sapphire interface. Then, a thin layer of high-performance silver paste was used to attach the exposed n-side of the GaN devices onto the flexible PET substrate. The structure was immersed into a hot acetone bath to selectively dissolve the adhesive layer, thus releasing the silicon carrier substrate. Finally, the samples were washed out adhesive residue by immersing in fresh acetone and alcohol baths. As a result, the GaN nanowire light emitters exhibited an excellent optical enhancement through bending of the substrate platform, which could pave a new approach for the development of flexible LED devices bearing remarkable performance based on various wide bandgap semiconductor nanowires.

3.3.2 Growing Nanowires on Graphene and Layered Material Substrates

For expanding the use of wide bandgap semiconductor nanowires in diverse applications of science and technology, it is highly essential to grow high-quality nanowires onto graphene and other layered material substrates. Compared with conventional substrates based on metals or semiconductors, graphene and layered material substrates exhibit several advantages for the development of high-performance electronic devices, such as a higher electrical and thermal conductivity, a smaller lattice mismatch, or a lower thermal expansion coefficient. Another merit is that the weak atomic bonding between successive layers of these layered materials could help in transferring the as-grown nanowires to other polymeric substrates [31]. Several research studies have reported the direct growth of wide bandgap nanowires on graphene films and subsequently transferred these nanowires onto other substrates for further characterizations and applications [32, 33]. For instance, Maboudian and coworkers reported the direct growth of SiC nanowires on graphene substrates as well as developed an effective method for transferring these nanowires from graphene to arbitrary substrates. In this study, graphene was firstly grown on copper using the well-known CVD technique. The as-grown graphene on copper then was used as a substrate for the growth of SiC nanowires via the VLS mechanism with the assistance of Ni catalyst in a low-pressure CVD furnace. Their experimental results confirmed an excellent contact between SiC nanowires and graphene as well as the ability to effectively transfer the SiC nanowires onto an arbitrary substrate by a simple sacrificial etch transfer [32]. In another study, Yi and coworkers reported the growth of high-areal-density GaN nanowires on large size graphene films using a Ni catalyst-assisted VLS mechanism. The as-grown GaN nanowires on graphene films were then transferred onto polymer substrates using a simple lift-off method for the fabrication of flexible photocatalyst [33]. More interestingly, graphene substrate can be employed to assist and control the growth of wide bandgap semiconductor nanowires in some cases [34, 35]. For example, Tchernycheva and coworkers recently developed a new approach to obtain selective area growth of GaN nanowires directly on graphene nanodots using a plasma-assisted MBE technique. In this study, CVD grown polycrystalline graphene patches were wet-transferred from their carrying copper foil to a Si wafer coated by 300 nm of SiO_2. Subsequently, these graphene patches were patterned to form highly ordered graphene nanodots using EBL technique. Finally, GaN nanowires can be grown on these graphene nanodomains with a perfect selectivity with respect to the SiO_2 surrounding surface. Figure 3.5 shows a series of SEM images of GaN nanowires selectively formed on graphene nanodots after the growth duration of four hours. These results unambiguously indicate that the GaN nanowires were only grown on the areas of graphene nanodots, and not on the areas of SiO_2 surfaces [34]. Furthermore, graphene can be used as a buffer layer to assist the growth of wide bandgap semiconductor nanowires on amorphous substrates. This is a huge advantage because it is quite difficult to directly grow wide bandgap semiconductor nanowires on amorphous substrates due to the lattice mismatch limits. For instance, Weman and coworkers developed a novel method to grow AlGaN nanowires on SiO_2/Si(100)

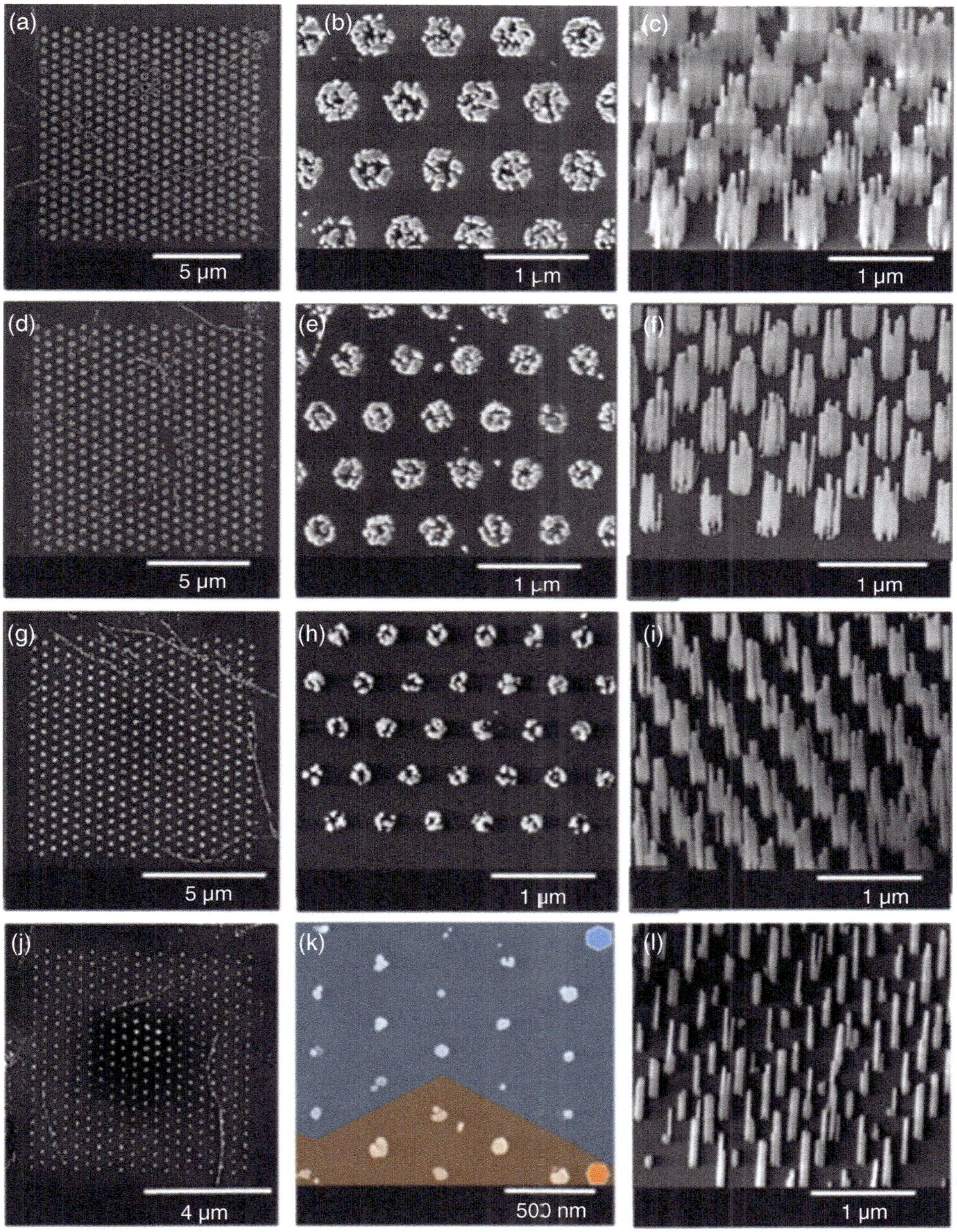

Figure 3.5 (a, b, d, e, g, h, j, k) Top view and (c, f, i, l) 45° tilted SEM images of GaN nanowires grown for four hours on patterned graphene. The rows correspond to dot radii R_{dot} = 200 nm in (a–c), 150 nm in (d–f), 100 nm in (g–i), and 50 nm in (j–l). *Source:* Reprinted from Morassi et al. [34].

substrates by employing graphene as a buffer layer. By optimizing the growth conditions, the authors were able to control the morphology of the AlGaN nanowires grown on graphene/ SiO_2/Si(111) substrates with an excellent optical properties [35].

Beside graphene, graphite and other layered materials have also been employed as substrates for the synthesis of wide bandgap semiconductor nanowires. For instance, Shi and coworkers reported a novel catalyst-free synthesis method to fabricate SiC nanowires/graphite hybrid system by carbothermal reduction of silicon powder at 1400–1600 °C [36]. In this work, they mixed silicon powder with ethyl cellulose and methanol to form a slurry, which was then uniformly coated on a graphite paper. Subsequently, the sample was kept in a vacuum furnace at different temperatures to form SiC nanowires driven by the VS mechanism. The authors found that increasing sintering temperature leads to significant impacts on the outcome of the synthesis process, such as more SiC nanowires formed, the diameter of the nanowires increased, and the morphology of the nanowires changed from flocculent to curvilinear and linear fashions. In some cases, wide bandgap semiconductor nanowires grown on graphite substrates exhibit better optical properties compared with the ones grown on conventional substrates as confirmed by the growth of GaN nanowires on graphite substrates using the MBE technique [37]. On the other hand, layered transition metal dichalcogenides can also be used as substrates for the growth of wide bandgap semiconductor nanowires. Recently, Ooi and coworkers reported the monolithic high-power, droop-free, and wavelength tunable InGaN/GaN nanowire light-emitting diodes on large-area MoS_2 layers by sulfurizing entire Mo substrates. The most important advantage in this approach is that MoS_2 serves as both a buffer layer for high-quality GaN nanowires growth and a sacrificial layer for epitaxy lift-off. The LED devices exhibited a low turn-on voltage of 2 V and light output power up to 1.5 mW emitting without an efficiency droop up to the current injection of 1 A. These interesting results could pave a new era for device design and applications in various emerging fields, ranging from energy harvesting to ultralow-profile optoelectronics [38].

3.4 Synthesis of Nanowire Heterostructures

3.4.1 Synthesis of One-Dimensional Heterostructures

With various wide bandgap semiconductor nanowires in hand, it is realized that for the broader applications of these nanowires in the fast-emerging fields of semiconductor technology, a controlled way to effectively enhance the overall performance of these nanowires is of utmost importance. With that in mind, researchers have demonstrated that the combination of dissimilar wide bandgap semiconductor nanowires into one-dimensional heterostructures could offer a new paradigm for generating new hybrid materials bearing a wide range of unprecedented properties that cannot be achieved by using individual nanowires alone. In this regard, synthesis methods play an important role in enabling sharp and clean interfaces as well as minimizing defect density in the nanowire heterostructures, and thus deciding the resulting performance of such hybrid materials [39–42]. Because of this, the bottom-up approaches normally offer a better flexibility and advantage compared with the top-down approaches to synthesize diverse nanowire heterostructures with a controlled heterointerfaces, chemical composition, and defect density. In general, the synthesis methods can combine different nanowires into two main different heterogeneous architectures, including axial and radical heterostructures as shown in Figure 3.6. An axial heterostructure is created by modulating the composition of the nanowire along the growth axis. In contrast, radical or core/shell nanowire heterostructures can be created by modulating the composition of the nanowire orthogonal to the long axis [42]. Compared with the bulk materials, the key merit of nanowire geometry for the formation of high quality heterostructures is the ability to minimize interface defects supported by the lateral relaxation of strain, even for systems with a large lattice mismatch [43, 44].

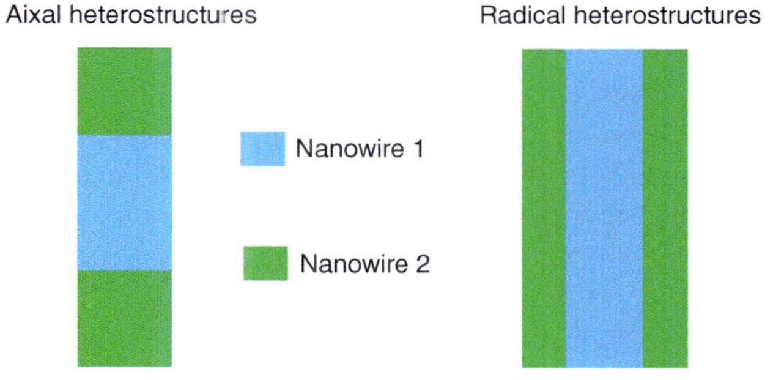

Figure 3.6 A cartoon illustrates two different architectures of nanowire heterostructures: axial (left) and radical heterostructures (right).

Among various growth techniques based on the bottom-up approach, CVD and MBE are the most common techniques for producing of a wide range of nanowire heterostructures due to their possibility to effectively control the outcome of the growth process via modulating different growth parameters, such as substrate conditions, temperature, pressure, growth duration, flux ratios, and precursors [45]. Employing CVD and MBE techniques, numerous research groups have devoted their efforts to controlling morphologies, defects, and interfaces of nanowire heterostructures that aim to enhance the overall performance of electronic devices consisting of such heterostructures [46–48]. For instance, Gogneau and coworkers demonstrated that the morphology of InGaN/GaN nanowire heterostructures critically depends on the In/Ga flux ratios affecting the local V/III ratio at the nanowire heterostructures growth front. By modulating this parameter, the authors were able to controllably tune morphologies of nanowire heterostructures from tapered island-like morphologies in locally N-rich conditions to disk-like morphologies in metal-rich conditions [46]. In some cases, new hybrid approaches based on the combination of both bottom-up and top-down techniques have also been used in order to promote the alignment and orientation of nanowire heterostructures grown on the chosen substrates. For example, Tyagi and coworkers reported the synthesis of highly aligned AlGaN/GaN axial nanowire heterostructures on sapphire using a combination of maskless reactive ion etching and metal organic vapor phase epitaxy. The as-grown nanowire heterostructures exhibited an excellent vertical alignment on sapphire substrates associated with their outstanding properties, which makes them as ideal platforms for the fabrication of high-performance electronic devices [49].

On the other hand, despite the recent achievements made on the heterostructure synthesis, there are still significant technical barriers to effectively control the axial selectivity for the two different growth modes. In term of growing axial nanowire heterostructures, one may simply claim that the continuation of the heterostructure elongation process can be achieved via switching the different vapor precursors. However, such selectivity is much more complicated because there is always a competition between anisotropic growth and conformal layer growth on the side facets of the growing nanowires for the formation of axial and radical heterostructures, respectively. In this context, careful tuning of growth parameters may offer appropriate solutions for overcoming this parasitic competing growth and thus, enhancing the growth mode selectivity [50, 51]. An excellent study reported by Myers and coworkers demonstrated that vertical or coaxial GaN/AlN nanowire heterostructures can be formed by adjusting the growth temperature using a two-step MBE technique. The most important point in this study is that the growth kinetics can be tuned to favor either vertical or radical growth by separating the nucleation and growth process into distinct steps [50]. In another interesting example, Lee and coworkers reported the controlled synthesis of InGaN/GaN multi quantum well uniaxial and coaxial nanowire heterostructures via MOCVD technique using Au catalysts with GaN seeded on Si(111) substrates. The authors proved that the *c*-plane and *m*-plane oriented nanowire heterostructures can be achieved by adjusting the growth temperature and working pressure as shown in Figure 3.7. Interestingly, both heterostructures exhibited sharp interfaces associated with superior electrical and optical properties, which represent huge advantages for developing high performance LEDs [52].

3.4.2 Synthesis of Mixed Dimensional Heterostructures

Inspired by huge achievements in the fabrication of nanowire heterostructures, mixed dimensional heterostructures constructed by various low-dimensional structures (such as 0D, 1D, and 2D) have recently attracted an increasing attention from researchers due to such

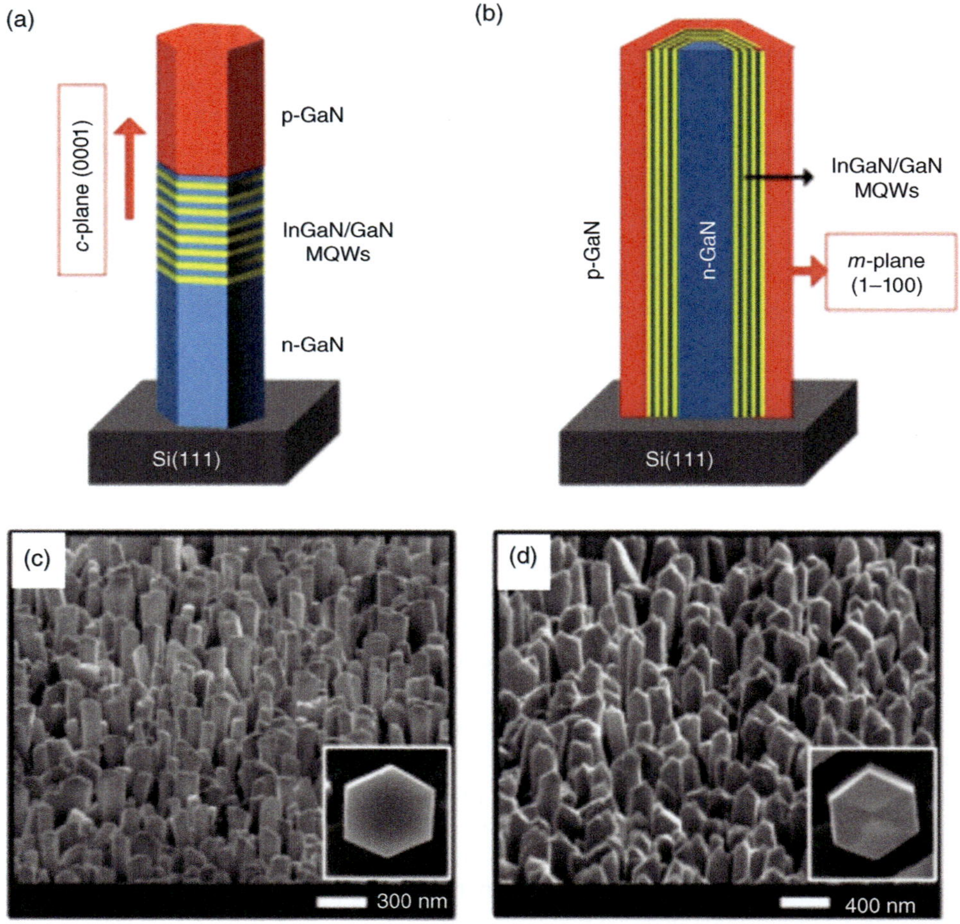

Figure 3.7 Schematic diagram of *c*-plane (uniaxial) oriented (a) and *m*-plane (coaxial) oriented multiple quantum-well heterostructures (b). Tilt-view FE-SEM images of *c*-plane oriented (c) and *m*-plane oriented (d) InGaN/GaN multiple quantum-well heterostructures grown on Si(111) substrate. The insets show the top-view of the uniaxial and coaxial NWs. *Source:* Reprinted from Ra et al. [52].

hybrid low-dimensional structures could maximize advantages from individual materials for the development of multifunctional electronic devices with unprecedented performance [53–55]. In this regard, graphene, a well-known 2D material made up of only a single atomic thickness of carbon atoms arranged into a honeycomb structure, is one of the most frequently used materials in the combination with various wide bandgap semiconductor nanowires for developing novel mixed dimensional materials with enhanced properties [56–59]. For instance, Kim and coworkers fabricated flexible ultraviolet photosensors based on GaN nanowires/graphene hybrid systems with different numbers of graphene layers. In this study, a light-absorbing layer of GaN nanowires was randomly embedded in a graphene sandwich structure, in which the number of bottom graphene layers was varied from zero to three while the top layer was fixed with graphene single layer. By optimizing the number of graphene layers, the authors found that the response time of the as-fabricated device at a given optical pulse was much shorter than other flexible photosensors fabricated using semiconductor nanowires alone. The structure of GaN/graphene devices and the corresponding SEM images of the nanowires are shown in Figure 3.8 [56]. In another work, Han and coworkers reported the synthesis of SiC nanowires/graphene hybrid materials with enhanced hydrophobicity and electromagnetic wave absorption property via chemical vapor infiltration. One of the advantages is that the growth of SiC nanowires dramatically improved

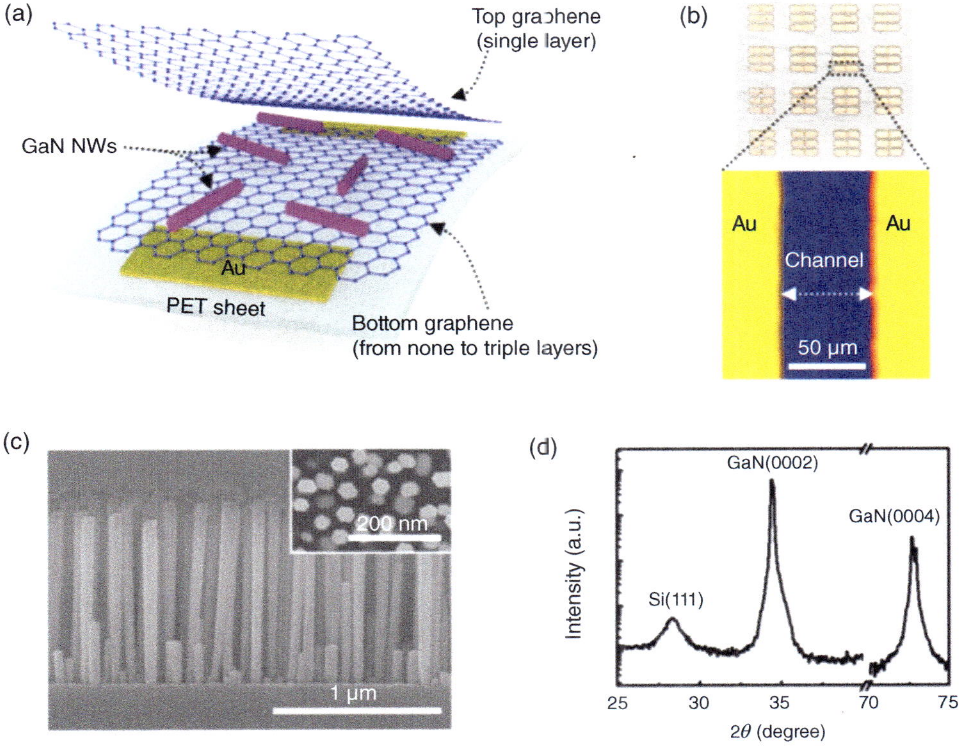

Figure 3.8 (a) Schematic of the process used to fabricate a flexible GaN-NW photosensor; (b) optical microscope image of the fabricated photosensors (top) and a magnified image of a single device (bottom); (c) cross-sectional FE-SEM image, where the inset is the plan-view image; (d) XRD curve for the GaN NWs formed on a Si(111) substrate. *Source:* Reprinted from Han et al. [56].

the yield strength of the hybrid material from 0.15 to 0.47 MPa in the linear elastic region. The SiC nanowires/graphene system had a wider effective absorbing bandwidth of 6.5 GHz in the frequency range of 9.3–15.8 GHz at a thickness of 2.0 nm [57]. Another example reported by Ha and coworkers where they prepared GaN nanowires/graphene systems for applications in hydrogen generation using MBE and CVD techniques. The most interesting point in this work is that the heterojunction formed between GaN nanowires and graphene enhanced the electron transfer between nanowires and reacted with protons on the active sites of graphene, and thus suppressing the recombination of charge carriers [58].

Beside graphene, other 2D materials from transition metal dichalcogenides have also been used to combine with many wide bandgap semiconductor nanowires for developing the next generations of hybrid semiconductor materials with superior properties. The Wang research group recently fabricated high-performance self-aligned MoS_2 field-effect transistors with enhanced photoresponsivity by the piezo-phototronic effect. The device was fabricated based on monolayer MoS_2 with a piezoelectric GaN nanowire as the local gate. The fabrication process in this study enabled the preservation of the intrinsic property of MoS_2 and suppresses the scattering center density in the MoS_2/GaN interface, which resulted in high electrical and photoelectric performances [60]. In another example, Wan and coworkers synthesized a 1D/2D heterostructures composed of SiC nanowires wrapped with MoS_2 nanosheets via the hydrothermal synthesis of MoS_2 on highly connected SiC nanowires. The authors found that SiC nanowires could reduce the activated water dissociation energy barrier as well as improve the active site exposure and charge transport, which enable efficient electrocatalytic hydrogen evolution [61].

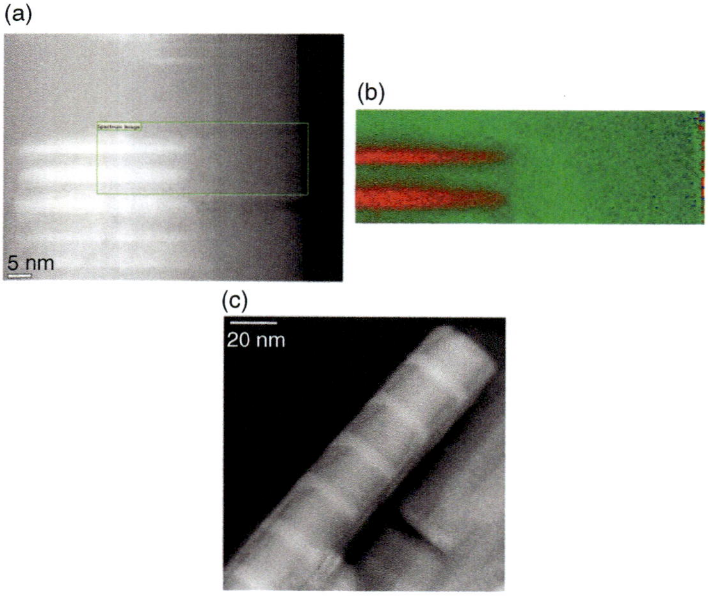

Figure 3.9 (a) High-resolution high-angle annular dark-field image showing the atomic number contrast between InGaN quantum dots and GaN and (b) electron energy loss spectrometry spectrum imaging (EELS-SI) performed in the selected region; (c) TEM image of GaN/$Al_{0.2}Ga_{0.8}N$ DINW shows sharp quantum well-like interfaces with no island formation. *Source:* Reprinted from Deshpande et al. [63].

Over the past years, 0D/1D heterostructures of dissimilar wide bandgap semiconductor nanowires have been rapidly emerging as promising materials for the fabrication of novel electronic devices having unique quantum performances. In this context, MBE has been widely employed to synthesize various 0D/1D heterostructures based on embedding of wide bandgap semiconductor quantum dots into wide bandgap semiconductor nanowires [62–64]. For example, Renard and coworkers reported the growth of unique heterostructures consisting of single GaN/AlN quantum dots embedded in GaN nanowires. The authors claimed that this unique quantum dot geometry could open-up new pathways toward the realization of vertically coupled identical quantum dots, which is very beneficial for potential applications such as UV emitters and room temperature single photo sources [62]. In another example, the Mi research group investigated the formation and nature of quantum confinement of InGaN disks incorporated in GaN nanowires grown on silicon substrates by MBE as shown in Figure 3.9. The authors successfully demonstrated the formation of self-organized InGaN quantum dots in the disk region of the GaN nanowires by recording of single photo emission from the single quantum dot [63].

References

1 Wacaser, B.A., Dick, K.A., Johansson, J. et al. (2009). Preferential interface nucleation: an expansion of the VLS growth mechanism for nanowires. *Advanced Materials* 21 (2): 153–165.

2 Wang, F., Dong, A., Sun, J. et al. (2006). Solution–liquid–solid growth of semiconductor nanowires. *Inorganic Chemistry* 45 (19): 7511–7521.

3 Wang, Y., Schmidt, V., Senz, S., and Gösele, U. (2006). Epitaxial growth of silicon nanowires using an aluminum catalyst. *Nature Nanotechnology* 1: 186–189.

4 Wu, Y. and Yang, P. (2001). Direct observation of vapor–liquid–solid nanowire growth. *Journal of the American Chemical Society* 123 (13): 3165–3166.

5 Kim, B.J., Tersoff, J., Wen, C.Y. et al. (2009). Determination of size effects during the phase transition of a nanoscale Au–Si eutectic. *Physical Review Letters* 103: 155701.

6 Kang, M.S., Lee, J.H., Bahng, W. et al. (2014). Top-down fabrication of 4H-SiC nano-channel field effect transistors. *Journal of Nanoscience and Nanotechnology* 14 (10): 7821–7823.

7 Liu, K., Avouris, P., Bucchignano, J. et al. (2002). Simple fabrication scheme for sub-10 nm electrode gaps using electron beam lithography. *Applied Physics Letters* 80: 865.

8 Juhasz, R., Elfström, N., and Linnros, J. (2005). Controlled fabrication of silicon nanowires by electron beam lithography and electrochemical size reduction. *Nano Letters* 5 (2): 275–280.

9 Feng, X.L., Matheny, M.H., Zorman, C.A. et al. (2010). Low voltage nanoelectrochemical switches based on silicon carbide nanowires. *Nano Letters* 10 (8): 2891–2896.

10 Cardinaud, C., Peignon, M.C., and Tessier, P.Y. (2000). Plasma etching: principles, mechanisms, application to micro- and nano-technologies. *Applied Surface Science* 164 (1): 72–83.

11 Kennedy, T., Bezuidenhout, M., Palaniappan, K. et al. (2015). Nanowire heterostructures comprising germanium stems and silicon branches as high-capacity Li-ion anodes with tunable rate capability. *Nano Letters* 9 (7): 7456–7465.

12 Lauhon, L.J., Gudiksen, M.S., and Lieber, C.M. (2004). Semiconductor nanowire heterostructures. *Philosophical Transactions of the Royal Society of London, Series A: Mathematical, Physical and Engineering Sciences* 362: 1247–1260.

13 Hersee, S.D., Sun, X., and Wang, X. (2006). The controlled growth of GaN nanowires. *Nano Letters* 6 (8): 1808–1811.

14 Schuster, F., Hetz, M., Weiszer, S. et al. (2015). Position-controlled growth of GaN nanowires and nanotubes on diamond by molecular beam epitaxy. *Nano Letters* 15 (3): 1773–1779.

15 Wang, X., Ocola, L.E., Divan, R.S., and Sumant, A.V. (2012). Nanopatterning of ultrananocrystalline diamond nanowires. *Nanotechnology* 23: 075301.

16 Fan, H.J., Werner, P., and Zacharias, M. (2006). Semiconductor nanowires: from self-organization to patterned growth. *Small* 2 (6): 700–717.

17 Tabassum, N., Kotha, M., Kaushik, V. et al. (2018). On-demand CMOS-compatible fabrication of ultrathin self-aligned SiC nanowire arrays. *Nanomaterials* 8 (11): 906.

18 Han, S., Jin, W., Tang, T. et al. (2003). Controlled growth of gallium nitride single crystal nanowires using a chemical vapor deposition method. *Journal of Materials Research* 18 (2): 245.

19 Nam, C.Y., Kim, J.Y., and Fischer, J.E. (2005). Focused ion beam platinum nanopatterning for GaN nanowires: Ohmic contacts and patterned growth. *Applied Physics Letters* 86: 193112.

20 Li, Y., Dorozhkin, P.S., Bando, Y., and Golberg, D. (2005). Controllable modification of SiC nanowires encapsulated in BN nanotubes. *Advanced Materials* 17 (5): 545–549.

21 McAlpine, M.C., Ahmad, H., Wang, D., and Heath, R.J. (2010). Highly ordered nanowire arrays on plastic substrates for ultrasensitive flexible chemical sensors. *Nature* 6: 379–384.

22 Ruck, L.G., Tsivion, D., Schvartzman, M. et al. (2014). Guided growth of horizontal GaN nanowires on quartz and their transfer to other substrates. *ACS Nano* 8 (3): 2838–2847.

23 Sun, Y. and Rogers, J.A. (2004). Fabricating semiconductor nano/microwires and transfer printing ordered arrays of them onto plastic substrates. *Nano Letters* 4 (10): 1953–1959.

24 Lee, C.H., Kim, D.R., and Zheng, X. (2010). Fabricating nanowire devices on diverse substrates by simple transfer-printing methods. *Proceedings of the National Academy of Sciences of the United States of America* 107 (22): 9950–9955.

25 Lee, K.N., Jung, S.K., Kim, W.H. et al. (2007). Well controlled assembly of silicon nanowires by nanowire transfer method. *Nanotechnology* 18: 445302.

26 Madaria, A.R., Kumar, A., Ishikawa, F.N., and Zhou, C. (2010). Uniform, highly conductive, and patented transparent films of a percolating silver nanowire network on rigid and flexible substrates using a dry transfer technique. *Nano Research* 3: 564–573.

27 Pham, T.A., Nguyen, T.K., Vadivelu, R.K. et al. (2020). A versatile sacrificial layer for transfer printing of wide bandgap materials for implantable and stretchable bioelectronics. *Advanced Functional Materials* 30 (40): 2004655.

28 Cui, Y., Chen, J., Di, Y. et al. (2017). High performance field emission of silicon carbide nanowires and their applications in flexible field emission displays. *AIP Advances* 7: 125219.

29 Asad, M., Wang, R., Ra, Y.H. et al. (2019). Optically invariant InGaN nanowire light-emitting diodes on flexible substrates under mechanical manipulation. *npj Flexible Electronics* 16: 2333.

30 Petroni, S., Tegola, C.L., Caretto, G. et al. (2011). Aluminum nitride piezo-MEMS on polyimide flexible substrates. *Microelectronic Engineering* 88 (8): 2372.

31 Zhao, C., Alfaraj, N., Subedi, R.C. et al. (2018). III-nitride nanowires on unconventional substrates: from materials to optoelectronic device applications. *Progress in Quantum Electronics* 61: 1–31.

32 Alper, J.P., Gutes, A., Carraro, C., and Maboudian, R. (2013). Semiconductor nanowires directly grown on graphene-towards wafer scale transferable nanowire arrays with improved electrical contact. *Nanoscale* 5: 4114–4118.

33 Park, J.B., Kim, N.J., Kim, Y.J. et al. (2014). Metal catalyst-assisted growth of GaN nanowires on graphene films for flexible photocatalyst applications. *Current Applied Physics* 14 (11): 1437–1442.

34 Morassi, M., Guan, N., Dubrovskii, V.G. et al. (2020). Selective area growth of GaN nanowires on graphene nanodots. *Crystal Growth & Design* 20 (2): 552–559.

35 Wang, Y., Dheeraj, D., Liu, Z. et al. (2019). AlGaN nanowires grown on SiO_2/Si(100) using graphene as a buffer layer. *Crystal Growth & Design* 19 (10): 5516–5522.

36 Hu, L., Zou, Y., Li, C.H. et al. (2020). Preparation of SiC nanowires on graphite paper with silicon powder. *Materials Letters* 269 (15): 127444.

37 Nakagawa, S., Tabata, T., Honda, Y. et al. (2013). GaN nanowires grown on a graphite substrate by radio frequency molecular beam epitaxy. *Japanese Journal of Applied Physics* 52: 08JE07.

38 Zhao, C., Ng, T.K., Tseng, C.C. et al. (2017). InGaN/GaN nanowires epitaxy on large-area MoS_2 for high-performance light-emitters. *RSC Advances* 7: 26665–26672.

39 Johansson, J. and Dick, K.A. (2011). Recent advances in semiconductor nanowire heterostructures. *CrystEngComm* 13: 7175–7184.

40 Hyun, J.K., Zhang, S., and Lauhon, L.J. (2013). Nanowire heterostructures. *Annual Review of Materials Research* 43: 451–479.

41 Hiruma, K., Tomioka, K., Mohan, P. et al. (2012). Semiconductor nanowires and nanotubes: from fundamentals to diverse applications. *Journal of Nanotechnology* 2012: 423879.

42 Jia, C., Lin, Z., Huang, Y., and Duan, X. (2019). Nanowire electronics: from nanoscale to macroscale. *Chemical Reviews* 119 (15): 9074–9135.

43 Bjork, M.T., Ohlsson, B.J., Sass, T. et al. (2002). One-dimensional steeplechase for electrons realized. *Nano Letters* 2: 87–89.

44 Gudiksen, M.S., Lauhon, L.J., Wang, J. et al. (2002). Growth of nanowire superlattice structures for nanoscale photonics and electronics. *Nature* 415: 617–620.

45 Ra, Y.H., Navamathavan, R., Park, J.H., and Lee, C.R. (2013). Coaxial InxGa$_1$_xN/GaN multiple quantum well nanowire arrays on Si(111) substrate for high-performance light-emitting diodes. *Nano Letters* 13 (8): 3506–3516.

46 Morassi, M., Largeau, L., Oehler, F. et al. (2018). Morphology tailoring and growth mechanism of indium-rich InGaN/GaN axial nanowire heterostructures by plasma-assisted molecular beam epitaxy. *Crystal Growth & Design* 18 (4): 2545–2554.

47 Ra, Y.H., Kang, S., and Lee, C.R. (2018). Ultraviolet light-emitting diode using nonpolar AlGaN core–shell nanowire heterostructures. *Advanced Optical Materials* 6 (14): 1701391.

48 Qian, F., Gradecak, S., Li, Y. et al. (2005). Core/multishell nanowire heterostructures as multicolor, high-efficiency light-emitting diodes. *Nano Letters* 5 (11): 2287–2291.

49 Lohani, J., Varshney, S., Rawal, D.S., and Tyagi, R. (2019). Vertically aligned nanowires comprising AlGaN/GaN axial heterostructure by convenient maskless reactive ion etching. *Materials Research Express* 6: 105001.

50 Carnevale, S.D., Yang, J., Philips, P.J. et al. (2011). Three-dimensional GaN/AlN nanowire heterostructures by separating nucleation and growth processes. *Nano Letters* 11 (2): 866–871.

51 Hiruma, K., Tomioka, K., Mohan, P. et al. (2012). Fabrication of axial and radical heterostructures for semiconductor nanowires by using selective-area metal-organic vapor phase epitaxy. *Journal of Nanotechnology* 2012: 169284.

52 Ra, Y.H., Navamathavan, R., Yoo, H.I., and Lee, C.R. (2014). Single nanowire light-emitting diodes using uniaxial and coaxial InGaN/GaN multiple quantum wells synthesized by metalorganic chemical vapor deposition. *Nano Letters* 14 (3): 1537–1545.

53 Sarau, G., Heilmann, M., Latzel, M. et al. (2019). GaN-based nanorods/graphene heterostructures for optoelectronic applications. *Physica Status Solidi B: Basic Solid State Physics* 256 (4): 1800454.

54 Kumanresan, V., Largeau, L., Madouri, A. et al. (2016). Epitaxy of GaN nanowires on graphene. *Nano Letters* 16 (8): 4895–4902.

55 Wang, D., Xue, C., Bai, H., and Jiang, N. (2015). Silicon carbide nanowires grown on graphene sheets. *Ceramics International* 41 (4): 5473–5477.

56 Han, S., Choi, I., Lee, C.R. et al. (2020). Fast response characteristics of flexible ultraviolet photosensors with GaN nanowires and graphene. *ACS Applied Materials & Interfaces* 12 (1): 970–979.

57 Cheng, Y., Tan, M., Hu, P. et al. (2018). Strong and thermostable SiC nanowires/graphene aerogel with enhanced hydrophobicity and electromagnetic wave absorption property. *Applied Surface Science* 448 (1): 138–144.

58 Bae, H., Rho, H., Min, J.W. et al. (2017). Improvement of efficiency in graphene/gallium nitride nanowire on silicon photoelectrode for overall water splitting. *Applied Surface Science* 422: 354–358.

59 Han, S., Lee, S.K., Choi, I. et al. (2018). Highly efficient and flexible photosensors with GaN nanowires horizontally embedded in a graphene sandwich channel. *ACS Applied Materials & Interfaces* 10 (44): 38173–38182.

60 Liu, X., Yang, X., Gao, G. et al. (2016). Enhancing photoresponsivity of self-aligned MoS_2 field-effect transistors by piezo phototronic effect from GaN nanowires. *ACS Nano* 10 (8): 7451–7457.

61 Peng, K., Zhou, J., Gao, H. et al. (2020). Emerging one-/two-dimensional heteronanostructure integrating SiC nanowires with MoS_2 nanosheets for efficient electrocatalytic hydrogen evolution. *ACS Applied Materials & Interfaces* 12 (17): 19519–19529.

62 Renard, J., Songmuang, R., Tourbot, G. et al. (2009). Evidence for quantum-confined Stark effect in GaN/AlN quantum dots in nanowires. *Phys. Rev.* B 80: 121305(R).

63 Deshpande, S., Frost, T., Yan, L. et al. (2015). Formation and nature of InGaN quantum dots in GaN nanowires. *Nano Letters* 15 (3): 1647–1653.

64 Deng, J., Hao, Z., Wang, L. et al. (2020). Studies on carrier recombination in GaN/AlN quantum dots in nanowires with a core–shell structure. *Nanomaterials* 10 (11): 2299.

4

Electrical Properties of Wide Bandgap Nanowires

4.1 Electrical Properties

The electrical properties of semiconductor nanowires (NWs) are generally characterized by the electrical conductivity or resistivity, which is dependent on several parameters, including carrier concentrations, mobility, ionization energy, and surface states [1–3]. The electrical resistivity of semiconductor NWs ρ depends on the carrier concentration n, which can be described by the following relationship:

$$\rho = \frac{1}{qn\mu} \tag{4.1}$$

where q is the electron charge and μ is the carrier mobility.

4.2 Measurement of Electrical Resistivity

In various semiconductor thin films or bulk materials, resistivity can be measured by four-point measurements that use a pair of electrodes for carrying currents and a pair of electrodes for measuring voltages. Different methods have been used to measure the electrical properties of NWs including four-point [4, 5] and three-point probes [6]. For example, Figure 4.1 shows the schematic sketch of four-point measurements to determine the conductivity/resistivity of NWs. A small current (I_{12}) is applied between two terminals/electrodes 1 and 2, while the voltage (V_{34}) between two terminals/electrodes 3 and 4 is measured. Therefore, the resistivity of the NW is calculated using the following equation [4]:

$$\rho = \frac{V_{34}}{I_{12}} \frac{\pi r^2}{L} \tag{4.2}$$

where L and r are the length and radius of the NW, respectively. Using the measured resistivity, the carrier concentration in NWs can be determined by calibration curves (also see Eq. (4.1)). In some cases, the measurement of electrical resistivity using the four-point measurements may lead to a different result for NWs compared with that of bulk materials with the same carrier concentration. For example, the resistivity of bulk silicon could be 1–2 orders of magnitude different to that of silicon nanowires (Si NWs) [3]. This difference is attributed

Wide Bandgap Nanowires: Synthesis, Properties, and Applications, First Edition. Tuan Anh Pham, Toan Dinh, Nam-Trung Nguyen, and Hoang-Phuong Phan

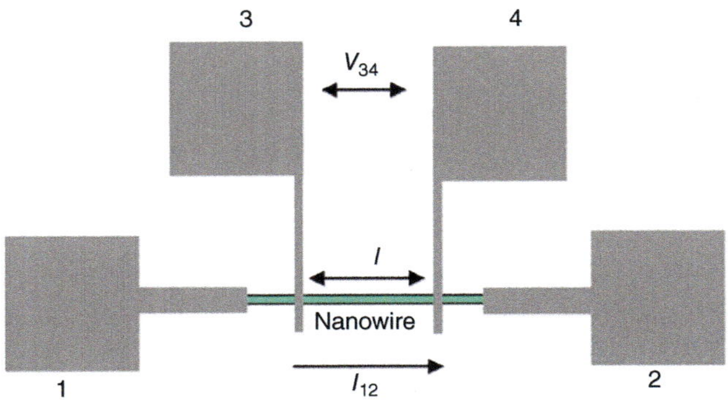

Figure 4.1 Schematic sketch of four-point measurements of the resistivity of a NW. A small current is applied between terminals 1 and 2, while the voltage between terminals 3 and 4 is measured. *Source:* Motayed et al. [4]/with permission of AIP Publishing.

to the quantum confinement effect that dominates the electrical transport of charge carriers in NWs.

Figure 4.2 shows the measurement of electrical properties of silicon carbide nanowires (SiC NWs) using a four-point-probe 3-ω method [5]. Figure 4.2a–d illustrates the scanning electron microscopy (SEM) images of SiC NWs with single (~60 nm diameter) and double (two identical NWs with ~60 nm diameter) structures. The 3-ω approach is used to characterize the electrical properties of the SiC NWs, Figure 4.2e. In this method, 3-ω signals are collected from a lock-in amplifier under the application of an alternating current (AC). Figure 4.2f,g shows the temperature dependence of the electrical properties of both single and double SiC NWs. The electrical resistance of the SiC NWs is measured in a small temperature range of 28.6–30.2 and 27–32.8 °C for single and double NWs, respectively. Therefore, the dependence of electrical properties of SiC NWs on temperature is evaluated using the temperature coefficient of resistance (TCR) as follows:

$$TCR = \frac{\Delta R}{R_o}\frac{1}{\Delta T} = \frac{R - R_o}{R_o}\frac{1}{T - T_o} \tag{4.3}$$

where R and R_0 are the electrical resistances measured at temperature T and reference temperature T_0, respectively. The resistance change of the single SiC NW is −10 Ω/K and that of the double SiC NW is −20 Ω/K. These resistance changes correspond to a TCR value of approximately −1000 ppm/K for both single and double SiC NWs. The electrical properties of wide bandgap (WBG) NWs depend on the excitation of charge carriers with increasing temperature and the change of carrier mobility. The detailed discussion on the generation of charge carriers and their mobility is presented in Section 4.3.

4.3 Fundamental Electrical Properties of Nanowires

Overall, the electrical conductivity of WBG semiconductor NWs depends on the carrier concentration and the mobility of carriers. The axial conductivity of a SiC NW can be determined from Ohm's law as [7]:

$$\sigma = \frac{4e^2}{m^* Sh}\tau\left(\frac{m^*}{2}\right)^{1/2}\int_0^\infty \varepsilon^{1/2}\frac{\partial f_0}{\partial \varepsilon}d\varepsilon = en\frac{e}{m^*}\tau \tag{4.4}$$

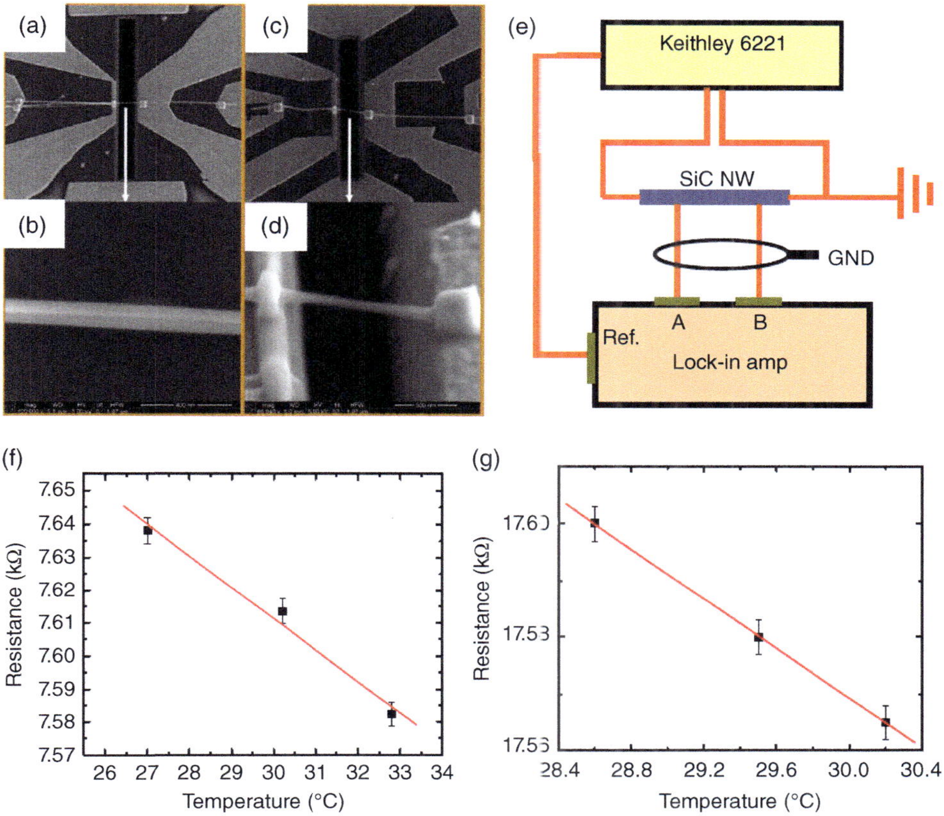

Figure 4.2 Measurement of electrical properties of SiC NWs using the four-point-probe 3-ω method. (a) SEM images of a double SiC NW on the four-point-probe electrodes pattern. (b) Magnified SEM image of the double SiC NW at its center. (c) SEM images of a single wire system. (d) Magnified SEM image of the single NW system. (e) Schematic diagram of the measurement setup. The third harmonic signal of the SiC NW is measured by using a lock-in amplifier and an AC current source. (f) Electrical resistance of the double SiC NW with temperature variation. (g) Electrical resistance of the single SiC NW with temperature variation. [5], Lee K, et. al., (2010) / with permission of IOP Publishing.

where n and μ are the electron concentration and electron mobility, respectively; T and S are the relaxation time and cross-sectional area of the NW; m^* denotes the effective mass. In general, the electrical conductivity of SiC NWs increases with increasing the diameters of the NWs and temperature. For example, Figure 4.3 shows an increase of the electrical conductivity of hydrogen passivated P-SiCNWss (Hyd.P-SiCNWs) from 50 to 60 S/cm when diameter increases from 0.75 to 1.45 nm [7]. Compared with Hyd.P-SiCNWs, phosphorus doped SiCNWss (P-SiC NWs) exhibits a stronger temperature dependence and weaker dependence on diameter. Different factors that influence the electrical conductivity of WBG NWs are discussed in the following sections.

4.3.1 Effect of Doping on Electrical Properties

Various dopants have been utilized to control the electrical resistivity of semiconductors, namely boron (B), aluminum (Al), and nitrogen (N). With impurities doped into a semiconductor NW, the Fermi level is shifted toward the conduction band of n-type semiconductors and the valance band of p-type semiconductors. This leads to a higher electrical

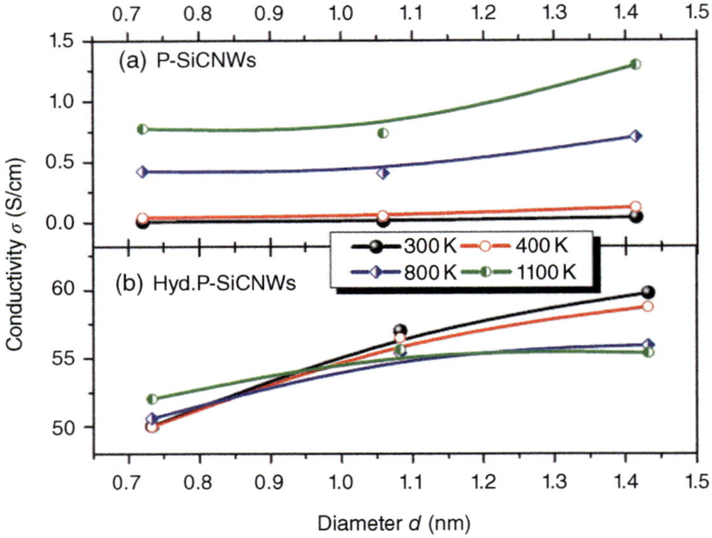

Figure 4.3 The dependence of SiC NWs conductivity on diameters. (a) P-SiCNWs. (b) Hyd.P-SiCNWs. *Source:* Li et al. [7]/with permission of Elsevier.

conductivity of semiconductors at room temperature due to the ionization of charge carriers. For example, when semiconductors are doped, the dopant atoms substitute semiconductor atoms in the lattice structures of the semiconductor. These dopants become electrically active donors/acceptors, leading to the increase in the electrical conductivity of the semiconductor. The carrier/dopant concentration in semiconductor NWs can be determined by counting the number of dopants per a unit volume using secondary-ion mass spectroscopy (SIMS) [8]. However, if the dopants at the NW surfaces are not properly substituted the semiconductor atoms or electrically inactive, the conductivity of semiconductor NWs would be different to the theoretically determined value. In addition, the increase of dopant concentration or dopant precursor will lead to the increases of the electrical conductivity or the decrease of the electrical resistivity of NWs. For example, Figure 4.4 shows the decrease of the electrical resistivity of Si NWs with increasing dopant precursor.

The electron concentration in NWs can be determined as follows [7].

$$\begin{aligned} n &= \frac{4e}{Sh}\left(\frac{m^*}{2}\right)^{1/2} \int_0^{E_F-E_c} E^{1/2}\frac{\partial f_0}{\partial E}dE \\ &= \frac{4e}{Sh}\bar{\tau}\left(\frac{m^*}{2}\right)^{1/2} E^{1/2}\left[1+\frac{\pi^2}{24}\left(\frac{k_BT}{E_c}\right)^2+\ldots\right] \end{aligned} \tag{4.5}$$

where E_c is the lowest energy of the conduction band. Equation (4.5) indicates a potential increase of carrier concentration with increasing temperature. For example, Figure 4.5 shows the dependence of electron concentration on temperature and diameter of the SiC NWs. Both P-SiCNWs and Hyd.P-SiCNWs exhibit the increase of the carrier concentration with increasing temperature or decreasing diameter.

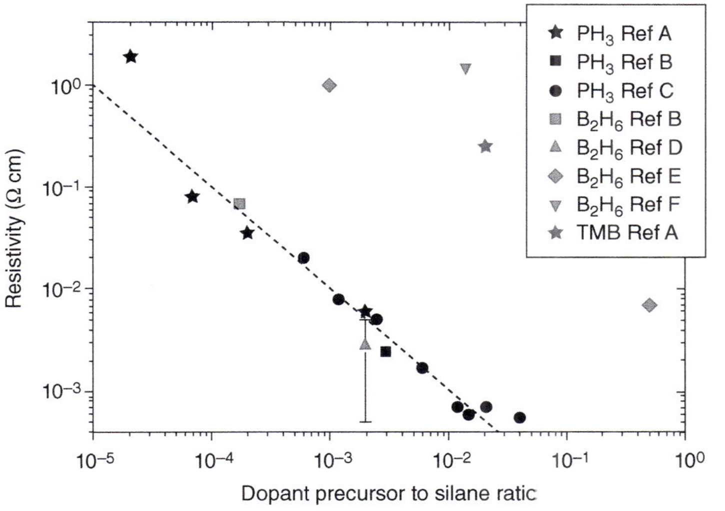

Figure 4.4 The dependence of Si NWs resistivity on dopant concentration. TMB = Trimethylborane. *Source:* Schmidt et al. [3]/with permission of John Wiley & Sons.

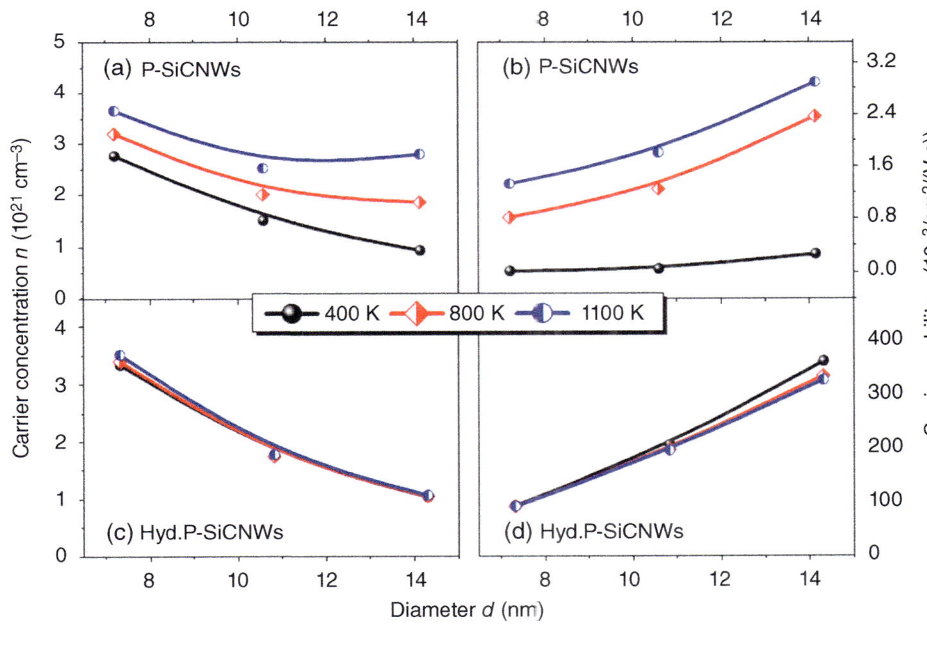

Figure 4.5 The dependence of electron concentration and mobility of SiC NWs on diameters. (a, b) P-SiCNWs. (c, d) Hyd.P-SiCNWs. *Source:* Li et al. [7]/with permission of Elsevier.

4.3.2 Mobility

In bulk semiconductors, the mobility of charge carriers is governed by the concentration of carriers, type of dopant, and temperature. The mobility of charge carriers typically decreases with increasing carrier concentration [9, 10]. This is due to the increase in scattering of ionized impurity atoms in the lattice structure. For example, n-type bulk silicon has a decrease in electron mobility of more than five times when the doping level increases from 10^{15} to 10^{18} cm^{-3} [10]. The mobility of hole is typically smaller than that of n-type semiconductors. In addition, the scattering effect is the primary factor which affects the mobility of charge carriers. Due to the high surface-to-volume ratio in semiconductor NWs, the considerable effect of surface scattering leads to a decrease in the mobility of the NWs. As such, the decrease of NW radius will result in the decrease of the NW volume or increase of the surface-to-volume ratio; this causes the decrease of the mobility [9, 10].

In addition, the surface dangling bonds of NWs can modify the electrical properties due to the discretization of the impurity energy band. For example, hydrogen atoms can be used to passivate surface dangling bonds of SiC NWs [7]. The mobility of SiC NWs depends on multiple scattering mechanisms which are described by the following equation [7]:

$$\frac{1}{\mu} = \frac{m^*}{e}\left(\frac{1}{\tau_{opt}} + \frac{1}{\tau_{ion}} + \frac{1}{\tau_{neu}}\right) \tag{4.6}$$

where τ_{opt}, τ_{ion}, and τ_{neu} represent the scattering caused by optical phonons, ionized impurities, and neutral impurities, respectively. In general, the mobility of NWs increases with increasing diameters due to their lower surface-to-volume ratio; the scattering at surface of NWs becomes less dominant. In addition, Hyd.P-SiC NWs have the decrease of carrier mobility with increasing temperature; P-SiC NWs show an opposite behavior [7]. This suggests the possibility to control the mobility of NWs using passivation which modifies the surface dangling bonds.

4.3.3 Activation/Ionization Energy

Activation/ionization energy refers to the energy that activates dopant atoms to ions. If the dopants are fully ionized, all dopant atoms become ions and charge carriers generated from the ionized atoms will be free to move in the lattice structure of semiconductor NWs. The level of ionization energy depends on temperature and electrostatic characteristics of the surrounding environments.

In general, the dependence of carrier/dopant concentration n in semiconductors on temperature T can be expressed as [11]:

$$n \propto T^{3/2} \exp\left(-\frac{E_a}{k_B T}\right) \tag{4.7}$$

where E_a is the activation/ionization energy and k_B is the Boltzmann constant. The increase of temperature will lead to the increase of carrier concentration. However, the carrier mobility decreases with increasing temperature as $\mu \sim 1/T\alpha$, where α is constant. The electrical resistivity of semiconductors can be described as:

$$\rho = \frac{1}{q\mu n} = T^{\alpha-3/2} \exp\left(\frac{E_a}{k_B T}\right) \tag{4.8}$$

Figure 4.6 shows the dependence of electrical resistance of SiC nanofilms on temperature; the activation energy and mobility constant are extracted from fitting curves of electrical resistance. Due to the high doping level of SiC (i.e. ~10^{19} cm^{-3}), a large number of dopants

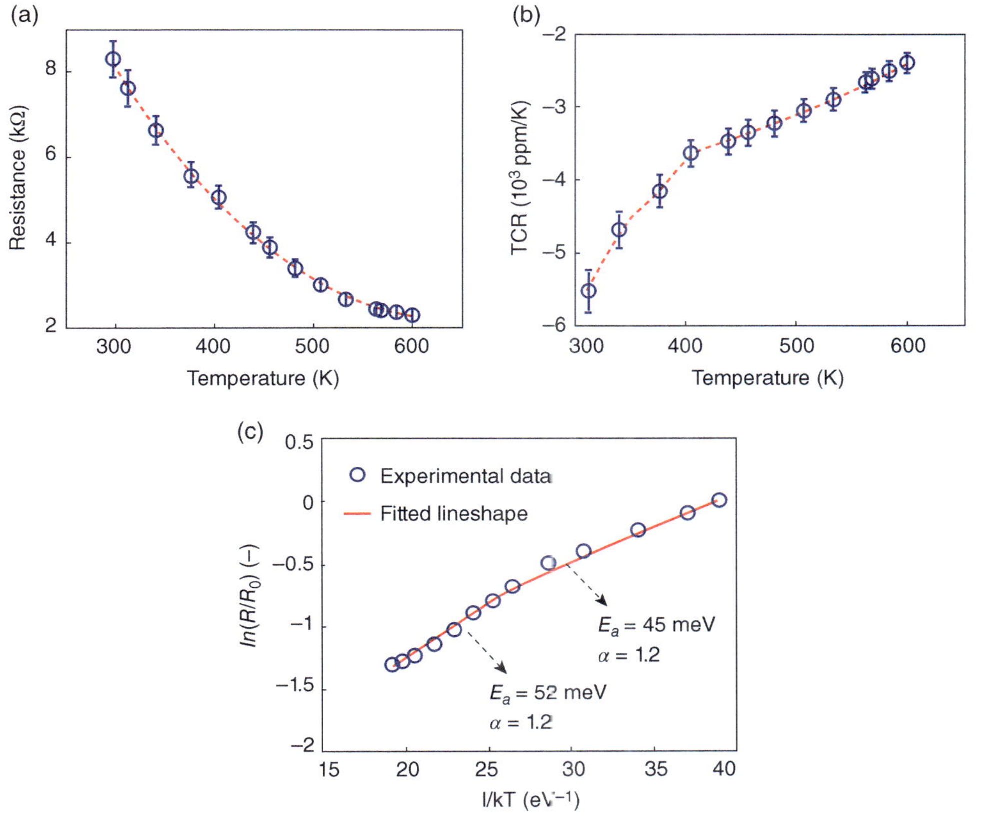

Figure 4.6 Activation energy of SiC. (a) The dependence of resistance on temperature. (b) Temperature coefficient of resistance (TCR) vs temperature. (c) Determination of activation energy using non-linear Arrhenius plot and its fitting curves. *Source:* Dinh et al. [11]/with permission of Royal Society of Chemistry.

are ionized at room temperature, leading to a relatively low activation energies of 45–52 meV (see Figure 4.6).

In GaN NWs, the activation energy can be determined from current–voltage (I–V) measurements while temperature is controlled. The source–drain current (I_{sd}) of GaN NW is determined as a function of temperature T:

$$I_{sd} \sim \exp\left(\frac{-E_a}{k_B T}\right) \tag{4.9}$$

Chen et al. [12] reported an activation energy of 18 meV based on the plot of $ln(I_{sd})$ vs $1/T$ for GaN. The activation energy (E_d) is associated to the difference in the conduction band energy E_C and donor energy E_D for n-type semiconductor NWs, which is determined as $E_d = E_C - E_D$. In addition, electron concentration is estimated as [4]:

$$n \cong \frac{1}{\sqrt{2}}\sqrt{N_D N_C}\exp\left(\frac{-E_d}{2k_B T}\right) \tag{4.10}$$

where N_D and N_C are the donor concentration and conduction band density of states, respectively. Using the relationship $\rho = 1/qn\mu$, the resistivity is calculated as [4]:

$$\rho \cong \frac{1}{\mu q}\sqrt{\frac{2}{N_D N_C}}\exp\left(\frac{E_d}{2k_B T}\right) \tag{4.11}$$

Figure 4.7 shows the fittings to determine the activation energy E_d of GaN NWs, indicating the increase of the activation energy with increasing temperature [4]. For example, the activation energy of GaN NWs is extracted to be 1.9 and 29.2 meV for the temperature below 200 K and above 200 K, respectively.

4.3.4 Dependence of Activation/Ionization Energy on NW Dimensions

In semiconductor NWs, the charge carriers shield electrostatic potential of ionized dopants, leading to the dependence of the ionization energy on the volume of NWs. In general, the activation/ionized energy increases with decreasing radius of NWs [3]. For example, when diameter of NWs increases, the ionized energy decreases using the following relationship:

$$E_I(r) = E_I^0 + \frac{2}{R\varepsilon_s}\frac{\varepsilon_s - \varepsilon_o}{\varepsilon_s + \varepsilon_o}F\left(\frac{\varepsilon_s}{\varepsilon_0}\right) \tag{4.12}$$

where E_I^0 is the ionization energy of dopant in bulk materials. ε_s and ε_o are relative permittivity of silicon and surrounding environment, respectively. Details on the function $F\left(\frac{\varepsilon_s}{\varepsilon_0}\right)$ can be found in Ref. [3].

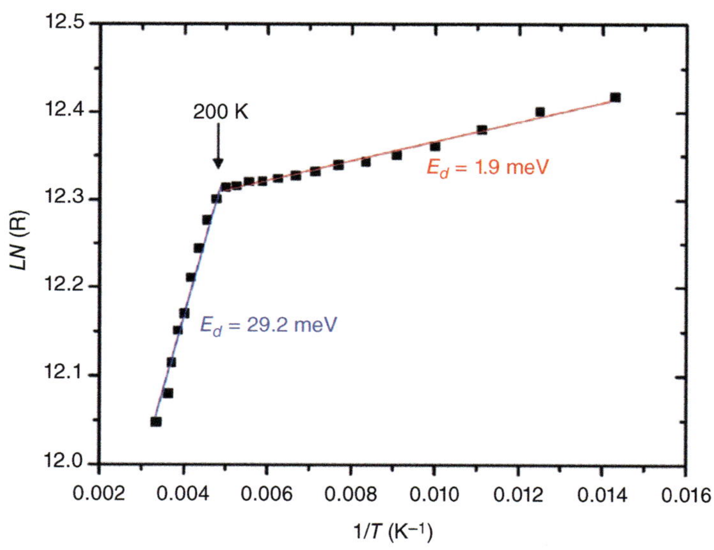

Figure 4.7 Activation energy of GaN nanowires. Red line is linear fit with E_d = 29.2 meV for a temperature range of 200–300 K and blue line shows the linear fit with E_d = 1.9 meV for the temperature range of 70–200 K. *Source:* Motayed et al. [4]/with permission of AIP Publishing.

4.4 Electrical Properties of Wide Bandgap Nanowire Based Devices

4.4.1 Single NW Electrical Sensing Devices

Single NWs are of interest for sensor applications due to their excellent electrical properties. The strong dependence of carrier mobility and surface modification on the input sensing signals such as stress/strain makes NWs a potential candidate for ultrasensitive sensors. Owing to the change of electrical properties, the sensing signals will be estimated or evaluated. The resistivity/conductivity of WBG NWs is required to be in a reasonable range for electrical measurement of sensing signals. For example, NWs with small diameter/cross-sectional areas could have low electrical conductivity of NWs, which makes great challenges for measurements and sensing applications. Another challenge is to form the non-Ohmic contact between the NWs and their electrodes. Non-Ohmic contact creates a barrier for reliable electrical measurements. Therefore, great effort has been paid to improve the synthesis process of NWs and understand the effect of physical sensing inputs to electrical properties of NWs. Figure 4.8a shows an SEM image of a three-probe method for measurement of electrical resistance of a ZnO NW under applied strain ϵ [6]. In this method, a current is applied and a voltage is measured. If the electrical contact between the NW and electrode is Ohmic, the resistance of the NW can be determined. In other word, the Ohmic contact is highly desirable for measuring the electrical properties of NWs. For example, Figure 4.8b shows the current–voltage characteristics of the ZnO NWs with deposition of Pt using electron-beam (e-beam) and ion-beam (i-beam). It is evident that the i-beam Pt deposition leads to the Ohmic contact between the ZnO NW and the electrode. This result could be due to the high energy Ga ions directly injected into the ZnO NWs, causing the increase of defect-assisted tunneling across the barrier between NW and metal electrode; hence the electrical contact is improved [6]. Figure 4.8c indicates the dependence of the electrical properties of ZnO NWs on environmental conditions. The resistance is stable in air while it decreases with increasing pressure or increasing of irradiation intensity. The dependence of the electrical resistance on environmental factors are attributed to the variation of depletion layer under the NW surface due to the absorption of vapor species and e-beam radiation. Figure 4.8d shows the dependence of the electrical resistivity on the diameter of ZnO NWs. In general, the resistivity decreases with increasing diameters or decreasing length of NWs. For example, the resistivity decreases by about 40% when diameters increase from 80 to 420 nm [6]. This is explained by the reduction of the cross-sectional area with decreasing diameter. In addition, hexagonal NWs have approximately 3 orders of magnitudes of resistivity greater than cylindrical NWs. This could be due to higher concentration of defects in the ZnO NWs with the hexagonal cross section [6].

4.4.2 Field-Effect Transistors (FETs)

Having excellent electrical properties, WBG NWs have attracted research interest for applications in switching devices or control of electrical signals, including field-effect transistors (FETs). NW based FETs utilize an electric field to regulate the electric current flowing in the NW. Different parameters have been characterized to evaluate the performance of WBG NW FETs. The figures-of-merits (FOMs) of WBG FETs include transconductance (g_m), carrier mobility (μ), carrier concentration (n), and the ratio between ON and OFF currents (I_{on-off}).

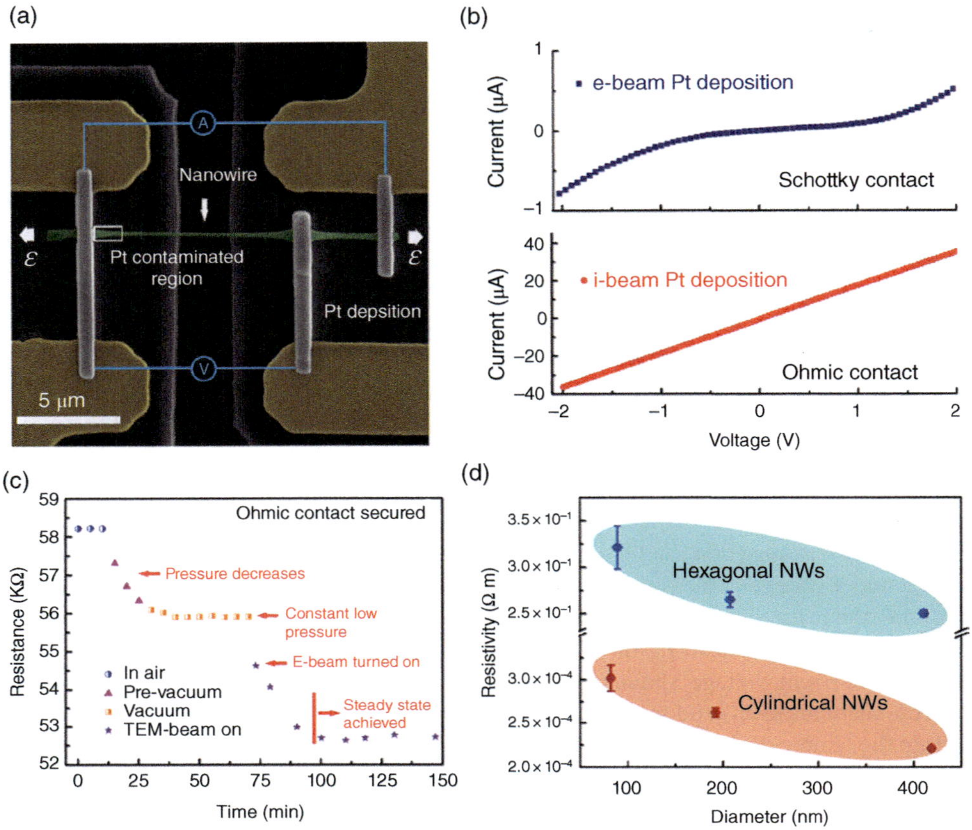

Figure 4.8 Electrical properties of ZnO NWs for strain sensors. (a) SEM image showing the three-point measurement of ZnO NWs. (b) The current–voltage (*I*–*V*) curves of a cylindrical ZnO NW with electrical contacts induced by e-beam (top) and i-beam (bottom), suggesting Schottky and Ohmic contact, respectively. (c) Electrical resistance of ZnO NWs with different environmental conditions. (d) The dependence of electrical resistivity on the diameter of ZnO NWs. TEM = Transmission electron microscopes. *Source:* [6], Wang X, et. al., (2015)/with permission of American Chemical Society.

These FOMs can be obtained from current–voltage (*C*–*V*) measurements and analysis of the FETs. High-performance NW-based FETs show high value of FOMs. Physics of conventional NW FETs based on Si material have been well documented. However, underlying physics of WBG semiconductor NW-based FETs have not been fully understood, and further investigation of electrical properties of FETs based on SiC, GaN, and diamond is required [13, 14].

4.4.3 SiC NW-Based FETs

SiC NWs have been employed to develop FETs with high performance. For example, Chen et al. [13] reported a high transconductance of 12 nS, a carrier density of 1.7×10^{17} cm^{-3}, a relatively low carrier mobility of 6.4 cm^2/(V s) for p-type aluminum doped 3C-SiC NWs. The ratio between ON and OFF currents (I_{on-off}) of SiC NWs was reported in the range from 1.5 to 1.34×10^4 [14–16]. Figure 4.9 shows the SEM image of SiC NWs and SiC NW based FET device with a high ON/OFF current ratio [14].

Various studies focused on the intrinsic n-type semiconductor behavior of SiC NWs. The n-type semiconducting properties of SiC NWs are attributed to the high density of donor states below the conduction band edge, which is resulted from structural defects, including stacking faults. For example, Rogdakis et al. characterized SiC NW-based FET device and the characterization results showed that it is a challenge to switch off the device with the Ohmic contact even at high gate bias voltages up to −40V. SiC NW-based FET devices could have a high concentration of charge carriers induced by unintentional doping [17]. Therefore, SiC NW-based FETs showed relatively poor device performance. To address this challenge, a number of studies have demonstrated the improved electrical properties of SiC NWs by surface modifications [18, 19] and control of doping concentrations and type of dopants [20, 21]. Figure 4.10 shows a typical relationship between the electrical current I_{DS} and the

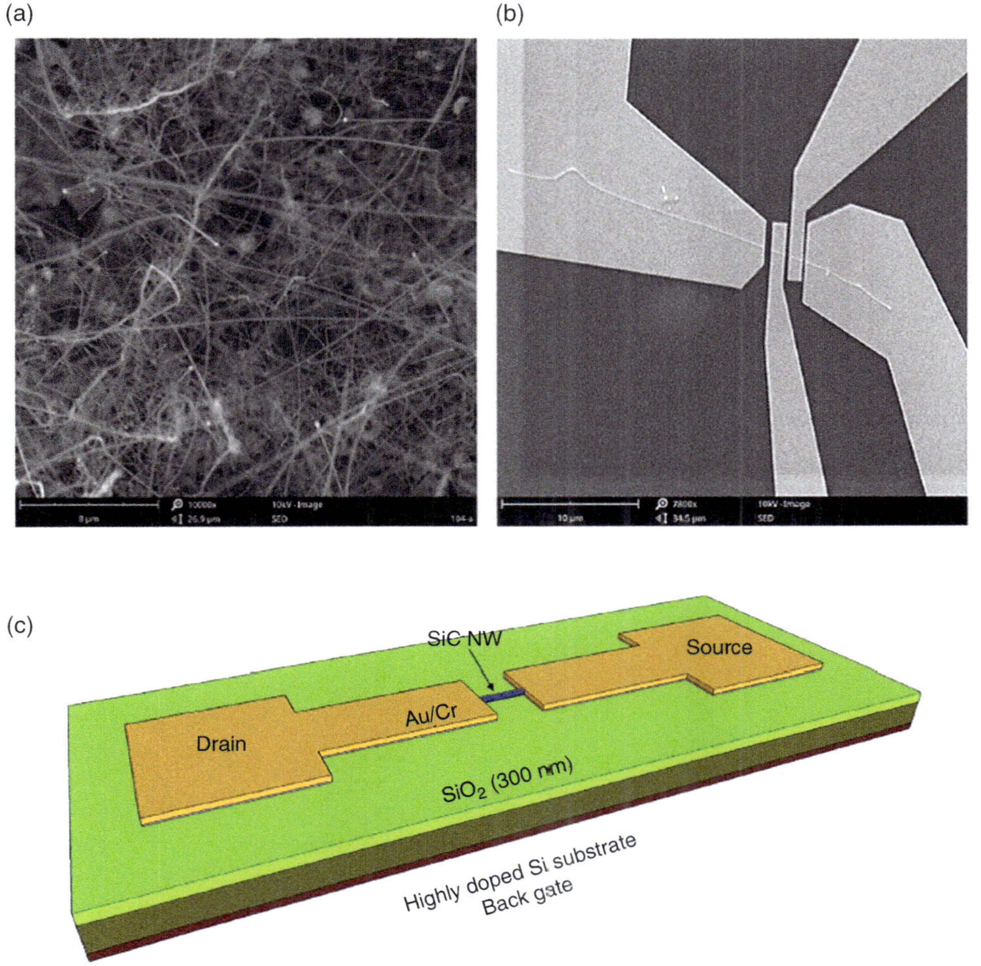

Figure 4.9 (a) An SEM image of p-type SiC NWs on SiO_2/Si substrate. (b) An SEM image of p-type SiC NWs with electrodes. (c) A schematic view of the SiC NWs FET device. *Source:* [14], Uzun A, et. al., (2019)/with permission of ELSEVIER.

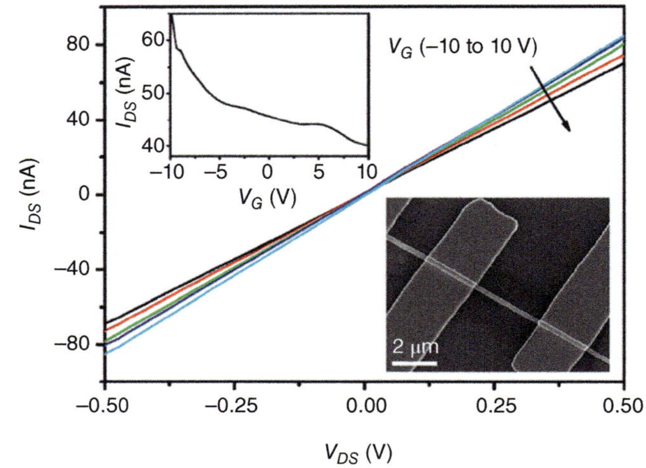

Figure 4.10 I_{DS}–V_{DS} characteristics of the Al-doped SiC NWs FET. Upper inset shows the I_{DS}–V_{DS} at V_{DS} = 0.5 V. Lower inset shows an SEM image of the device. *Source:* [13], Chen Y, et. al., (2011)/with permission from Royal Society of Chemistry.

drain–source voltage V_{DS} with applied gate voltages from −10 to 10 V. The resistance of SiC NWs is determined by:

$$R = \rho \frac{L}{\pi r^2} \tag{4.13}$$

where L and r are the length and radius of the SiC NW, respectively. The resistance can also be determined from current–voltage (C–V) measurements as: $R = V_{DS}/I_{DS}$. Therefore, the resistivity ρ is calculated as:

$$\rho = \frac{V_{DS}\pi r^2}{I_{DS}L} \tag{4.14}$$

The hole mobility can be calculated using [4, 13]:

$$\mu = g_m \frac{L^2 V_{DS} \ln(2h/r)}{2\pi\varepsilon_0\varepsilon} \tag{4.15}$$

where $g_m = \frac{\partial I_{DS}}{\partial V_g}$, determined by the slope of the current–voltage ($I_{DS} - V_g$) curve, is the transconductance of NWs; h is the dielectric thickness of the SiO_2 layer, ε_0 is the permittivity of free space, and ε is the permittivity of silicon dioxide. Therfore, the carrier concentration can be determined as: $n = \frac{I}{q\mu\rho}$.

4.4.4 GaN NW-Based FETs

Research on GaN NW-based FETs has largely focused on fabrication and characterization of unintentionally doped GaN NWs. These GaN NWs typically show n-type semiconducting properties as a result of the intrinsic residual nitrogen. The electrical properties of GaN NWs are characterized by the I–V characteristics [4, 22, 23]. Several approaches have been utilized to improve the electrical properties of GaN NWs for high-performance electronic devices. For example, the electrical properties of GaN NWs can be enhanced by doping [12, 24], modification of NW surfaces [25, 26], and formation of heterostructures [27]. In an early work, Chen et al. characterized Mn-doped GaN NWs that exhibit enormous improvements of electrical conductivity compared with intrinsic GaN NW [12]. Figure 4.11 shows 2 orders of magnitude of conductivity enhancements based on the I–V measurement of intrinsic and Mn-doped GaN NWs [12]. The formation of the heterostructures using dissimilar group III-nitride semiconductors in NWs is an effective strategy to control the transport of charge carriers, providing superior and controllable electrical properties [27–29].

A pioneering work on WBG NWs for high electron mobility transistors was reported by the Lieber group using dopant-free GaN/AlN/AlGaN radial NW heterostructures [30]. Figure 4.12a shows the high-angle annular darkfield scanning transmission electron microscopy (HAADF-STEM) image of the cross-sectional area of a GaN/AlN/AlGaN NW which was fabricated by sequential shell growth and following NW elongation using metal–organic chemical vapor deposition (MOCVD). Figure 4.12b shows the STEM image and the X-ray spectroscopy (EDX) elemental mapping of the GaN/AlN/AlGaN NW. The NW showed a high electron mobility of 3100 and 21 000 $cm^2/(V s)$ at room temperature and 5 K, respectively. This research showed significant difference of the electrical properties of FETs fabricated from the GaN/AlN/$Al_{0.25}Ga_{0.75}N$ and undoped GaN NWs. The I_{DS}–V_{DS} and I_{DS}–V_{GS} characteristics shown in Figure 4.12c,d indicated that the GaN/AlN/AlGaN NW device acts as depletion mode FET and reveal the accumulation of electron carriers. In addition, the GaN/AlN/$A_{0.25}Ga_{0.75}N$ heterostructure displays 50-time smaller transconductance and very high resistivity at $V_{GS} = 0$, compared to undoped GaN NW-based FETs. Since both heterostructures and GaN NWs are intrinsic, the enormously enhanced electrical conductivity of the GaN/AlN/AlGaN FET is attributed to the formation of a confined free electron gas by modifying the band structure of the radical NW heterostructure. Therefore, it is the capable

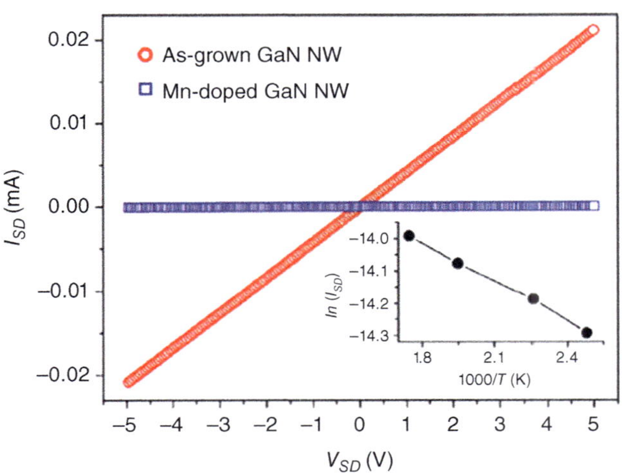

Figure 4.11 Current–voltage (I–V) characteristics of intrinsic GaN nanowire and Mn-doped GaN nanowire. *Source:* Chen et al. [12]/with permission of AIP Publishing.

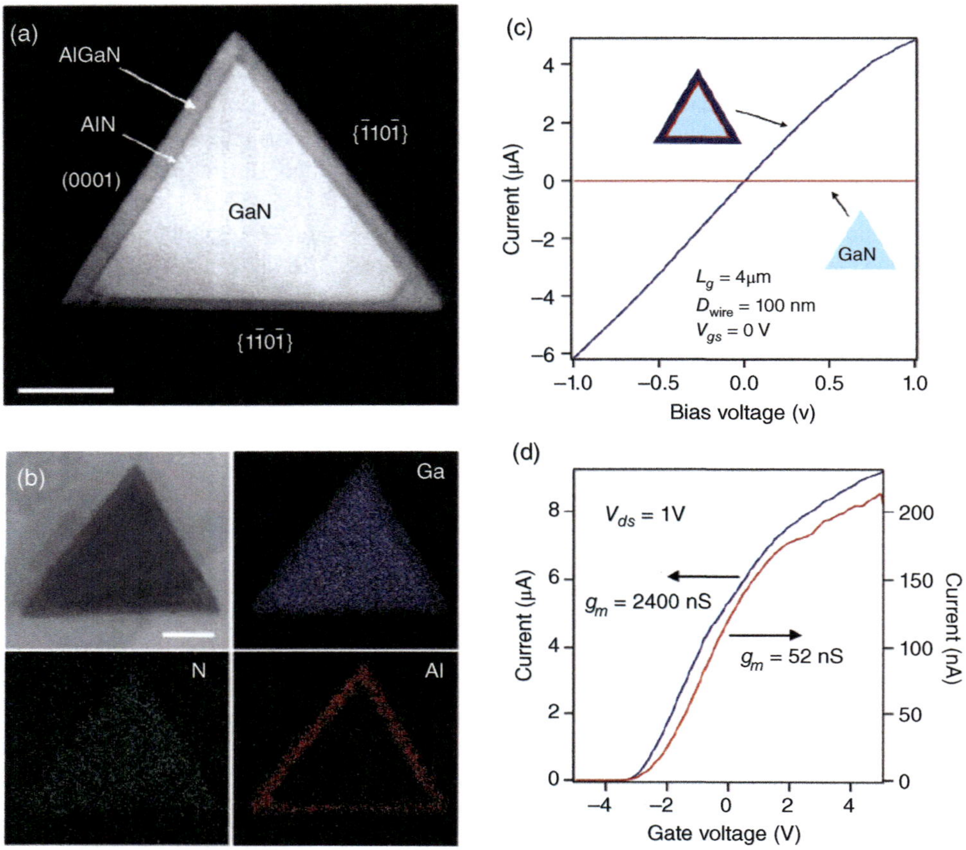

Figure 4.12 (a) HAADF-STEM image of the cross-sectional area of a GaN/AlN/AlGaN nanowire. Scale bar is 50 nm. (b) STEM image and the EDX elemental mapping of the GaN/AlN/AlGaN nanowire. (c) I_{DS}–V_{DS} characteristics recorded on 100-nm-diameter GaN/AlN/$Al_{0.25}Ga_{0.75}N$ (blue) and GaN (red) nanowires. (d) I_{DS}–V_{GS} characteristics of the same GaN/AlN/$Al_{0.25}Ga_{0.75}N$ (blue) and GaN (red) nanowires for V_{DS} = 1 V. *Source:* [30], Li Y, et. al., (2006)/with permission of American Chemical Society.

of controlling the electrical conductivity of NWs by designing band structure in III-nitride radial NW heterostructures.

4.4.5 ZnO NW-Based FETs

The ZnO NW-based FET has attracted considerable interest for sensors and electronic devices [31]. The control of density, alignments, and electrical properties of ZnO NWs could lead to the enhancement of the performance of various ZnO NW based sensing devices. For example, Boubenia et al. [31] reported on the growth and electrical characterization of a ZnO NW based FET for sensing applications. Figure 4.13a shows the schematic sketch of a ZnO NW constructed on silicon oxide layer with a thickness of 170 nm. Figure 4.13b shows the atomic force microscopy (AFM) image of the NW-FET device, indicating the perfect cover of the NW by electrodes. Figure 4.13c,d shows the electrical properties of NW-FET which is characterized by current–voltage measurements. The NW-FET exhibits an increase of the drain–source current (I_{DS}) with increasing gate–source voltage (V_{GS}). The output scans

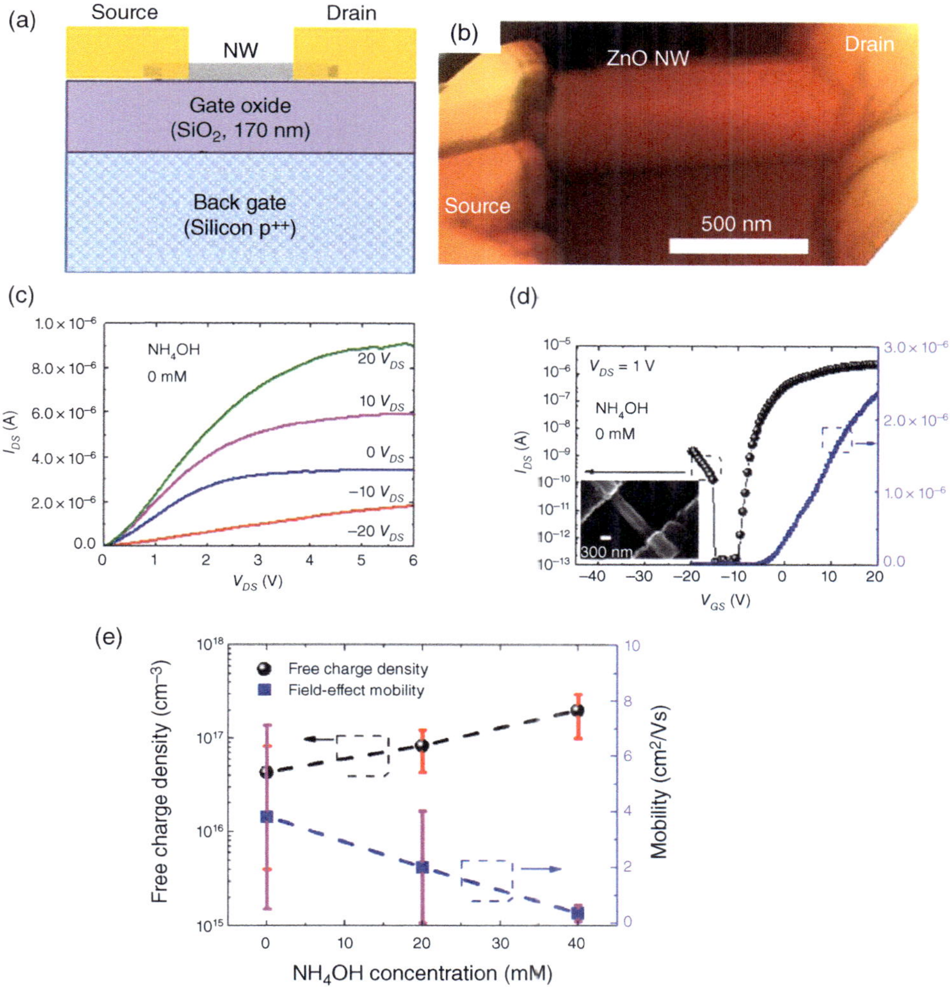

Figure 4.13 Electrical properties of ZnO NW based transistors. (a) Schematic sketch of a ZnO NW-FET. (b) AFM image of the ZnO NW-FET. (c, d) Electrical characterization results for ZnO NW-FETs with zero NH_4OH concentration. (e) Mobility and free charge density with the addition of NH_4OH concentration. *Source:* [31], Boubenia S, et. al., (2017)/with permission of Springer Nature.

show a linear dependence of I_{DS} on increasing drain–source voltage (V_{DS}) (at $V_{DS} \leq 0.1$ V). The drain–source current (I_{DS}) reached the saturating region at a higher V_{DS} range. In addition, ZnO NWs show the excellent chemical sensing properties. For example, an increase of the drain–source current (I_{DS}) with increasing gate–source voltage (V_{GS}) is observed for ZnO NW FETs when ammonium hydroxide is applied. The increase in V_{GS} toward a positive value will lead to the increase of I_{DS}, which indicates that ZnO NWs act as the n-channel accumulation type FET. The intrinsic semiconducting properties of ZnO NWs can be determined as follows [31]:

Field effect mobility,

$$\mu_{FE} = \frac{g_{m \times L^2}}{V_{DS} \times C_{NW}} \tag{4.16}$$

Free charge density,

$$n_e = \frac{C_{NW} V_{TH}}{2\mu r^2 qL} \tag{4.17}$$

where the gate capacitance C_{NW} is determined as:

$$C_{NW} = \frac{2\pi\varepsilon\varepsilon_0 L}{cosh^{-1}\left(\frac{r_{NW} + t_{ox}}{r_{NW}}\right)} \tag{4.18}$$

where r_{NW} and t_{ox} are the radius of the NW and thickness of silicon oxide layer, respectively; ε and ε_0 are the absolute permittivity and permittivity of the silicon oxide layer, respectively. Figure 4.13e shows the calculated free charge density and field-effect mobility of the ZnO NWs. The field effect mobility decreases almost 10 times when the concentration of ammonium hydroxide is 40 mM, while the free charge density increases approximately 20 times. However, due to the chemical reactions between ZnO NW surface and the metal contact, the extracted mobility and carrier density vary significantly (see Figure 4.13e). This emphasizes the importance of high-quality surface of NWs and electrical contact in determination of electrical properties of NW-based devices.

References

1 Zekentes, K. and Rogdakis, K. (2011). SiC nanowires: material and devices. *Journal of Physics D: Applied Physics* 44 (13): 133001.

2 Schmidt, V., Wittemann, J., and Gosele, U. (2010). Growth, thermodynamics, and electrical properties of silicon nanowires. *Chemical Reviews* 110 (1): 361–388.

3 Schmidt, V., Wittemann, J.V., Senz, S., and Gösele, U. (2009). Silicon nanowires: a review on aspects of their growth and their electrical properties. *Advanced Materials* 21 (25, 26): 2681–2702.

4 Motayed, A., Davydov, A.V., Mohammad, S., and Melngailis, J. (2008). Experimental investigation of electron transport properties of gallium nitride nanowires. *Journal of Applied Physics* 104 (2): 024302.

5 Lee, K., Choi, T., Lee, S., and Poulikakos, D. (2010). Focused ion beam-assisted manipulation of single and double β-SiC nanowires and their thermal conductivity measurements by the four-point-probe 3-ω method. *Nanotechnology* 21 (12): 125301.

6 Wang, X., Chen, K., Zhang, Y. et al. (2015). Growth conditions control the elastic and electrical properties of ZnO nanowires. *Nano Letters* 15 (12): 7886–7892.

7 Li, Y.-J., Li, S.-L., Gong, P. et al. (2018). Effect of surface dangling bonds on transport properties of phosphorous doped SiC nanowires. *Physica E: Low-Dimensional Systems and Nanostructures* 104: 247–253.

8 Benninghoven, A. (1973). Surface investigation of solids by the statical method of secondary ion mass spectroscopy (SIMS). *Surface Science* 35: 427–457.

9 Sze, S.M., Li, Y., and Ng, K.K. (2021). *Physics of Semiconductor Devices*. Wiley.

10 Kasap, S.O. (2006). *Principles of Electronic Materials and Devices*. New York: McGraw Hill.

11 Dinh, T., Phan, H.-P., Kozeki, T. et al. (2015). Thermoresistive properties of p-type 3C-SiC nanoscale thin films for high-temperature MEMS thermal-based sensors. *RSC Advances* 5 (128): 106083–106086.

12 Chen, X., Lee, S.J., and Moskovits, M. (2007). Modification of the electronic properties of GaN nanowires by Mn doping. *Applied Physics Letters* 91 (8): 082109.

13 Chen, Y., Zhang, X., Zhao, Q. et al. (2011). p-Type 3C-SiC nanowires and their optical and electrical transport properties. *Chemical Communications* 47 (22): 6398–6400.

14 Uzun, A. and Teker, K. (2019). Silicon carbide nanowire field effect transistors with high on/off current ratio. *Microelectronic Engineering* 205: 59–62.

15 Seong, H.K., Lee, S.Y., Choi, H.J. et al. (ed.) (2006). Fabrication and electrical transport properties of CVD grown silicon carbide nanowires (SiC NWs) for field effect transistor. In: *Materials Science Forum*, 771–774. Trans Tech Publications Ltd.

16 Zhou, W., Fang, F., Hou, Z. et al. (2006). Field-effect transistor based on β-SiC nanowire. *IEEE Electron Device Letters* 27 (6): 463–465.

17 Rogdakis, K., Lee, S.Y., Bescond, M. et al. (2008). 3C-silicon carbide nanowire FET: an experimental and theoretical approach. *IEEE Transactions on Electron Devices* 55 (8): 1970–1976.

18 Jang, C.-O., Kim, T.-H., Lee, S.-Y. et al. (2008). Low-resistance ohmic contacts to SiC nanowires and their applications to field-effect transistors. *Nanotechnology* 19 (34): 345203.

19 Li, Y., Dorozhkin, P.S., Bando, Y., and Golberg, D. (2005). Controllable modification of SiC nanowires encapsulated in BN nanotubes. *Advanced Materials* 17 (5): 545–549.

20 Rogdakis, K., Bano, E., Montes, L. et al. (2010). Field effect transistors based on catalyst-free grown 3C-SiC nanowires. In: *Materials Science Forum*, vol. 645–648 (ed. A.J. Bauer, P. Friedrichs, M. Krieger, et al.), 1235–1238. https://doi.org/10.4028/www.scientific.net/MSF.645-648.1235.

21 Mathur, S., Barth, S., Shen, H. et al. (2005). Size-dependent photoconductance in SnO_2 nanowires. *Small* 1 (7): 713–717.

22 Ham, M.-H., Choi, J.-H., Hwang, W. et al. (2006). Contact characteristics in GaN nanowire devices. *Nanotechnology* 17 (9): 2203.

23 Wu, H., Cha, H.-Y., Chandrashekhar, M. et al. (2006). High-yield GaN nanowire synthesis and field-effect transistor fabrication. *Journal of Electronic Materials* 35 (4): 670–674.

24 Richter, T., Meijers, H.L.R., Calarco, R., and Marso, M. (2008). Doping concentration of GaN nanowires determined by opto-electrical measurements. *Nano Letters* 8 (9): 3056–3059.

25 Calarco, R., Stoica, T., Brandt, O., and Geelhaar, L. (2011). Surface-induced effects in GaN nanowires. *Journal of Materials Research* 26 (17): 2157.

26 Dobrokhotov, V., McIlroy, D., Norton, M.G. et al. (2006). Principles and mechanisms of gas sensing by GaN nanowires functionalized with gold nanoparticles. *Journal of Applied Physics* 99 (10): 104302.

27 Songmuang, R., Katsaros, G., Monroy, E. et al. (2010). Quantum transport in GaN/AlN double-barrier heterostructure nanowires. *Nano Letters* 10 (9): 3545–3550.

28 Roul, B., Kumar, M., Rajpalke, M.K. et al. (2015). Binary group III-nitride based heterostructures: band offsets and transport properties. *Journal of Physics D: Applied Physics* 48 (42): 423001.

29 Rizzi, A. (2002). Surface and interface electronic properties of group III-nitride heterostructures. *Applied Surface Science* 190 (1–4): 311–317.

30 Li, Y., Xiang, J., Qian, F. et al. (2006). Dopant-free GaN/AlN/AlGaN radial nanowire heterostructures as high electron mobility transistors. *Nano Letters* 6 (7): 1468–1473.

31 Boubenia, S., Dahiya, A.S., Poulin-Vittrant, G. et al. (2017). A facile hydrothermal approach for the density tunable growth of ZnO nanowires and their electrical characterizations. *Scientific Reports* 7 (1): 1–10.

5

Mechanical Properties of Wide Bandgap Nanowires

5.1 Characterization Techniques

Various methods have been used to characterize the mechanical properties of nanowires (NWs), including bending, indenting, and vibrating. The main challenge in mechanical testing of NWs is the manipulation/handling of nanoscale structures. The generation and control of micro-/nanonewton forces and measurement of micro-/nanometer displacements can be implemented using different advanced micro-/nano-characterization methodologies such as atomic force microscopy (AFM) testing, nano-indenting, *in situ* electron microscope testing, and vibration testing.

5.1.1 Bending and Buckling Methods

Using micro-/nano-characterization techniques, several methods have been developed to examine the mechanical properties of wide bandgap (WBG) semiconductor nanowires [1]. Nanoscale bending approach, which uses *in situ* scanning electron microscopy (SEM) to generate nanoforces, is commonly employed to deform nanowires. For example, Wong et al. [2] deployed AFM to characterize the mechanical properties of SiC nanorods protruding from a SiO pad. Figure 5.1a,b shows the schematic sketches of the bending beam method used for charactering mechanical properties of the nanowires/nanotubes [2]. Using the data of force and displacement, the mechanical properties, e.g. modulus of elasticity or Young's modulus, can be calculated. Theoretically, Young's modulus E of the NW can be deduced from the following relationship [2]:

$$EI\frac{d^4y}{dx^4} = -f + P\delta\left(x-a\right) \tag{5.1}$$

where f is the beam-surface friction force and I is the moment of inertia. The friction force is small and therefore negligible, the nanobeam force constant $k(x)$ is determined as [2]:

$$k\left(x\right) = \frac{3I}{x^3}E \tag{5.2}$$

If $k(x)$ is determined for different values of x, the measurement of Young's modulus E would be accurate and reliable without the need of exact point for applying nanoforce. Using

Wide Bandgap Nanowires: Synthesis, Properties, and Applications, First Edition. Tuan Anh Pham, Toan Dinh, Nam-Trung Nguyen, and Hoang-Phuong Phan.

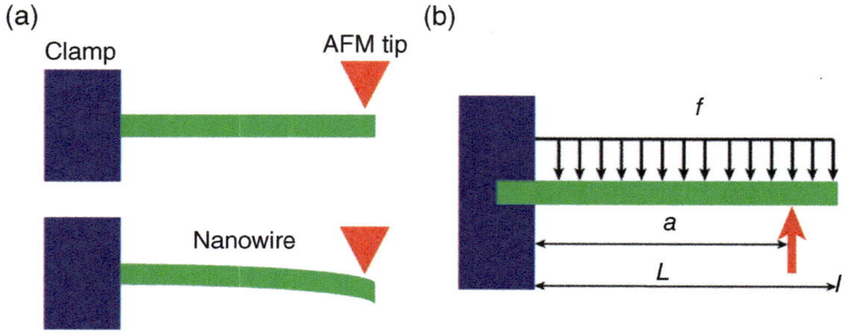

Figure 5.1 Characterization of mechanical properties of nanotubes/nanowires by bending. (a) Schematic sketches of beam bending [2]. The AFM tip (red triangle) creates the bending force to the nanowire. (b) Schematic sketch of a clamped nanowire with a free end. The beam with the length L undergoes a point load P and a distributed friction force f.

the plot of $k(x)$ vs x from experimental data, Young's modulus E can be calculated. As such, the theoretical value of E is calculated using [1]:

$$E = \frac{Pa^3}{3I\delta} \tag{5.3}$$

where P and δ are the loading force and deflection of the beam. Wong et al. [2] determined the range of Young's modulus from 610 to 660 GPa, which was in solid agreement with the theoretical value of ~600 GPa for a single SiC crystal along the ⟨111⟩ direction. In addition, three-point bending was used to characterize Young's modulus of nanotubes and nanowires [3, 4]. The common configuration of three-point bending uses a doubly clamped beam with a load (e.g. AFM tip) applying to the middle of the beam.

Nanoscale buckling testing of NWs can be conducted by applying compression using nanomanipulator and AFM cantilever. Figure 5.2a,b shows the SEM images of an NW in buckling experiments [5]. The effective length L_e is half actual length of the nanowire. Figure 5.2c shows the dependence of separation distance between two AFM probes on the applied load. Young's modulus of NWs from the buckling tests can be determined as [5]:

$$E = P_{cr} L_e^2 / \pi^2 I \tag{5.4}$$

where I is the moment of inertia and P_{cr} is the buckling force.

Other methods were also utilized to investigate the mechanical properties of nanowires, including SEM [6], field emission microscopy (FEM) [7], and transmission electron microscopy (TEM) [8]. The integration of micro-electromechanical systems (MEMS) actuators and AFM in TEM system has been deployed as a powerful tool for nano/atomic scale characterization of mechanical properties of nanowires. Under applied voltage, suspended MEMS based thermal actuators can generate micro/nanoforces which are used to deform nanowires [9]. The elastic modulus of NWs can be calculated from the force–displacement relationship. Huang et al. [10] reported the deformation, facture mechanisms, and strengths of GaN nanowires. The GaN nanowires were characterized in real-time using TEM combined with scanning probe microscope (SPM). Figure 5.3a shows the relationship between force/load and displacement of the GaN nanowires. The elastic behavior is observed before fracture without plastic deformation. Therefore, GaN nanowires display the brittle nature. In addition, Figure 5.3b shows the facture strength of GaN nanowires, indicating the increase of fracture strength with decreasing diameters of the nanowires. For example, the

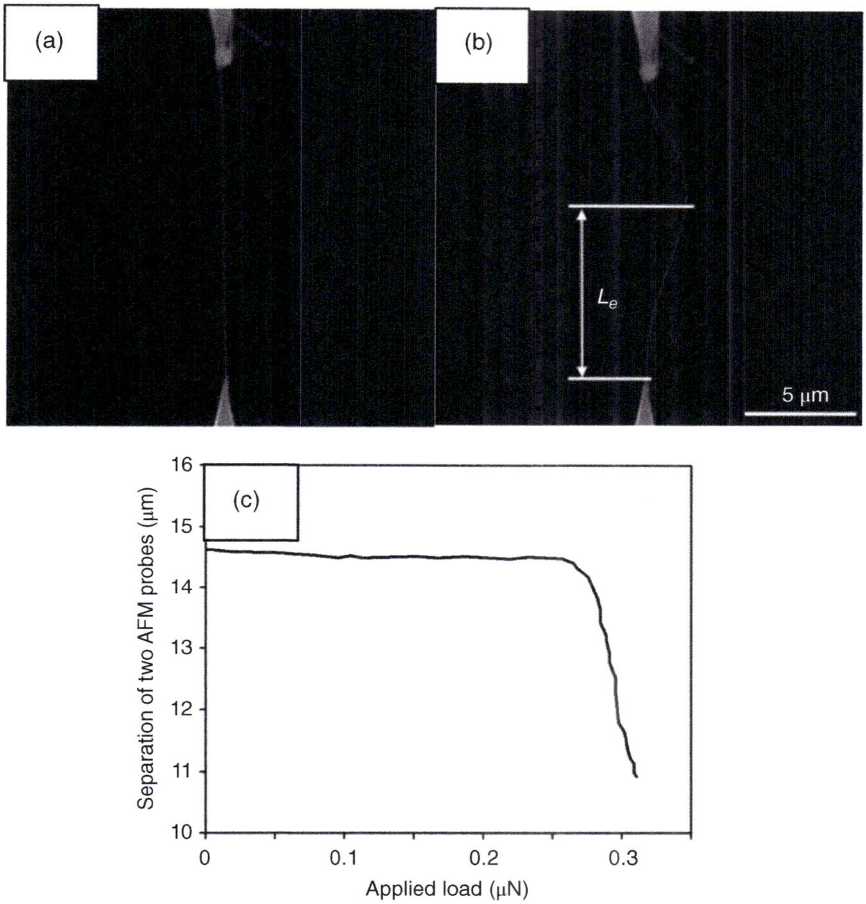

Figure 5.2 Buckling method. (a, b) SEM images of buckling experiments on NWs. (c) The dependence of separation distance between two AFM probes on applied load. *Source:* [5], Lin CH, et. al., (2010)/ with permission of JOHN WILEY & SONS, INC.

fracture strength of 0.21–1.76 GPa was observed for GaN nanowires with diameters ranging from 150 to 700 nm. The size effect on mechanical properties of GaN nanowires is attributed to the strain gradient induced by bending during compression testing.

Zhang et al. characterized one-dimensional disordered (ODD) structure of SiC NWs, showing an ultrahigh tensile strength [11]. The *in situ* tensile tests of single SiC NWs with different ODD were conducted in SEM (see Figure 5.4a). Figure 5.4b shows the typical stress–strain curve of SiC NWs with the tensile strength of 13.7 GPa and high elastic strain of 12%. Molecular dynamics (MD) simulation in Figure 5.4c shows the facture cleavage mode of the SiC NWs with ODD structure, which is in solid agreement with experimental results. Therefore, SiC NWs exhibit the nature of brittle properties, similar to GaN NWs.

5.1.2 Nano Indenting Method

Nanoindentation is commonly utilized to directly measure the mechanical properties of semiconductor nanowires. During the nano indentation process, a nanoscale probe induces

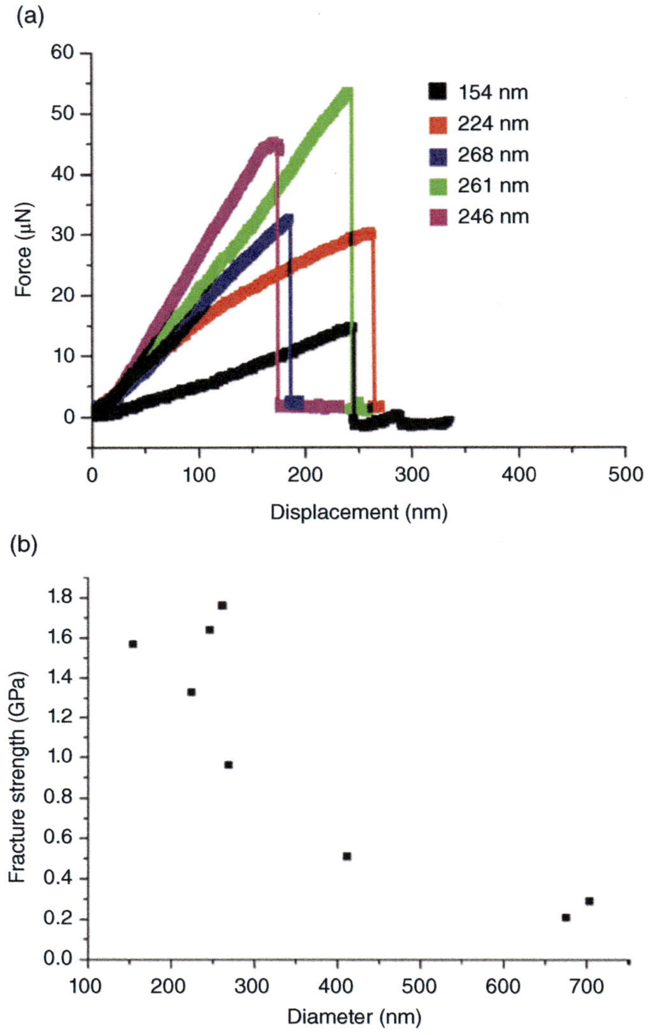

Figure 5.3 Mechanical properties of GaN nanowires. (a) Force vs displacement. (b) Fracture strength measured with different diameters. *Source:* Huang et al. [10]/with permission of John Wiley & Sons.

an impression area on the surface of a nanowire, which is fixed on its substrate. Nano indenter is commonly integrated with SEM, TEM, and Raman characterization facility. For example, Feng et al. studied the mechanical properties of nanowire using nanoindentation [12]. The hardness of GaN nanowires was found to be 46.7 ± 5.6 GPa. Figure 5.5a shows the schematic sketch of indentation of a nanowire on a substrate where R_1 and R_2 are the diameters of indenter and nanowire, respectively. Since the contacts of nanoindenter and nanowire are in series, the equivalent stiffness of the system is determined as [12]:

$$\frac{1}{S} = \frac{1}{S_1} + \frac{1}{S_2} \tag{5.5}$$

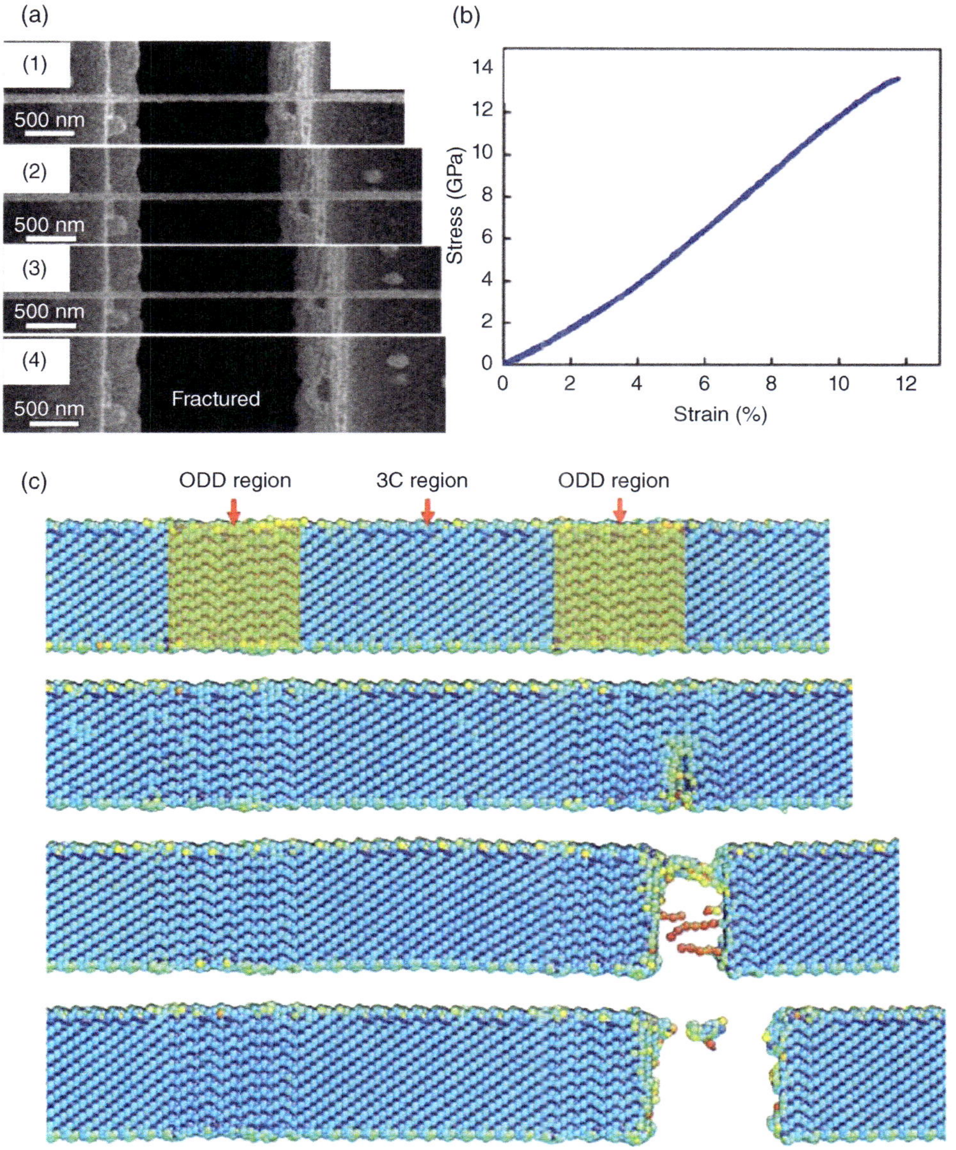

Figure 5.4 *In situ* tensile test. (a) A tensile process of a single SiC NW; (b) stress–strain curve of the single SiC NW. (c) MD simulation result of the SiC nanowire with periodical ODD structure, showing the fracture propagation. *Source:* [11], Zhang X, et. al., (2020)/Springer Nature/CC BY 4.0.

where S_1 and S_2 are the contact stiffness for contacts 1 and 2, respectively. Figure 5.5b shows the scanning force microscopy (SFM) results before and after indentation on the GaN nanowire. Based on the height of the SFM image, the radius of the GaN nanowire is estimated. The analysis results can also provide detail information on the mechanical properties of the GaN nanowire.

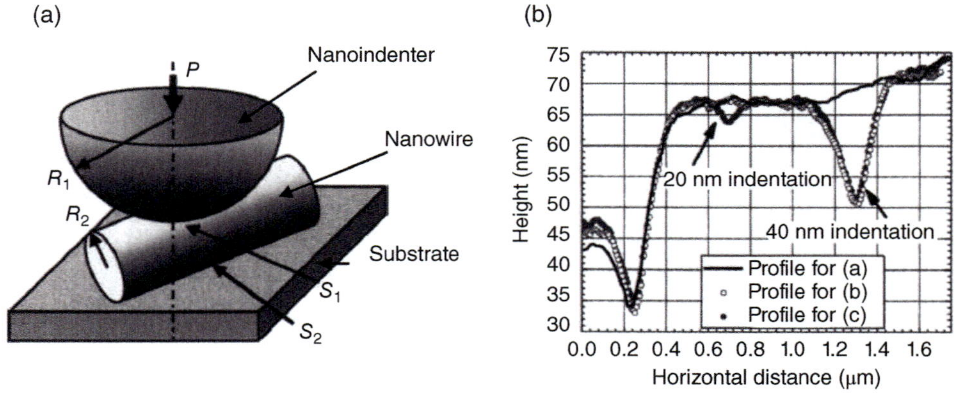

Figure 5.5 Nanoindentation of nanowires. (a) Schematic sketch of the indentation of a nanowire on substrate. (b) Topographical profiles along the GaN nanowire, showing the indentation of 20–40 nm. *Source:* Feng et al. [12]/with permission of AIP Publishing.

5.1.3 Resonance Testing Method

The AFM vibration measurements were used to determine the surface modulus and elasticity of nanowires. In this resonance method, AFM scans the surface of the nanowire and its substrate. The frequency changes are measured and the contact stiffness between substrate and nanowire is calculated, and subsequently the modulus of elasticity is determined.

Another approach is to drive the NW into resonance. For example, an *ac* electrical signal is applied and swept across a range of frequency. The vibration of NW can be measured by the SEM technique. For example, Figure 5.6a shows the schematic sketch of an experimental setup using piezoelectric sheet to drive nanowires into resonance [13]. Figure 5.6b shows the SEM image of a stationary ZnO NW, which is clamped at one end on the tip of the tungsten probe. Figure 5.6c indicates the vibration of ZnO NW measured by SEM, respectively. Figure 5.6d shows the amplitude–frequency response of the first mode with the fitting curve. The measurements are extended to different deposition widths of the clamp system [13]. Therefore, the resonant frequency f can be extracted from the fitting curve. Based on beam theory, Young's modulus E of nanowires can be determined using the following equation [14]:

$$E = \frac{64\pi^2 \rho}{\beta_n^{\ 4}} \frac{L^4}{D^2} f_n^{\ 2} \tag{5.6}$$

where ρ, L, and D is the density, length, and diameter of the nanowire, respectively; f_n is the nth mode resonance frequency. To detect the motion of nanowire, the scanning laser Doppler is also utilized to measure the change in velocity of nanowire when it vibrates. For this method, the external excitation such as piezoelectric actuation is required and the damping of nanowire needs to be considered for measurement of the resonant frequency. With the nanocharacterization laser system, the motion can be detected precisely with a resolution of laser spot down to 1 pm, enabling the measurement of a nanowire with diameter of below 50 nm [15].

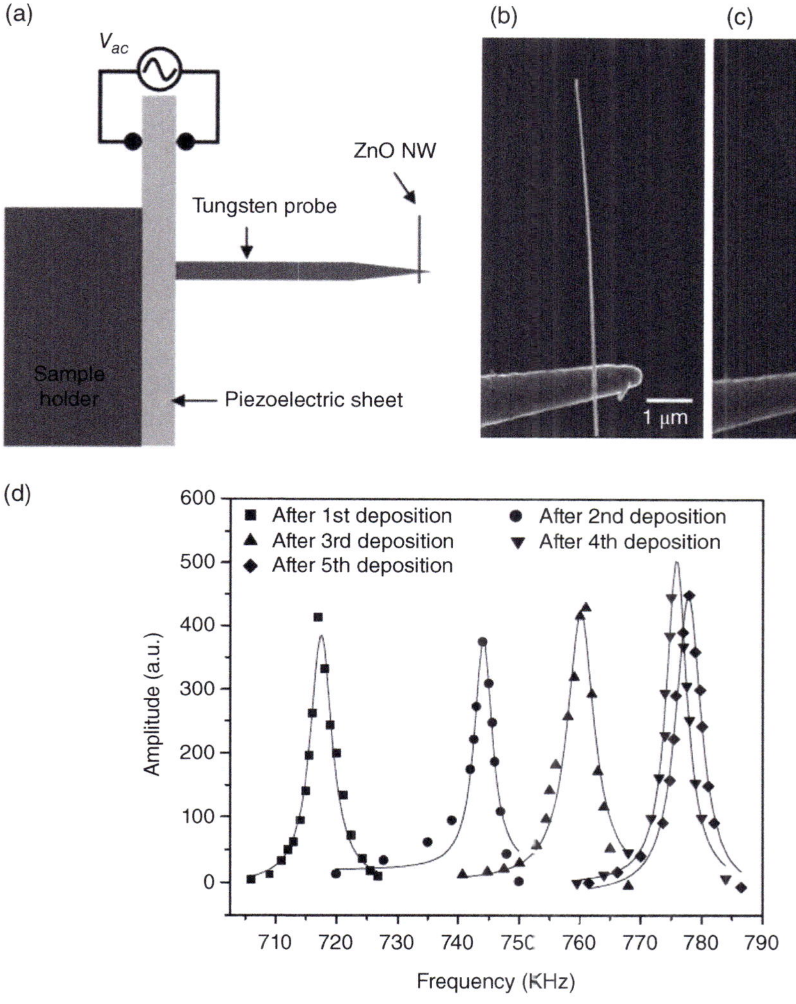

Figure 5.6 Resonance measurement of NWs. (a) Schematic sketch of experimental setup for resonance test using piezoelectric sheet. (b) SEM image of a stationary ZnO NW with one end clamped on the tip of the tungsten probe. (c) SEM image of the ZnO NW under resonance at the fundamental harmonic mode. (d) Amplitude–frequency response curve of the first resonance mode of the ZnO nanowire with Lorentzian fitting (solid line). *Source:* [13], Qin Q, et. al., (2012)/with permission of JOHN WILEY & SONS, INC.

5.2 Impact of Defects and Microstructures on Mechanical Properties of NWs

Controlling/tuning the mechanical properties of nanowires is important for the design and fabrication of nanowire-based devices. It is desirable to understand the key parameters that influence the mechanical properties of nanowires such as Young's modulus, elastic strain, and strength. Several influencing parameters during material growth and nanomachining processes include but are not limited to defects, size/dimensions, temperature, doping levels, growth orientations, and surface coating [9, 16–19].

5.2.1 Defects

Basically, point and line defects can exist in the atomic structure of crystalline materials [2, 20]. Point defects could lead to the increase in average bond length and decrease Young's modulus [1]. Vacancy, a type of point defect, can significantly decrease the modulus of elasticity of SiC NWs. For example, Si and C vacancies are known as intrinsic point defects in SiC structure. MD simulation has intensively been utilized to study the impact of defects on mechanical properties of WBG nanowires. MD simulation at atomic level has shown that atomic disorders could change the bonding at grain boundaries and affect Young's modulus of nanowires. The increase of grain boundaries could lead to the significant decrease in the modulus of elasticity while the stacking defect does not have a significant impact of Young's modulus.

Figure 5.7a shows the stress–strain curves of various SiC NWs with different thickness of stacking defects in comparison to single crystalline (SC) 3C-SiC NWs [21]. The mechanical responses show brittle properties of SiC NWs. Young's modulus decreases approximately 4% from 283.9 to 272.8 GPa, while the tensile strength drops up to 20% from 25.2 to 20.1 GPa (Figure 5.7b). Figure 5.7c shows the tensile strength as functions of thickness of stacking defects. As such, the increase in thickness leads to the decrease of tensile strength. Figure 5.7d reveals an intragranular facture of the SiC NWs with stacking defects at 19.47°.

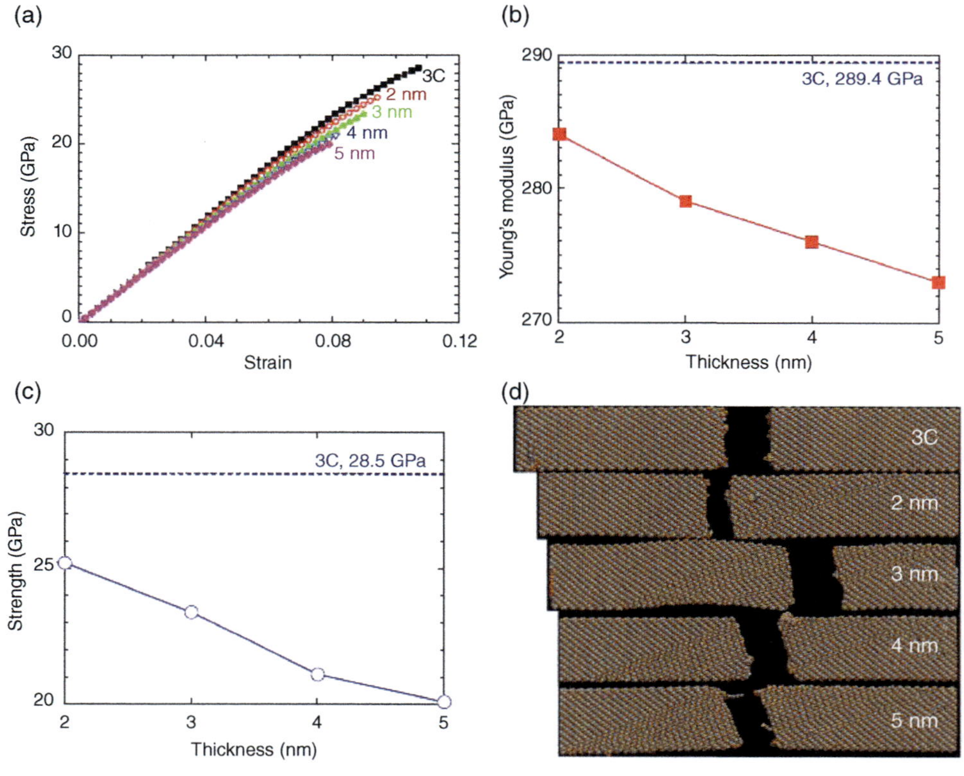

Figure 5.7 Impact of defects on mechanical properties of SiC NWs (stacking defects at 19.47°). (a) Stress–strain curves of SiC NWs with thicknesses of stacking defects from 2 to 5 nm. (b, c) Young's modulus and tensile strength as functions of thickness of stacking defects. (d) Patterns after fracture. *Source:* [21], Wang J, et. al., (2011)/with permission of IOP Publishing.

The existence of local point defects and surface defects has strong impact on the mechanical properties of GaN nanowires [22]. Shi-Feng et al. reported the stress–strain curves of GaN nanowires with different random defect concentrations from 1% to 10%. GaN nanowires showed the linear response before facture, indicating their brittle properties [22]. The tensile strength decreases from 40 to 18 GPa when the defect concentration increases from 1% to 10%. The defects also have significant impact on Young's modulus of GaN nanowires. For example, the decrease of Young's modulus is typically observed with increase of defect concentrations. With random defects, the value of Young's modulus E decreases from 272 to 156 GPa, while E value is from 282 to 244 GPa for the surface defects [22]. The decrease of modulus of elasticity is attributed to the formation of darling bonds or weak bonds or wrong bonds in the GaN lattice.

In addition, considerable research interest has been paid to understand the impact of defects on the mechanical properties of ZnO NWs. For example, Young's modulus of ZnO NWs was reported to be approximately 140 GPa with variation strongly depending on surface effects, structural defects, testing approaches, and environmental factors [23–25]. Furthermore, the facture strength of ZnO NWs is also associated with surface and atomic-vacancy defects which are directly relevant to the size/diameter of the nanowires [26, 27]. Figure 5.8a shows the TEM images of a broken ZnO NW with two segments [28]. The fracture appears to be associated with surface defects which cause the weakening of ZnO NWs strength. It was proved that ZnO NWs do not exhibit plastic deformation before fracture. In another word, ZnO NWs show the brittle characteristics. Figure 5.8b shows the increase of the fracture strength from 2.9 to 8.1 GPa with the decrease of diameter from 316 to 114 nm. Overall, the fracture strength depends on the size of surface defects, e.g. larger ZnO NWs showing higher defects on the nanowire surfaces and smaller fracture strengths.

It has been reported that defects also significantly reduce the elastic strain of materials. Elastic strain refers to the maximum strain value of elastic deformation at fracture of brittle materials, or plastic deformation at fracture of ductile materials. Theoretically, the elastic strain of crystalline materials could achieve a value of 10% [29]. However, due to the presence of defects, crystalline materials have a limit tensile strain and elastic shear from 0.2% to 0.3%. Nanowires could be utilized as ideal specimens for evaluating the intrinsic elastic strain of materials due to their nearly defect-free structures. For example, SiC nanowires exhibited an elastic strain of 2% in TEM and 10% in AFM bending tests. Other testing of SiC NWs also indicated an elastic strain of 4.5–7% depending on types of materials, nature of tests and techniques. These results are in solid agreement with theoretical calculation from first-principle and MD [1].

5.2.2 Effect of Structures, Dimensions, and Temperatures

Microstructures play an important role in tuning mechanical properties of nanowires. For example, Dai et al. [16] conducted the mechanical testing with different GaN nanowire structures, including single crystalline (SC) and obtuse-angle twin (OT), shown in inset Figure 5.9a. Figure 5.9a shows Young's modulus of the SC GaN nanowires and OT GaN nanowires. The GaN nanowires with SC structure and diameter of 92–11 nm showed a Young's modulus of 271–388 GPa. However, the GaN nanowires with OT structure and diameter of 89–171 nm exhibited a low Young's modulus of average 66 GPa, which is approximately five times lower than that of the GaN nanowires with SC structure. It is important to note that, Young's modulus also varies with diameters of GaN nanowires.

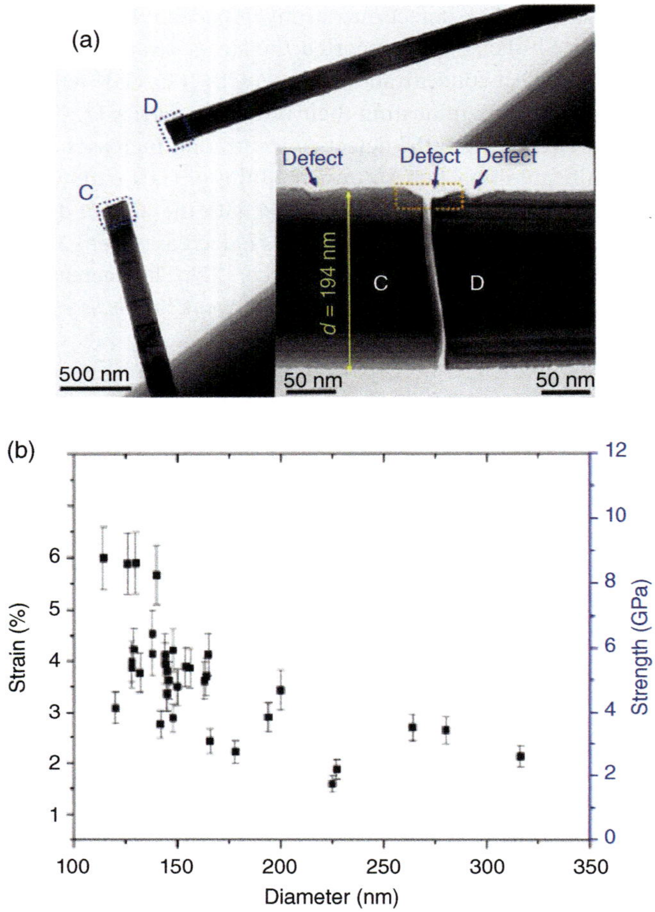

Figure 5.8 Effect of defects on the mechanical properties of ZnO NWs. (a) TEM images of the two fracture ends, C and D. The inset is the enlarged TEM images of the fracture ends, indicating the fracture initiated at a relatively large surface defect. (b) Fracture strain (%) and strength plotted as a function of the diameters. *Source:* [28], Roy A, et. al., (2017)/with permission of Springer Nature.

Figure 5.9b shows the typical stress–strain curve of OT GaN NWs. The linear characteristics of stress–strain before failure indicate the brittle behavior of OT GaN NWs with a Young's modulus of 124 GPa.

Mechanical properties of nanowires depend upon their dimensions or diameters. In general, dimension changes could affect the defect density and surface-to-volume ratio. Subsequently, size effect on strength and modulus of elasticity has been investigated toward tuning/controlling mechanical properties of various nanowires. Cheng et al. investigated the size effects on the fracture strength of SiC NWs with diameters ranging from 17 to 45 nm [9]. The NWs showed linear elastic behavior before brittle fracture. Figure 5.10a shows TEM image of a SiC NW with the growth direction of ⟨111⟩. The 3C-SiC structure has inclined stacking faults (SF), high defects, and pure 3C single crystalline, which are marked by I, II, and III, respectively. Figure 5.10b shows the high-resolution transmission electron microscopy (HRTEM) of a SiC NW with a 19.47° SF (segment A) and high-defective (HD)

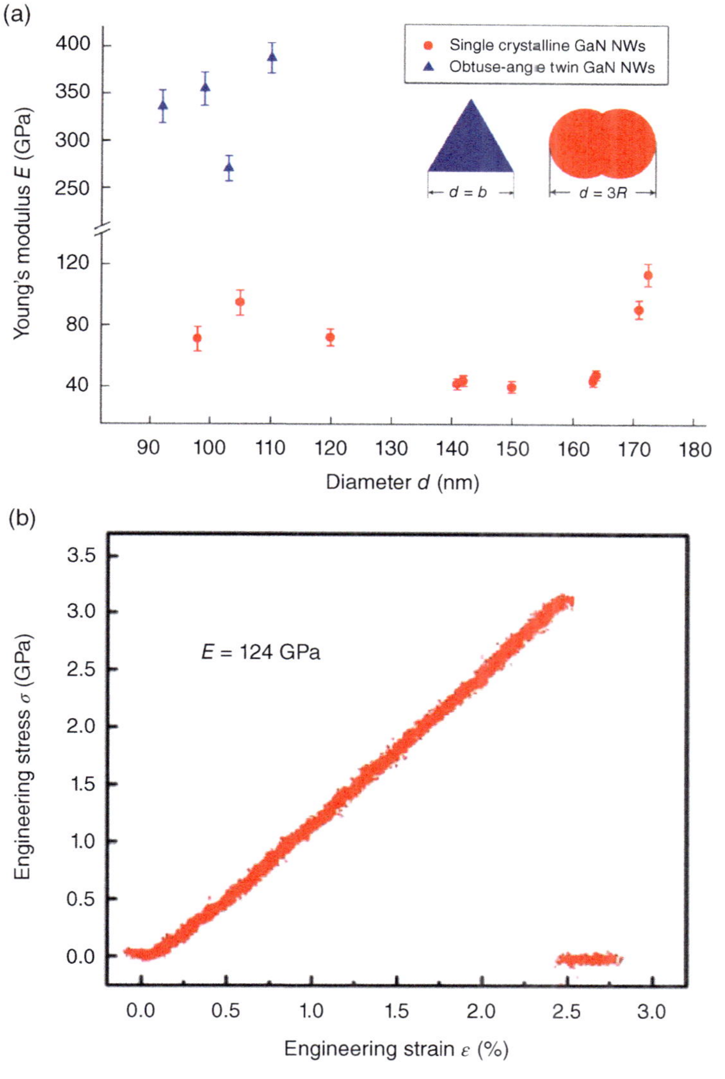

Figure 5.9 Effect of nanowire dimensions on Young's modulus of GaN NWs. (a) Young's modulus E with diameter d of GaN NWs determined from resonance experiments. (b) Stress–strain curve of GaN NW. The value of Young's modulus is 124 GPa. *Source:* Dai et al. [16]/with permission of American Chemical Society.

structure (segment B). Figure 5.10c shows the fracture strength of NWs, which depend on diameters. The fracture strength of SiC NWs increases from 8.1 to 25.3 GPa when the diameter decreases from 45 to 17 nm [9]. The increase of fracture strength with decreasing diameter is also observed in other single crystalline semiconductor nanowires. In general, the small nanowires show higher strength. This is attributed to the impact of the surface effect and defect density on the mechanical properties of nanowires. The value of Young's modulus of SiC NWs was reported in a wide range from 166 to 1270 GPa, while the value for bulk SiC materials is from 500 to 600 GPa [9]. The dependence of strength of SiC NWs on diameter is attributed to the change in defect density. Smaller diameter nanowires have

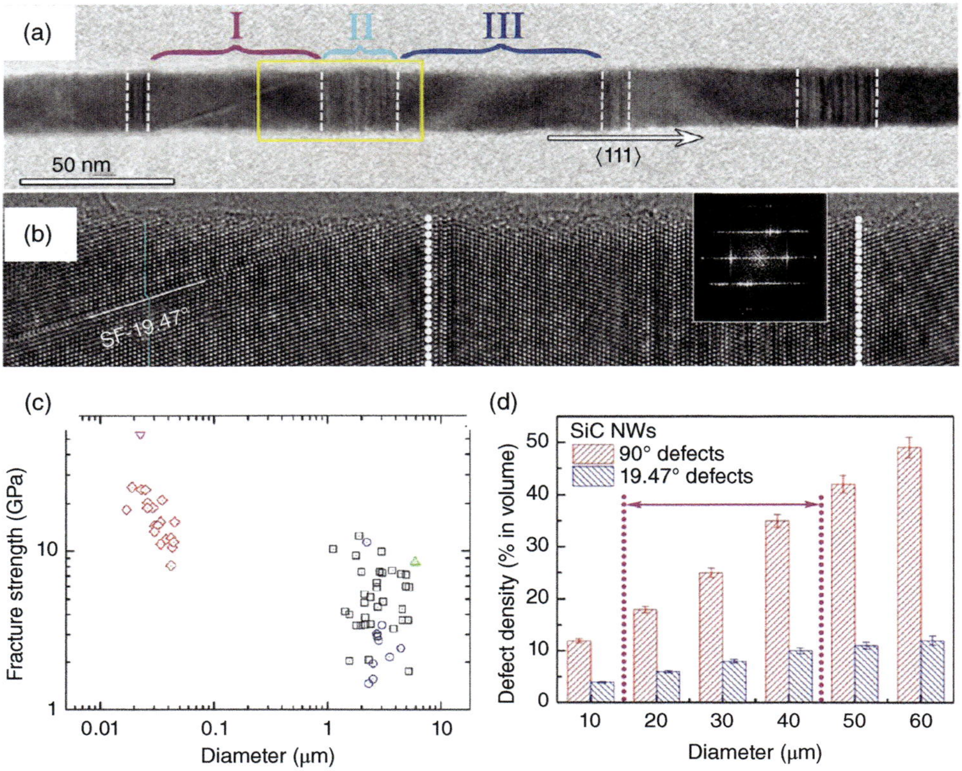

Figure 5.10 (a) TEM image of 3C-SiC NW ⟨111⟩, the 3C structure with inclined stacking faults (I), highly defective structure (II), and pure cubic structure (III). (b) HRTEM observation of the NW containing 3C structure with a 19.47° SF (segment A) and high-defective structures (segment B). (c) Fracture strength of SiC NWs as a function of diameter. (d) Defect density as a function of NW diameter. 90° and 19.47° defects refer to highly defective structures and 3C structures with a 19.47° stacking faults, respectively. *Source:* [9], Cheng G, et. al., (2014)/with permission of American Chemical Society.

lower defect density. Figure 5.10d shows the dependence of two types of the defect density on the diameter. The SiC NWs have 3C structures with a 19.47° stacking faults and high-defective structures with 90° stacking faults. MD simulation and experimental results indicated that single crystalline 3C-SiC has highest strength, while the SiC NWs with 19.47° stacking faults have lowest strength. This is attributed to the size-dependent defect density which is dominant in SiC NWs with 19.47° stacking faults.

Surface effect also plays an important role in tuning mechanical properties of nanowires. In material growth and synthesis, the surface of nanowire could be affected by uncontrollable amorphous silicon dioxide sheath [17]. For example, Ma et al. studied the SiC NWs with diameters ranging from 130 to 150 nm and covered by SiO_2 layer with a thickness of 5–14 nm [17]. Figure 5.11a shows a HRTEM image of an SiO_2-coated SiC (SiO_2@SiC) nanowire. Figure 5.11b shows the X-ray diffraction (XRD) pattern measured for the SiC nanowires with the wave crest at low diffraction angle of 15–30° indicating the presence of amorphous-state SiO_2 shells. Figure 5.11c shows the relationship between force and indentation depth from bending experiments for the SiO_2@SiC nanowires. Based on this relationship, Young's modulus can be determined by Euler–Bernoulli beam theory.

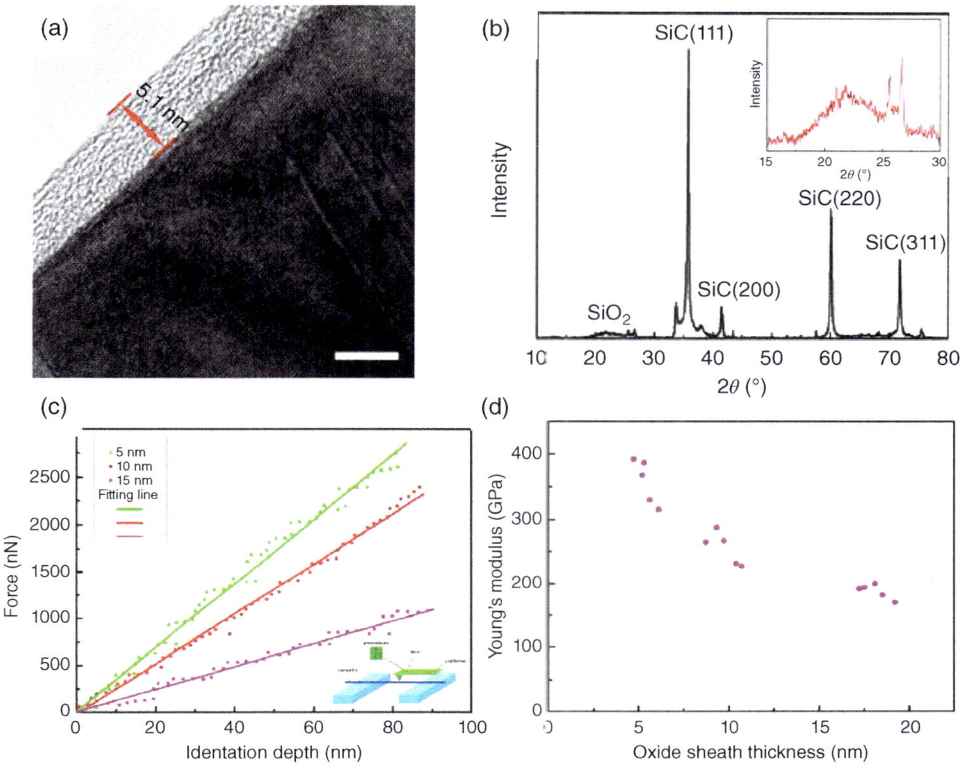

Figure 5.11 Mechanical properties of SiC nanowire with SiO_2. (a) TEM image of SiO_2@SiC nanowires with different shell thicknesses. The scale bar is 5 nm. (b) X-ray diffraction pattern. (c) Force–displacement curves and Young's modulus of SiC NWs with different SiO_2 thicknesses. (d) Young's modulus calculated at different sheath thicknesses. *Source:* [17], Ma J, et. al., (2016)/with permission of Springer Nature.

Figure 5.11d shows Young's modulus of SiO_2@SiC nanowire, which decreases with increasing thickness of the SiO_2 layer. As such, Young's modulus of SiO_2@SiC nanowire decreases from approximately 400 to 250 GPa when the thickness of SiO_2 increases from 5 to 20 nm. This value is a composite modulus, which combines the modulus of elasticity of SiC core and SiO_2 shell. Using the core–shell theory, the effective Young modulus E_e is calculated as [17]:

$$E_e = E_s + \left(E_c - E_s\right)\left(\frac{d_c}{d_t}\right)^4 \tag{5.7}$$

where $E_s = 73$ GPa is Young's modulus of SiO_2; $E_c = 603$ GPa is Young's modulus of SiC core; d_c and d_t are the core diameter and total diameter, respectively. Both diameter and temperature have strong effects on the mechanical properties of SiC nanowires [19]. In general, the higher the temperature is, the lower the strength and the lower Young's modulus of SiC nanowires are. MD results show the increase in the strength and Young's modulus with increasing diameter of the nanowires [19].

Similar to SiC and GaN NWs, ZnO NWs show brittle characteristics [28]. The mechanical properties of ZnO NWs are different from the bulk material. The dependence of mechanical

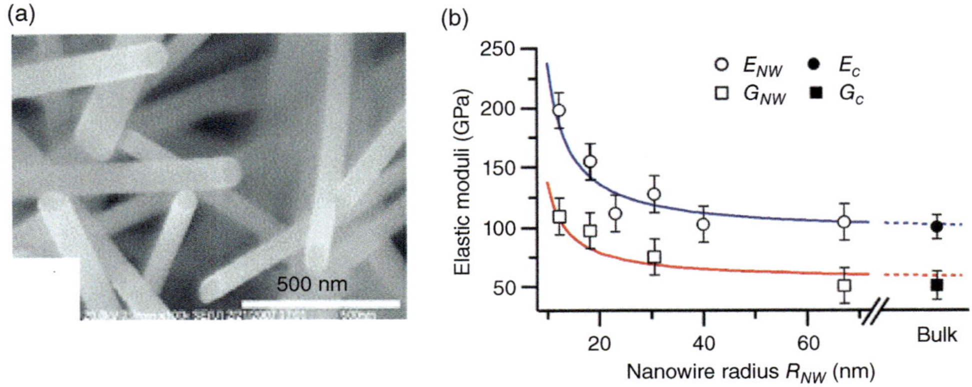

Figure 5.12 (a) SEM image of ZnO nanowires. (b) Elastic moduli E_{NW} and G_{NW} obtained from measurements (data points) and fitted as functions for E_{NW} (blue curve) and G_{NW} (red curve). *Source:* [30], Stan G, et. al., (2017) / with permission of American Chemical Society.

properties on the diameter of ZnO NWs has been known as a controversial topic in various studies. For example, some research demonstrated the independence of Young's modulus (approximately 140 GPa) on the diameter, while other studies revealed an increase of Young's modulus with decreasing diameter. For example, Stan et al. demonstrated the diameter dependent radial (E_{NW}) and tangential (G_{NW}) elastic moduli of ZnO NWs [30]. Figure 5.12a shows the SEM image of ZnO NWs grown on Au catalyst layers deposited on SiO_2 substrates using thermal evaporation. Figure 5.12b indicates an increase of both radial and tangential elastic moduli with decreasing diameters of ZnO NWs. The increase of elastic moduli is attributed to the surface stiffening effect. This effect is commonly described by a core–shell model where the bulk core and a (stiffer) surface shell contribute to the elastic properties of ZnO NWs with the dominant effect from the shell at nanoscale levels [30].

5.3 Anelasticity and Plasticity Properties

5.3.1 Anelasticity

Anelasticity or viscoelasticity is also an important mechanical behavior of nanowires. Anelasticity refers to the time-dependent elastic response of nanowire under loading and unloading conditions. Due to the presence of defects, nanowires exhibit more significant time-dependent in response to applied stress/strain. For example, GaAs NWs exhibited remarkable anelasticity of 0.5–1% for a diameter of 25 nm and no stacking faults [31]. The anelasticity reduces to 0.3–0.6% for GaAs NWs with the same diameter but with stacking faults. This is attributed to the impact of amorphous/crystalline interfaces at the surface of the NWs [31]. In addition, it is believed that the anelasticity of NWs is associated with size and surface effects. However, much work needs to be done for clarifying the anelasticity of WBG nanowires [1].

5.3.2 Plasticity

Plasticity refers to the permanent deformation of nanowire materials under large applied strain. Plasticity is associated with dislocation movements and the interaction of dislocations

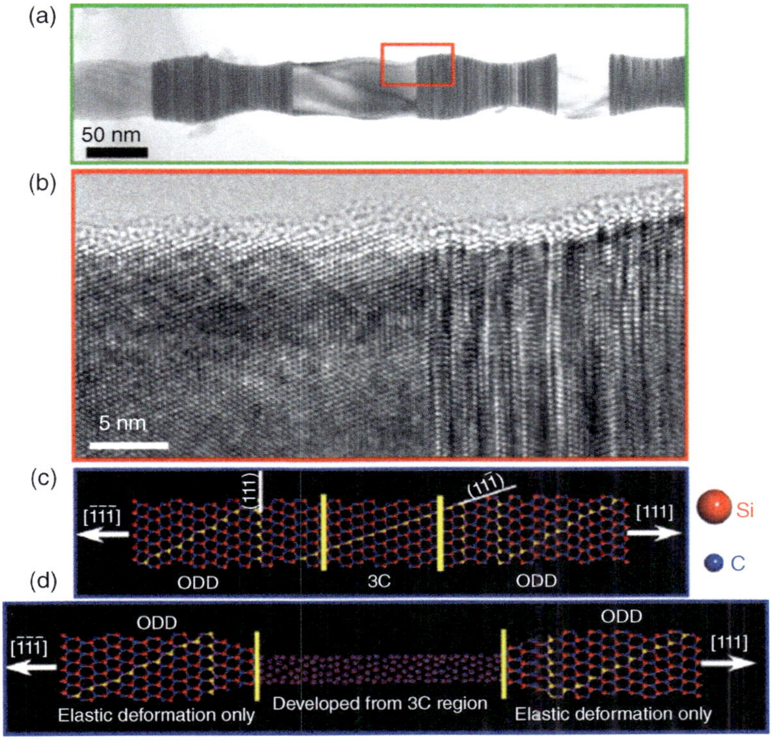

Figure 5.13 Super-plasticity of β-SiC nanowires. (a) A TEM image of a SiC NW. (b) The HRTEM image of the 3C and ODD structured segments. (c) Two-dimensional atomic projection model of a SiC NW with two ODD/HD segments and one 3C segment in the middle. (d) Plastic deformation process in 3C structured segment. *Source:* [6], Zhang Y, et. al., (2007)/with permission of John Wiley & Sons, Inc.

and the grain boundaries. Nanowires could behave different to bulk materials due to the high surface-to-volume ratio where the surface, interfaces, and size effects play the main role in the plastic behavior of the nanowire materials [32]. For example, Zhang et al. studied the super-plasticity of β-SiC nanowires at low temperatures, indicating that single crystalline β-SiC [11] nanowires exhibited super-plasticity close to 220% elongation [6]. Figure 5.13a shows a TEM image of the SiC NW. Figure 5.13b indicates ODD or high defective (HD) structure formed on the SiC NW.

In the SiC NWs, the ODD/HD structural parts only exhibited elastic deformation, Figure 5.13c,d. The atomic slip occurs on the type of (11−1) plane of the NW region when the load is applied along [111] or [−1−1−1], Figure 5.13c. Figure 5.13d shows the super-plastic flow of the 3C-structured region, which is confined within the two yellow-blockers. The plastic deformation, followed by fracture, is described by quadra-step process, including dislocation → amorphization → super-plastic amorphous → fracture [6]. Transformation of crystalline to amorphous structures contributes to the super-plasticity of SiC NWs.

5.3.3 Brittle to Ductile Transition

For the use of WBG NWs for practical applications, it is essential to examine the mechanical properties of NWs. In addition, understanding of their dynamic responses, such as elastic–plastic response [6, 32, 33] and brittle–ductile transition [33–35], is also important.

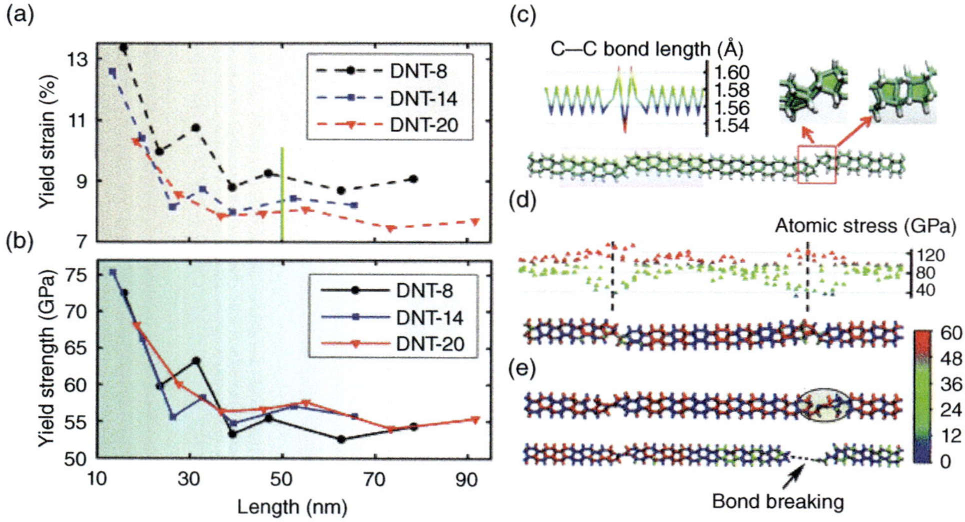

Figure 5.14 Brittle to ductile transition in diamond nanothread (DNT). (a) Estimation of yield strain. (b) Yield strength of the DNT constructed by DNT-8, DNT-14, and DNT-20 units. (c) The distribution of C—C bond in the DNT. The insets show the pentagons and hexagon that realized the Stone–Wales (SW) defect. (d) Stress concentration at the SW defect region. (e) Bond rupture at the pentagon (11.2% strain), initiating the failure of the DNT from defect (13.4% strain). *Source:* Zhan et al. [36] / with permission of Royal Society of Chemistry.

Investigation of mechanical responses of WBG NWs to strain rates and temperature plays an important role in design and fabrication of high-performance nanowire-based devices with tunable mechanical properties. Recently, Zhan et al. deployed large-scale MD approaches to study the transition from brittle to ductile behavior of the sp^3 bonded diamond nanothread (DNT), which varies with the length of the poly-benzene sections [36]. Figure 5.14a,b shows the relationship between the yield strain/strength and the lengths of DNT samples. Both yield strain and yield strength decrease with the increase of the length of DNT samples. To explain the dependence of yield strain and strength on the length of samples, atomic simulations (Figure 5.14c–e) present the stress distribution along the direction of the length of samples. The defect regions exhibit stress concentration or higher stress value during tensile tests. The initial bonding breaking exists in the pentagons and causes the failure of the nanostructure with increasing elongation [36]. The brittle-to-ductile transition is resulted from the hardening process of the Stone–Wales transformation under applied tensile stresses. The Stone–Wales defect plays an important role as grain boundaries that interrupt the flow of the poly-benzene rings in the DNT structure [36]. The dynamic transition in nanowires could open new opportunities to manipulate their mechanical properties at the nanoscale level, which is not achievable with the bulk platform.

References

1 Wang, S., Shan, Z., and Huang, H. (2017). The mechanical properties of nanowires. *Advanced Science* 4 (4): 1600332.

2 Wong, E.W., Sheehan, P.E., and Lieber, C.M. (1997). Nanobeam mechanics: elasticity, strength, and toughness of nanorods and nanotubes. *Science* 277 (5334): 1971–1975.

3 Salvetat, J.-P., Briggs, G.A.D., Bonard, J.-M. et al. (1999). Elastic and shear moduli of single-walled carbon nanotube ropes. *Physical Review Letters* 82 (5): 944.

4 Ni, H., Li, X., Cheng, G., and Klie, R. (2006). Elastic modulus of single-crystal GaN nanowires. *Journal of Materials Research* 21 (11): 2882–2887.

5 Lin, C.H., Ni, H., Wang, X. et al. (2010). In situ nanomechanical characterization of single-crystalline boron nanowires by buckling. *Small* 6 (8): 927–931.

6 Zhang, Y., Han, X., Zheng, K. et al. (2007). Direct observation of super-plasticity of beta-SiC nanowires at low temperature. *Advanced Functional Materials* 17 (17): 3435–3440.

7 Perisanu, S., Gouttenoire, V., Vincent, P. et al. (2008). Mechanical properties of SiC nanowires determined by scanning electron and field emission microscopies. *Physical Review B* 77 (16): 165434.

8 Han, X., Zhang, Y., Zheng, K. et al. (2007). Low-temperature in situ large strain plasticity of ceramic SiC nanowires and its atomic-scale mechanism. *Nano Letters* 7 (2): 452–457.

9 Cheng, G., Chang, T.-H., Qin, Q. et al. (2014). Mechanical properties of silicon carbide nanowires: effect of size-dependent defect density. *Nano Letters* 14 (2): 754–758.

10 Huang, J.Y., Zheng, H., Mao, S. et al. (2011). In situ nanomechanics of GaN nanowires. *Nano Letters* 11 (4): 1618–1622.

11 Zhang, X., Wang, J., Yang, Z. et al. (2020). Strong structural occupation ratio effect on mechanical properties of silicon carbide nanowires. *Scientific Reports* 10 (1): 1–8.

12 Feng, G., Nix, W.D., Yoon, Y., and Lee, C.J. (2006). A study of the mechanical properties of nanowires using nanoindentation. *Journal of Applied Physics* 99 (7): 074304.

13 Qin, Q., Xu, F., Cao, Y. et al. (2012). Measuring true Young's modulus of a cantilevered nanowire: effect of clamping on resonance frequency. *Small* 8 (16): 2571–2576.

14 Zhang, H., Ding, W., and Aidun, D.K. (2015). Mechanical properties of crystalline silicon carbide nanowires. *Journal of Nanoscience and Nanotechnology* 15 (2): 1660–1668.

15 Gil-Santos, E., Ramos, D., Martínez, J. et al. (2010). Nanomechanical mass sensing and stiffness spectrometry based on two-dimensional vibrations of resonant nanowires. *Nature Nanotechnology* 5 (9): 641–645.

16 Dai, S., Zhao, J., He, M.-R. et al. (2015). Elastic properties of GaN nanowires: revealing the influence of planar defects on Young's modulus at nanoscale. *Nano Letters* 15 (1): 8–15.

17 Ma, J., Liu, Y., Hao, P. et al. (2016). Effect of different oxide thickness on the bending Young's modulus of SiO_2@SiC nanowires. *Scientific Reports* 6 (1): 1–7.

18 Bernal, R.A., Agrawal, R., Peng, B. et al. (2011). Effect of growth orientation and diameter on the elasticity of GaN nanowires. A combined in situ TEM and atomistic modeling investigation. *Nano Letters* 11 (2): 548–555.

19 Ma, L., Hao, T., and Hossain, Z.M. (2019). Size-dependent toughness and strength in defective 3C-SiC nanowires. *Journal of Applied Physics* 126 (16): 164303.

20 Xia, Y., Yang, P., Sun, Y. et al. (2003). One-dimensional nanostructures: synthesis, characterization, and applications. *Advanced Materials* 15 (5): 353–389.

21 Wang, J., Lu, C., Wang, Q. et al. (2011). Influence of microstructures on mechanical behaviours of SiC nanowires: a molecular dynamics study. *Nanotechnology* 23 (2): 025703.

22 Shi-Feng, X., Shang-Da, C., and Ai-Kah, S. (2011) The effect of atomic vacancies and grain boundaries on mechanical properties of GaN nanowires. *Chinese Physics Letters* 28 (6): 066201.

23 Asthana, A., Momeni, K., Prasad, A. et al. (2011). *In situ* observation of size-scale effects on the mechanical properties of ZnO nanowires. *Nanotechnology* 22 (26): 265712.

24 Xu, F., Qin, Q., Mishra, A. et al. (2010). Mechanical properties of ZnO nanowires under different loading modes. *Nano Research* 3 (4): 271–280.

25 Wang, X., Chen, K., Zhang, Y. et al. (2015). Growth conditions control the elastic and electrical properties of ZnO nanowires. *Nano Letters* 15 (12): 7886–7892.

26 Agrawal, R., Peng, B., and Espinosa, H.D. (2009). Experimental-computational investigation of ZnO nanowires strength and fracture. *Nano Letters* 9 (12): 4177–4183.

27 He, M.-R., Xiao, P., Zhao, J. et al. (2011). Quantifying the defect-dominated size effect of fracture strain in single crystalline ZnO nanowires. *Journal of Applied Physics* 109 (12): 123504.

28 Roy, A., Mead, J., Wang, S., and Huang, H. (2017). Effects of surface defects on the mechanical properties of ZnO nanowires. *Scientific Reports* 7 (1): 1–8.

29 Latanison, R. and Pickens, J. (2012). *Atomistics of Fracture*. Springer Science & Business Media.

30 Stan, G., Ciobanu, C., Parthangal, P.M., and Cook, R.F. (2007). Diameter-dependent radial and tangential elastic moduli of ZnO nanowires. *Nano Letters* 7 (12): 3691–3697.

31 Chen, B., Gao, Q., Wang, Y. et al. (2013). Anelastic behavior in GaAs semiconductor nanowires. *Nano Letters* 13 (7): 3169–3172.

32 Kaganer, V.M., Jenichen, B., and Brandt, O. (2016). Elastic versus plastic strain relaxation in coalesced GaN nanowires: an X-ray diffraction study. *Physical Review Applied* 6 (6): 064023.

33 Wang, R., Wang, C., Feng, Y., and Tang, C. (2018). Mechanical responses of *a*-axis GaN nanowires under axial loads. *Nanotechnology* 29 (9): 095707.

34 Wang, Z., Zu, X., Gao, F., and Weber, W.J. (2006). Atomistic simulation of brittle to ductile transition in GaN nanotubes. *Applied Physics Letters* 89 (24): 243123.

35 Wang, Z., Zu, X., Yang, L. et al. (2007). Atomistic simulations of the size, orientation, and temperature dependence of tensile behavior in GaN nanowires. *Physical Review B* 76 (4): 045310.

36 Zhan, H., Zhang, G., Tan, V.B. et al. (2016). From brittle to ductile: a structure dependent ductility of diamond nanothread. *Nanoscale* 8 (21): 11177–11184.

6

Optical Properties of Wide Bandgap Nanowires

6.1 Optical Properties of WBG NWs

The optical properties of wide bandgap (WBG) semiconductor nanowires (NWs) have attracted great research attention in the last 10 years due to their tremendous advantages for the development of ultraviolet (UV) sensors and optoelectronic devices.

6.1.1 Photoluminescence Characterization of NWs

Photoluminescence (PL) measurements are utilized as a popular approach to study the optical properties of nanowires, including emission intensity, photon confinement, and quantum efficiency. Nanowires made by high-quality crystalline materials typically show spectral peaks near the band edges. Optical properties of WBG materials depend upon the quantum-confinement effects when their diameters are shrunk below a critical value (the Bohr radius). Nanowires that have a high surface area will have their optical properties depending on the material compositions, structures, and dimensions. In general, the emission and absorption of nanowires are polarized in their axis direction. Nanowires typically show redshift in PL peaks, which can be explained based on the quantum confinement phenomenon observed when the diameter is shrunk down [1].

In the early work, Liang et al. conducted PL characterization of SiC nanowires for optoelectronic devices using WBG nanowires [2]. The authors reported two PL emission peaks at the wavelength of 340 and 440 nm, and the blueshift in the emission peak is attributed to the quantum confinement effect, which shows different properties to the PL results observed in bulk porous SiC [3]. In another work, Wu et al. [4] reported the quantum confinement effect in nanocrystalline 3C-SiC nanowires with diameters ranging from 1 to 6 nm in aqueous solution. Figure 6.1a shows the PL of SiC NWs under excitation of different wavelengths at room temperature. It is evident that the emission band is from 240 to 500 nm. The redshift of PL peaks is observed with increasing excitation wavelengths, indicating the quantum confinement effect. Figure 6.1b shows the PL peak position of five SiC NW samples as a function of excitation wavelengths. Overall, the PL peak wavelength increases with increasing excitation wavelengths. Figure 6.1c shows the PL peak intensity as a function of excitation wavelengths. One SiC NW exhibited low intensity due to low crystallite concentration. Figure 6.1d shows the emission photos with emission spots of blue (~450 nm), bluegreen (~510 nm), and green-yellow (~545 nm) (from left to right), respectively.

Wide Bandgap Nanowires: Synthesis, Properties, and Applications, First Edition. Tuan Anh Pham, Toan Dinh, Nam-Trung Nguyen, and Hoang-Phuong Phan.

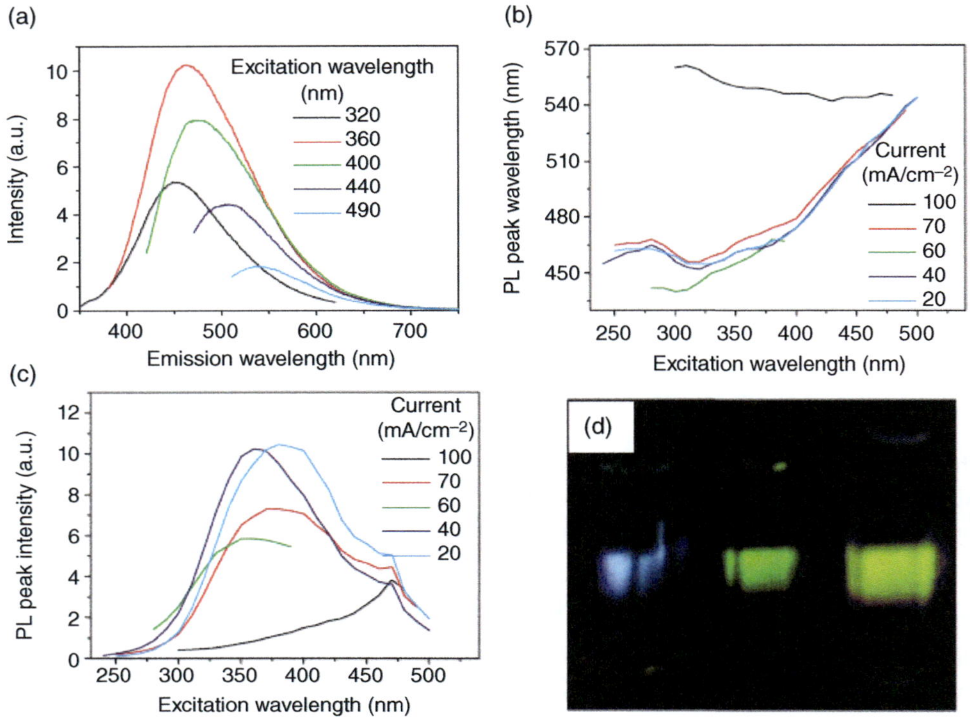

Figure 6.1 Optical properties of SiC NWs. (a) Photoluminescence spectra of SiC NWs under five different excitation wavelengths ranging from 320 to 490 nm. (b) Photoluminescence peak position vs the excitation wavelength for five samples. (c) Photoluminescence peak intensity vs the excitation wavelength for five samples. (d) Emission photos of SiC NWs. The emission wavelengths can be identified to be blue (~450 nm), bluegreen (~510 nm), and green-yellow (~545 nm) (from left to right), respectively. *Source:* [4], Wu X, et. al., (2005)/with permission of American Physical Society.

Several primary factors govern the optical properties of WBG semiconductor nanowires, including material morphologies, microstructures, size, defects, strain, doping, and growth orientations [4–10]. These factors will be discussed in Sections 6.1.2–6.1.5.

6.1.2 Size-Dependent Optical Properties

In semiconductor, the widening of energy bandgap E^* could come from quantum confinements of electron–hole when the diameter of nanowire decreases to a critical value d. The size dependence of energy bandgap can be described as [4]:

$$E^* = E_g + \frac{h^2}{8\mu r^2} - 1.8\frac{e^2}{8\pi\varepsilon\varepsilon_o r} \tag{6.1}$$

where $E_g = 2.2\,\text{eV}$ is the bandgap of SiC materials; h is the Planck constant; $r = d/2$ is the radius of the SiC NWs; μ is the deduced mass of the excitation; and ϵ is the dielectric constant. The requirement of crystallite size with quantum confinement effect is $d/R < 2$, where R is the Bohr radius. For example, it was determined that with a nanocrystallite size of 3.9 nm, SiC NWs showed redshift in PL spectrum and quantum confinement effect [4].

6.1.3 Shape/Morphology-Dependent Optical Properties

Wu et al. [5] demonstrated SiC nanowires in different morphologies such as nanocylinders, nanoprisms, and nanobamboos by controlling reaction temperature from 1470, 1550 to 1630 °C. Figure 6.2 shows the PL of SiC NWs with different morphologies. It is evident that SiC nanocylinders, nanoprisms, and nanobamboos exhibited blue light emission at 422 nm (or 2.94 eV in photon energy), a UV-blue emission at 406 nm (or 3.05 eV), and a sharp blue light emission peak at 440 nm (2.82 eV), respectively. Since bulk 3C-SiC has an energy bandgap of 2.2 eV, the results indicated a blueshift, which could be explained based on the shape dependence or size dependence in corresponding to the quantum confinement effect [4].

6.1.4 Effect of Crystal Orientation

Chin et al. investigated the crystal-orientation dependence of PL in GaN nanowires [11]. Figure 6.3 shows the PL measurement results for *c*-axis GaN nanowires on glass and *a*-axis GaN nanowires on graphite. The blueshift of approximately 140 meV between *c*-axis peak and *a*-axis peak was observed for the GaN nanowires. This blueshift is attributed to the surface states can trap photoexcited carriers, which cause the blueshifted PL spectrum of *a*-axis GaN nanowires [11].

The development of optoelectronic and photonic devices with tunable optical properties has been achieved using different methods including modifying surfaces of NWs, constructing heterostructures, and controlling detect density [12–16].

In WBG NWs including III nitrides, the surface charge properties and doping have strong influence on the optoelectronic performance. For instance, engineering the surfaces of nanowires by tuning Fermi level of GaInN nanowire surfaces can improve the photoconductivity of III-nitride photodetectors [17]. With regards to doping, GaN are typically doped with magnesium (Mg) and silicon (Si) to form p-type and n-type semiconducting nanowires,

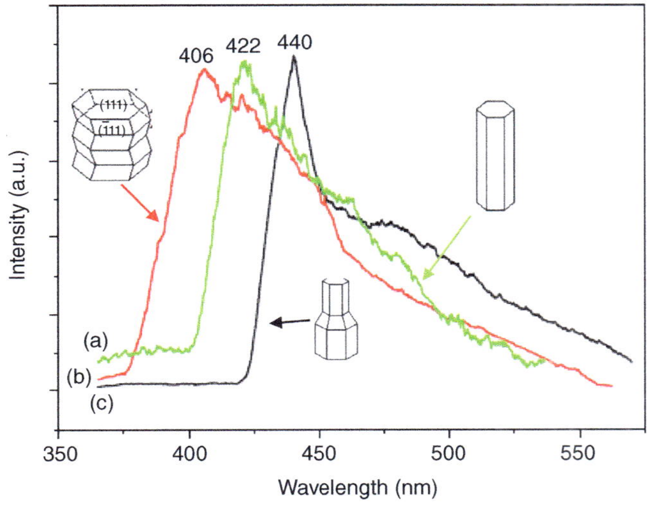

Figure 6.2 PL spectra of the different morphologies of 3C-SiC nanowires: (a) nanocylinders, (b) nanoprisms, and (c) nanobamboos, respectively. *Source:* Wu et al. [5]/with permission of IOP Publishing.

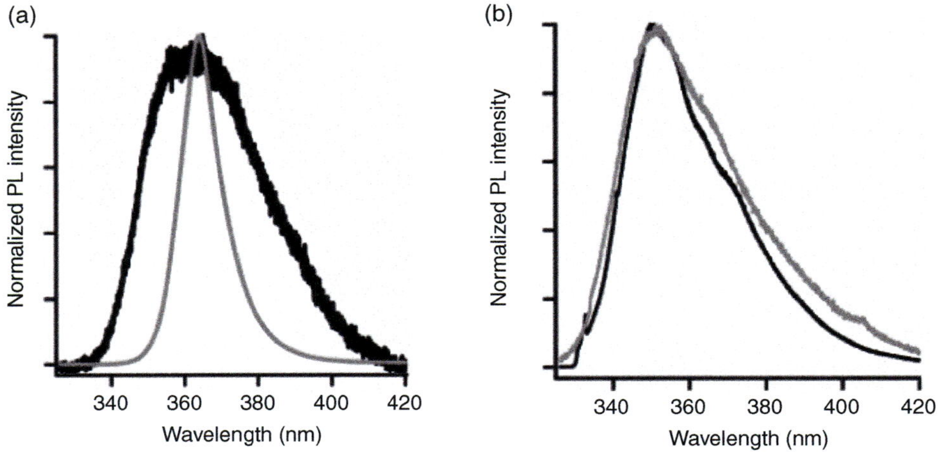

Figure 6.3 Impact of crystal orientation on optical properties of GaN nanowires. (a) Photoluminescence of *c*-axis GaN nanowires (black) and polycrystalline GaN (gray) at low CW power. (b) Photoluminescence of *a*-axis GaN nanowires using pulsed laser excitation (gray) and low CW power laser excitation (black). *Source:* Chin et al. [11]/with permission of American Chemical Society.

respectively. Selection of suitable doping level and type of dopants can significantly improve the charge transport and performance of optoelectronic devices [18].

6.1.5 Tuning Optical Properties of NWs

Qian et al. fabricated single crystal n-GaN/$In_xGa_{1-x}N$/GaN/p-AlGaN/p-GaN nanowire heterostructures with an n-GaN core and $In_xGa_{1-x}N$/GaN/p-AlGaN/p-GaN shells [12]. This nanowire heterostructure has well-defined radical modulation of doping, composition, and thickness. Figure 6.4a shows the cross-sectional view of the GaN-based core/multishell structure with n- and p-GaN utilized as electron and hole injection areas, respectively. Low injection current leads to the localization of carriers at the tip of the nanowire, while high injection current results in the dominance of carriers in other parts. This enables the tuning of emission color from green to blue by the applied injection current. The authors conducted electroluminescence (EL) measurements under different excitation wavelengths [12]. Figure 6.4b shows the normalized EL spectra of five GaN-based core/multishell structures. The tenability of EL peaks was observed with maxima at 367, 412, 459, 510, and 577 nm, which is in solid agreement with band-edge emission from $In_xGa_{1-x}N$ with In compositions of 1%, 10%, 20%, 25%, and 35%, respectively [12, 19]. Holmes et al. proposed single photon emitters for on-chip quantum communication using GaN core, AlGaN shell, and GaN quantum dot (QD) in a core–shell nanowire structure [13]. The authors have concluded that WBG materials including GaN would enable excitation confinement at high temperatures. Figure 6.4c shows the schematic sketch of the structure of core–shell GaN/AlGaN structure with embedded GaN QD. Figure 6.4d–f shows the micro-PL spectrum of single device of GaN-QD GaN/AlGaN nanowire measured at different temperatures 3.9, 150, and 300 K. PL peak is observed at 4.34 eV with full width at half maximum of 1.45 meV. As temperature increased to 300 K, the redshift of 87 meV with a broaden width of 36 meV. This is attributed to the decrease of the energy bandgap of materials with increasing temperature and the interaction of acoustic phonons [13, 20, 21].

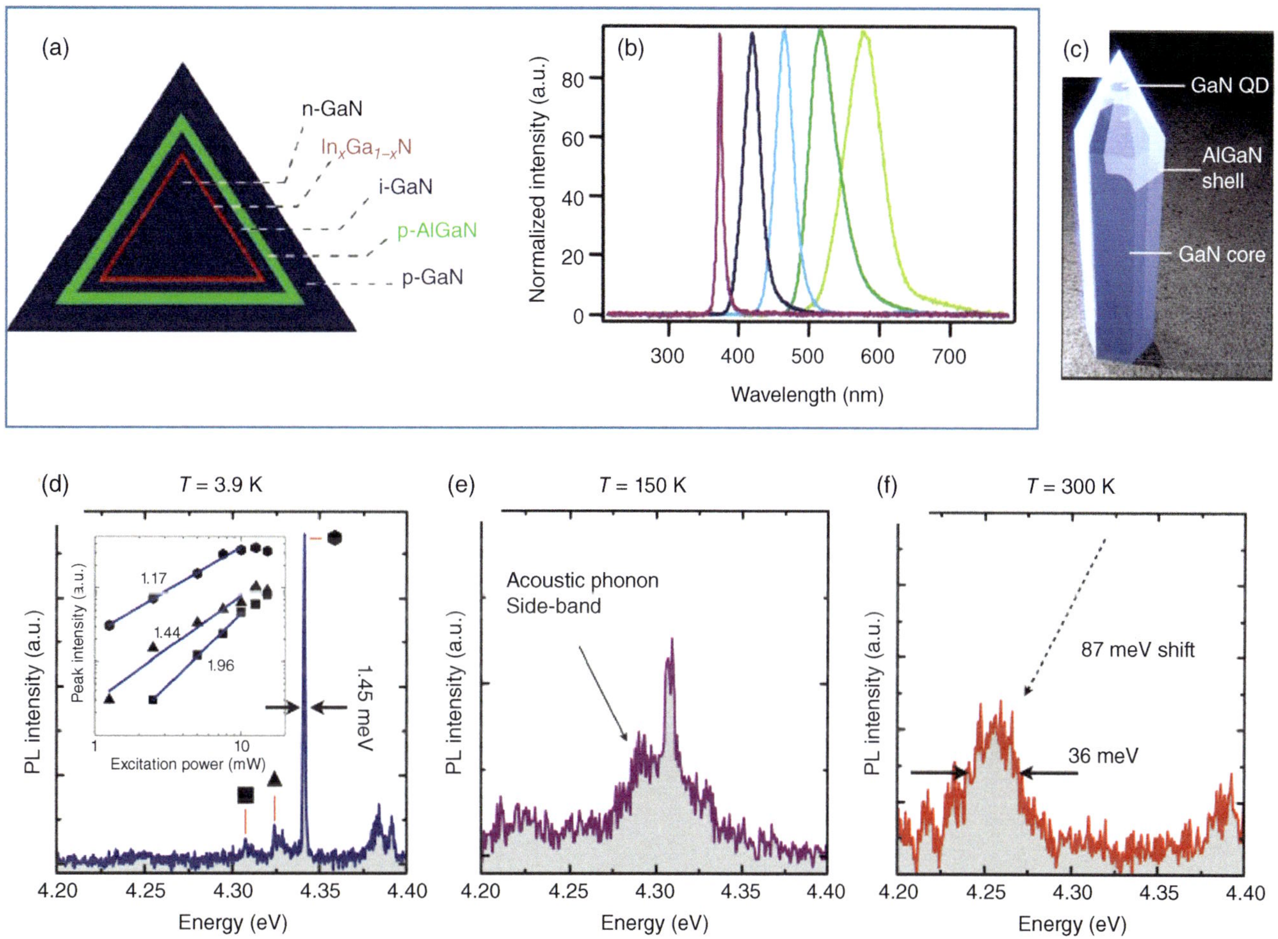

Figure 6.4 Design of core–shell nanowire heterostructure for tuning optical properties. (a) Cross-sectional view of the nanowire structure with a n-GaN core and $In_xGa_{1-x}N$/GaN/p-AlGaN/p-GaN shells. (b) EL spectra of five representative forward-biased multicolor GaN-based core/multishell nanowire LEDs. *Source*: Qian et al. [12]/with permission of American Chemical Society. (c) Schematic of a GaN nanowire containing a single quantum dot (QD). (d) Emission spectra from a single GaN nanowire-QD measured at 3.9 K. (e) Emission spectra at 150 K. (f) Emission spectra at 300 K. *Source:* [13], Holmes MJ, et. al., (2014)/with permission of John Wiley & Sons, Inc.

The optical properties of WBG nanowires and nanotubes can be used to develop robust light sources for emitting phonons in a wide range of applications such as security and communication [14, 22, 23]. Babinec et al. reported the development of a single-photon source using a nitrogen-vacancy center in a diamond nanowire, exhibiting 10 times greater flux than bulk diamond devices with 10 times lower power consumption [14]. Figure 6.5a presents scanning electron microscope (SEM) image of diamond nanowires with ~200 nm diameter and ~2 μm long. To tune the electrical properties of nanowires, the nitrogen-vacancy centers are randomly formed during the growth and etching processes. Figure 6.5b shows the microscope image of nanowire device array. The photon count rate is highest at red area where the nitrogen-vacancy centers are embedded. Figure 6.5c presents the PL spectrum of photons showing zero-phonon line at 637 nm and phonon sideband from 640 to 780 nm. In another work, Nguyen et al. developed light-emitting diodes (LEDs) using InGaN/GaN dot-in-a-wire nanoheterostructure grown on Si(111) [15], Figure 6.5d. This structure was demonstrated to have strong emission in whole visible wavelength range by controlling the dot composition. Figure 6.5e presents the tunable emission spectra of the nanowire measured at 300 K. Figure 6.5f shows the high quantum efficiency of above 20% was achieved with high current density of 200 A/cm^2 [15]. The internal quantum efficiency (IQE), the ratio of the number of electron–hole (e–h) pairs generated to the number of photons absorbed, was calculated by comparing the PL intensities measured at 300 and 10 K using a blue laser with a wavelength of 405 nm. The IQE is estimated using the following equation [24]:

$$IQE = \frac{q * P_{\text{int}}}{I * hv} \tag{6.2}$$

where P_{int} and I are the emitted optical power and injected current, respectively. hv is the energy of the emitted photons and q is the electron charge. The light extraction efficiency is known as the ratio of number of photons emitted into free space to the number of phonons emitted from active region. The external quantum efficiency (EQE) is determined as [25]:

$$EQE = \frac{q * P_{out}}{I * hv} \tag{6.3}$$

where P_{out} is the measured optical power.

The integration of multiple nanowires in a single device enables the enhancement of photon transport and device performance. InGaN/GaN quantum-disks-in-nanowire LEDs emitting at ~710 nm was demonstrated with stable light output and capability of sustaining up to 600 mA electrical current over an eight hours period without catastrophic failure [16]. This indicates the potential of using WBG nanowire for integration of optical devices using in harsh environments. In addition, the development of large-scale nanowire-based devices would contribute to a wide range of low cost high-performance applications such as nano-optoelectronic devices and nanosensors arrays. For example, Wang et al. demonstrated large scale growth and ZnO nanorod/nanowire array, which are aligned vertically on the substrate with uniform shape and length [26]. Figure 6.6a–d shows the SEM images of aligned ZnO nanorods viewed at different angles. The properties of the ZnO nanorod arrays are characterized by PL. Figure 6.6e shows strong dependence of the PL intensity on the angles of input laser. The PL intensity increases with decreasing input angles from 45°, 30°, 15°, to 0°. This result indicates the strong luminescence emission along the axis of the ZnO nanorods. The

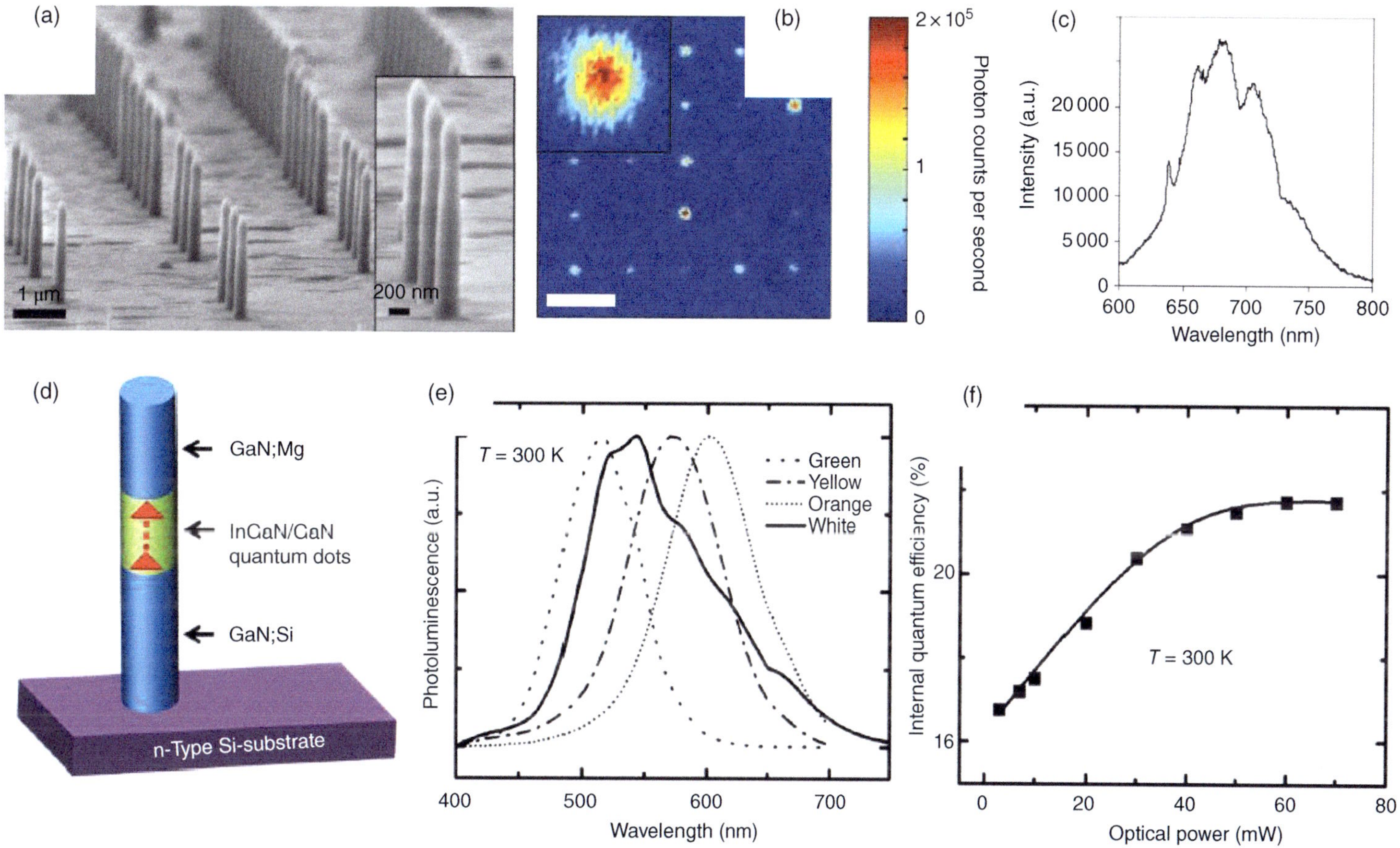

Figure 6.5 Optical properties of diamond nanowires. (a) SEM image showing an array of the200-nm-diameter diamond nanowires. (b) Confocal microscope image of nanowires. (c) Photoluminescence spectrum of photons. *Source:* [14], Babinec et al. (2010)/with permission of Springer Nature. (d) Schematic sketch of InGaN/GaN dot-in-a-wire nanoheterostructure grown on Si(111). (e) PL spectra of InGaN/GaN dot-in-a-wire heterostructures grown on Si(111) measured at room temperature. (f) Internal quantum efficiency with pumping power. *Source:* Nguyen et al. [15]/with permission of IOP Publishing.

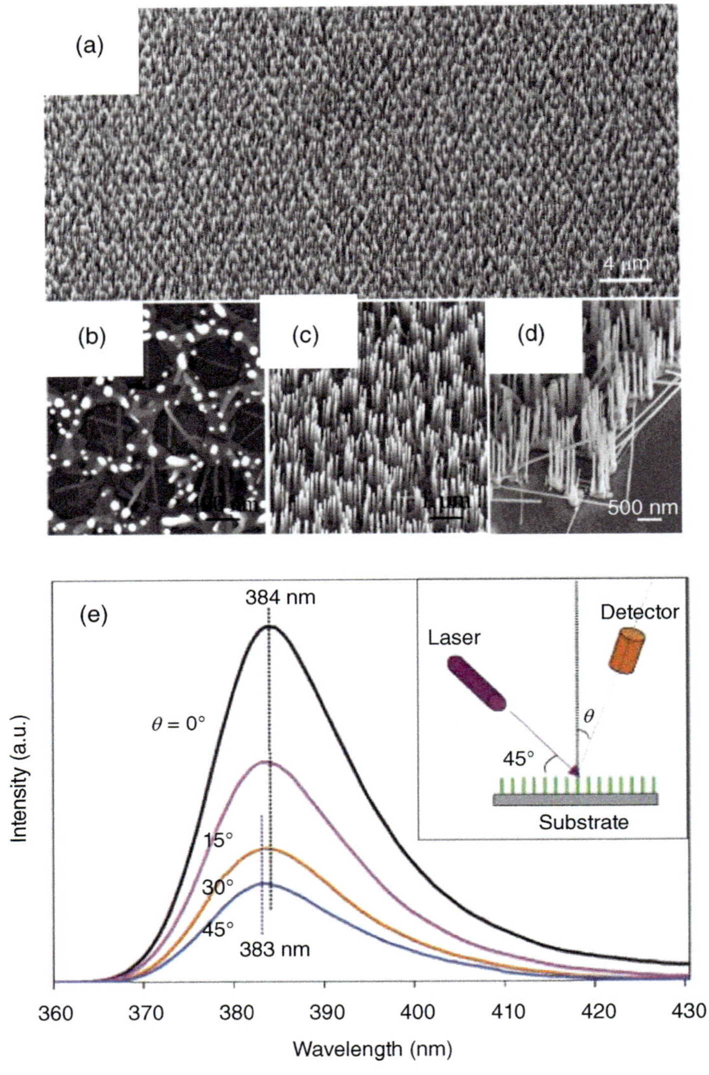

Figure 6.6 The optical properties of aligned ZnO nanorods/nanowires. (a) Side view of the aligned ZnO nanorods at an angle of 30°. (b) Top view of aligned ZnO nanorods. (c) 30° view of aligned ZnO nanorods. (d) Aligned ZnO nanorods at the edge. (e) Photoluminescence (PL) spectra of aligned ZnO nanorod array. The inset shows the experimental set up. *Source:* [26], Wang X, et. al., (2004)/with permission of American Chemical Society.

luminescence peak is observed to be relative stable at 387 nm. In addition, the luminescent peak could also have a very light shift from 384 to 383 nm when the detection angle increase from 0° to 40°.

6.2 Wide Bandgap Nanowire Visible Light-Emitting Diodes (LEDs)

Optoelectronics of nanowires refer to the electronic devices that source and control lights and optical inputs. In the past few years, WBG nanowire-based optoelectronic devices have been successfully fabricated, such as solar cells, photodetectors, and LEDs. This section discusses the development of WBG nanowire-based electronic devices used in a wide spectral range from UV to visible wavelengths. The performance of nanowire optoelectronic shows strong dependence on the surface charge properties of nanowires due to their high surface-to-volume ratio. Owing to a large surface-to-volume ratio, nanowires and nanostructures have lattice strain relaxation on the surface, dislocation-free in bulk region, and effective/uniform doping, which make WBG semiconductor nanowires (e.g. GaN and AlGaN) a potential candidate for high performance optical devices.

6.2.1 GaN Nanowire-Based LEDs

6.2.1.1 Core–Shell/Axial Nanowires

Core–shell/axial nanowires include a core region covering by coaxial shell layers and active regions [18]. Research on AlGaN/GaN core–shells has focused on investigation of fundamental optical properties with different structural designs; InGaN/GaN core–shell nanowire heterostructures have been demonstrated for light emitting applications [27]. Figure 6.4a shows the design of core–shell nanowire heterostructure with an n-GaN core and $In_xGa_{1-x}N$/GaN/p-AlGaN/p-GaN shells. The p-GaN outer shell and the n-GaN core shell act as hole and electron injection sources, respectively. By controlling the composition of the active regions (e.g. InGaN quantum wells), the emission wavelengths from the nanowires can be manipulated. The special design and structure of core–shell/axial nanowires reduce polarization and associated quantum confined effects, enabling the radiative recombination of the active regions and high light extraction efficiency. The large active area between the core and shell contributes to the enhanced performance of GaN nanowire LEDs. The quantum efficiency of radial nanowire LEDs is typically in the range from 20% to 40%.

The speed response and performance of conventional GaN LEDs are limited by spontaneous and piezoelectric polarization causing long carrier lifetimes. Using GaN/InGaN core–shell nanowire LEDs, the high modulation speed and large modulation bandwidth of over 1 GHz have been demonstrated [27]. Figure 6.7a shows the array of p-GaN/InGaN NWs grown on highly n-type conductive Si(111) substrate, which is used as the LED cathode [27]. The nanowire structure is covered with spin-on-glass and Ni/Au contacts. The sketch in Figure 6.7b illustrates the overall structure of the GaN/InGaN nanowire LED. Figure 6.7c shows the EL measurements of the LED at room temperature (e.g. 300 K). Figure 6.7d presents the setup for measurement of frequency response of EL with direct current (DC) offset and square pulse from generator. Figure 6.7e indicates the extracted 10–90% rise and fall times of 210 and 220 ps, respectively, for the on/off modulation of 1.1 GHz, and Figure 6.7f proves the high frequency limit of 1.3 GHz with a rise- and fall-times of 170 and 310 ps, respectively. The disadvantage of using array of nanowires is that light emitted from one nanowire can be absorbed by surrounding ones, which reduces the quantum efficiency of LEDs. The requirements of high-temperature growth of the core–shell nanowires, which utilize metal–organic chemical vapor deposition (MOCVD), have raised challenges in the development of core–shell nanowire LEDs for a wide range of colors such as red, yellow, and orange.

6.2.1.2 Axial Nanowire LEDs

Compared with planar LEDs, nanowire LEDs could exhibit higher output power and higher extraction efficiency due to the low residual stress/strain and reduced polarization. Different growth methods have been used to form axial nanowire LEDs, including bottom-up and top-down approaches. Among them, selective growth methods can be utilized to tune the

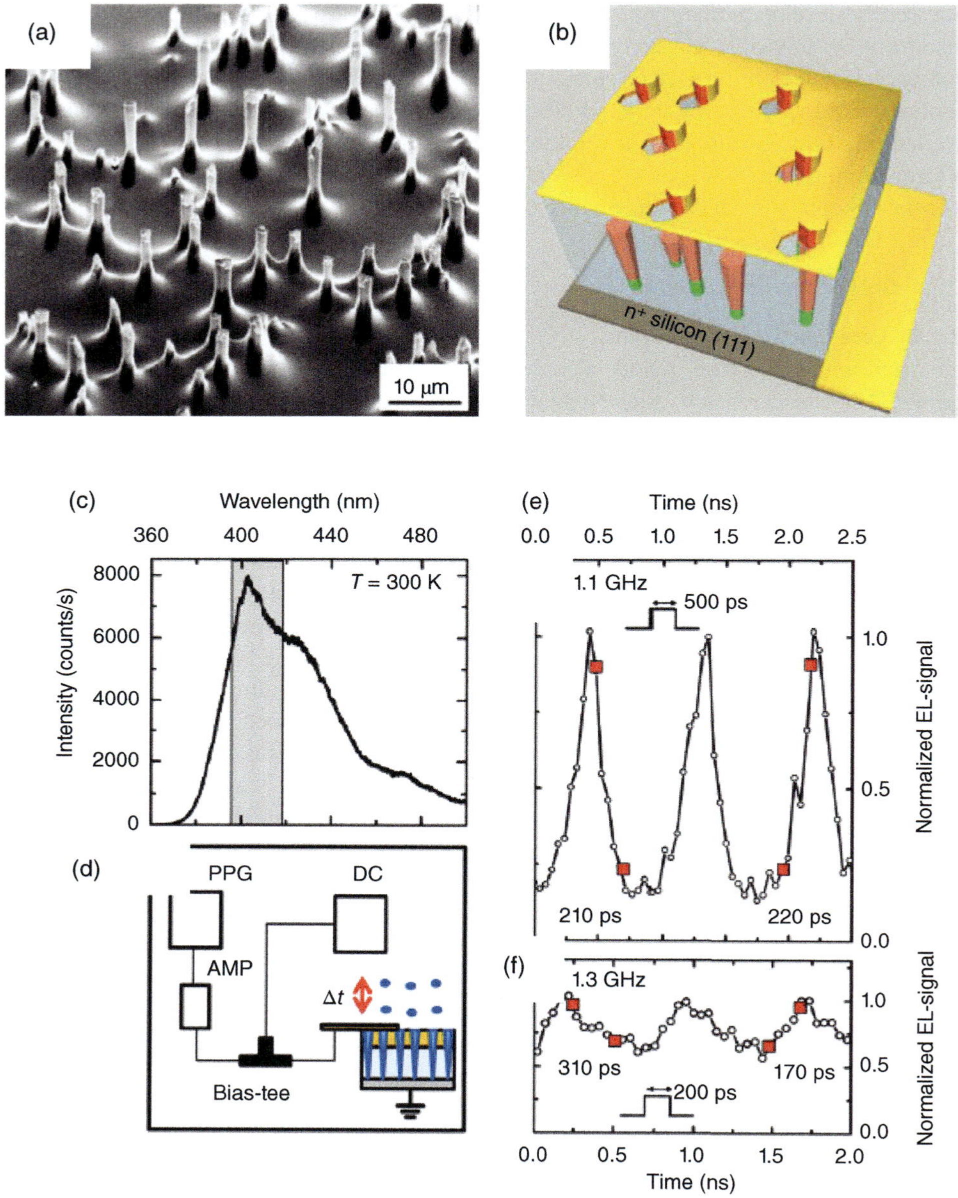

Figure 6.7 GaN/InGaN nanowire array light-emitting diode (LED). (a) Array NWs covered with spin-on-glass and Ni/Au contacts. (b) Sketch of the nanowire array LED. (c) Electroluminescence (EL) vs wavelength. (d) High-speed measurement setup. (e) Electroluminescence signal showing a frequency of 1.1 GHz. (f) A pulse width of 200 ps at a frequency of 1.3 GHz. The red points represent 90% to 10% rise- and fall-times. *Source:* [27], Koester R, et. al., (2015)/with permission of American Chemical Society.

emission properties of LEDs, thanks to the well-defined dimension, position, and controllable surface morphology of QDs. For example, Nguyen et al. demonstrated InGaN/GaN axial nanowire LEDs grown on Si(111) [28]. Figure 6.8a shows a 45° tilted SEM image of the InGaN/GaN nanowire with a focus on the QD area. The authors utilized the p-type modulation doping in a dot-in-a-wire structure to achieve an IQE of up to 56.8%, which is attributed to the 3D carrier confinement formed in the heterostructure with strain and defect-free GaN nanowires. Figure 6.8b indicates the tenability of PL spectra, indicating the influence of material composition and size, which are achieved via tuning the growth conditions such as growth duration and substrate temperature. In another work, Guo et al. reported a double heterostructures of n-Ga/InGaN/p-GaN, which are achieved with Mg and Si dopant species [29]. The color tenability was demonstrated from red to UV by changing the composition of In element. The IQE can be up to 20–35%. The advantage of using QDs is the capability of embedding them in the center region of nanowires using the strain induced alignment.

In addition to the QD structures, quantum well nanowires can also enhance the performance of LEDs via carrier confinement. For example, InGaN/GaN quantum disk nanowire LEDs grown on Si(111) were demonstrated, with EL measured through semitransparent electrodes showing tunable emission colors from green to red [30]. GaN/AlGaN nanocolumn LEDs grown on Si(111) was demonstrated for UV wavelengths at room temperature [31]. Bryan et al. demonstrated high IQE on AlGaN multiple quantum wells grown on AlN substrates [32]. The authors reported a record high IQE of up to 80% achieved at 258 nm at a carrier density of 10^{18} cm^{-3}. The high quantum efficiency of AlGaN on AlN is attributed to the decrease of the non-radiative coefficient and a reduction of midgap defect luminescence. The challenges of developing high-performance quantum well LEDs are the large dislocation densities resulted from substrate materials and the high activation energy (e.g. approximately 500 mV for Mg dopant) of p-type GaN or AlGaN to activate the injection current in LEDs.

Recent work has focused on the development of multiple quantum well micro-/nanowire LEDs for flexible and stretchable applications, in which the LEDs can function well under applied stress/strain with high-efficiency current injection [33, 34].

Figure 6.9a shows the bending model of a nanowire LED under compressive and tensile strains. The strain value ε is determined from $\varepsilon = \frac{H}{2R}$, where H is the thickness of the substrate and R is the bending radius. Figure 6.9b shows the change of PL intensity of the InGaN/GaN multiple-quantum-well microwire-based flexible LED under compressive

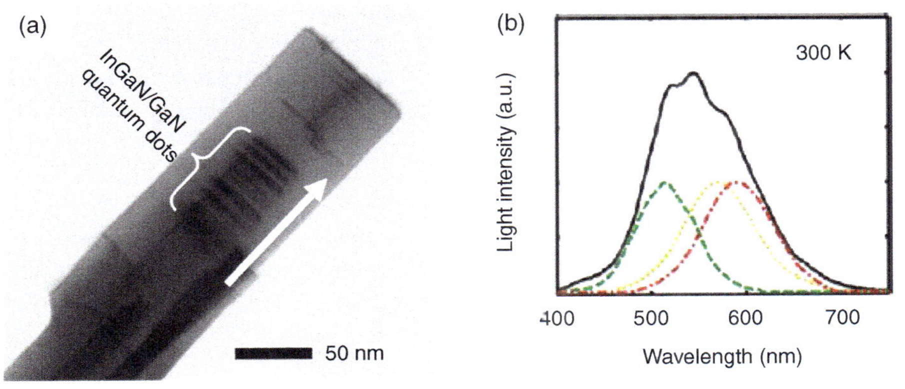

Figure 6.8 InGaN/GaN dot-in-a-wire heterostructures grown on a Si(111) substrate. (a) A 45° tilted SEM image of the InGaN/GaN nanowire. (b) Room temperature photoluminescence spectra of InGaN/GaN nanowires grown under different conditions. *Source:* [28], Nguyen HPT, et. al., (2011)/with permission of American Chemical Society.

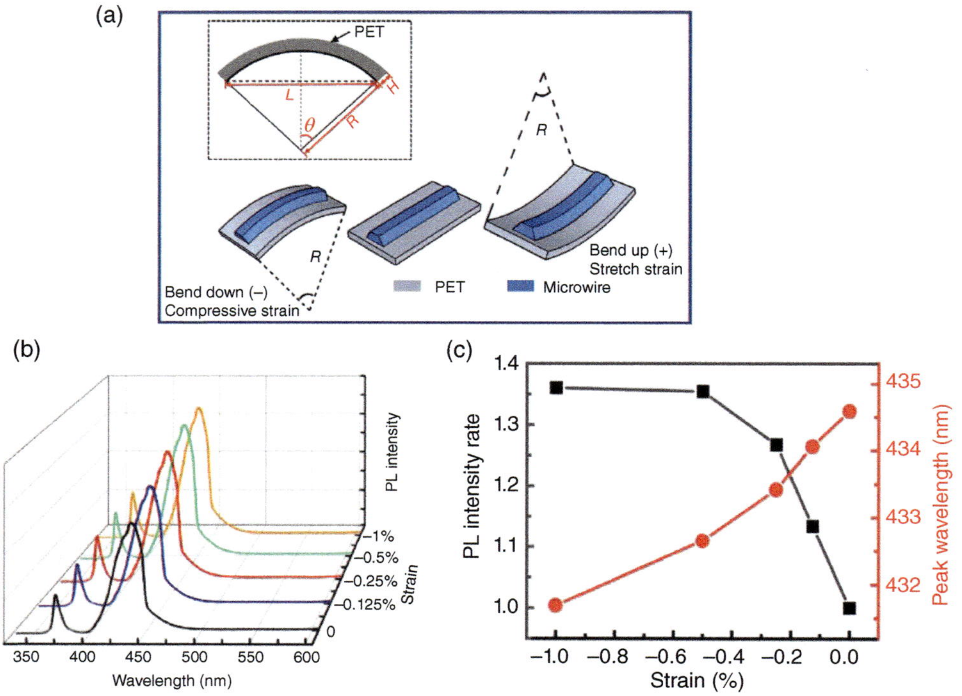

Figure 6.9 Flexible quantum well InGaN/GaN micro/nanowire LEDs. (a) Model of a single micro/nanowire under bending conditions. (b) PL spectra of the micro/nanowire LED under compressive strains. (c) The change of peak wavelength and intensity rate under compressive strain. *Source:* Luo et al. [34]/with permission of John Wiley & Sons.

strain/stress. The data on the change of PL is presented in Figure 6.9c, indicating the decrease of the peak wavelengths and increase of PL intensity rate with increasing compressive strain. As such, the peak wavelength shows blueshifts from 434.6 to 431.7 nm and the intensity rate from 1.0 to 1.36 when strain is applied from 0% to −1% [34]. Compared with the strain-free conditions, the LEDs achieved better performance with increasing compressive strain, while the quantum efficiency increases by 14.2%. The effect of strain on the performance on the InGaN/GaN multiple-quantum-well microwire-based flexible LED is attributed to the strain-induced piezoelectric polarization at the interface. It is expected that research on tuning performance of LEDs by applied strain/stress and development of flexible/stretchable LEDs will continue receiving considerable interest.

6.2.2 GaN Nanowire UV LEDs

AlGaN nanowire-on-Si platforms have attracted considerable interest for UV LEDs [35–37]. AlGaN holds a large, direct, and tunable bandgap energy of 3.4–6 eV, which is suitable for UV and deep UV LEDs with the wavelengths from 207 to 364 nm [38]. Different substrates have been utilized to grown AlGaN nanowire UV LEDs, including silicon (Si), metal, and graphene with the best LED performance found for AlGaN nanowires grown on Si. Performance LEDs can be improved using high doping concentration of NWs by surface modification, which can enhance the current injection to LEDs. Figure 6.10a shows

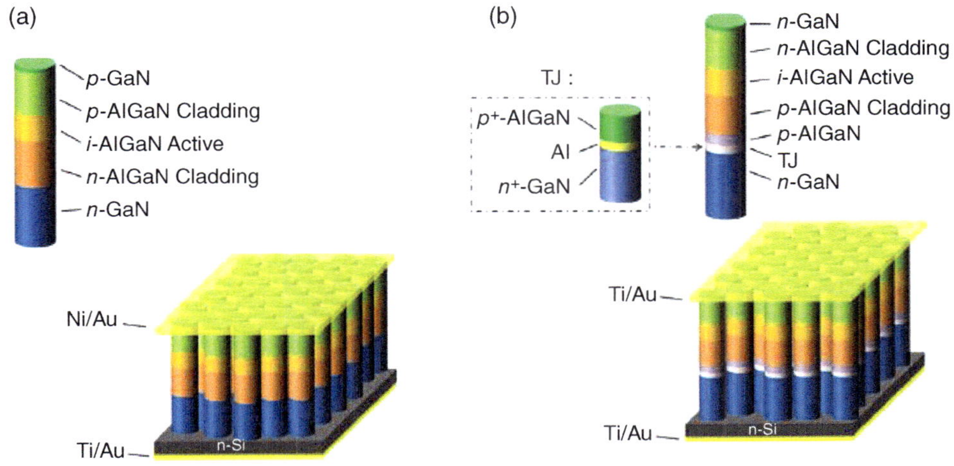

Figure 6.10 AlGaN nanowire UV LEDs. (a) AlGaN nanowire based standard UV LED structure. (b) AlGaN nanowire tunnel junction (TJ) UV LED structure. *Source:* Zhao et al. [35]/with permission of AIP Publishing.

schematic of the AlGaN nanowire standard UV LED structure and Figure 6.10b illustrates the structure of a tunnel junction (TJ) UV LED based on AlGaN nanowires. Compared with the standard structure, the device with tunnel junction exhibited an improvement of light intensity of more than 2 orders of magnitude under electrical injection [35]. The improvement is attributed to the carrier injection to the active region (Figure 6.10b). Therefore, the formation of tunnel junction has been utilized to enhance the performance of UV LEDs thanks to the incorporation of dopants in the nanowire. Based on the EL spectra, all emission components in UV LEDs can be investigated. The EL spectra of TJ UV LED were studied under continuous wave (CW) biasing up to 60 mA [35]. The EL spectra were observed up to 650 nm under an injection of 20 mA. The data indicated the high the emission peak around 242 nm. Currently, research on UV LEDs using GaN is still facing challenges with efficiency droop which could be explained based on Auger recombination effect, and coupling effect of polarization with carrier transport [37, 39]. The challenges also include the difficulty in enhancing the performance of AlGaN UV LEDs to a close point compared to InGaN quantum well devices [39]. In addition, LEDs working in deep UV range have low efficiency if the LED structures exhibit high defect density [36].

While AlGaN nanowire UV LEDs constructed on Si have been demonstrated highest light output power and low efficiency droop, other substrates have also been utilized including metal foils and metal coated substrates owing to their excellent electrical properties and possibility of use as electrical contact. However, the use of metals results in the delamination effect caused by the difference in thermal expansion between the nanowire and its substrate, and the electrical leakage, which reduce the performance of LEDs [38, 40]. Another issue of using metal contact is the lack of transparency for LEDs and optoelectronic devices. Recent research has successfully demonstrated the use of nanomaterials (e.g. graphene) as transparent electrodes and electrical contacts owing to their excellent electrical conductivity and transparency to visible and UV wavelengths [25]. Figure 6.11a shows the GaN/AlGaN nanocolumns LED device with a bias applied between the two contacts, and light emitted through the graphene/silica glass substrate (see [25] for details on fabrication and material

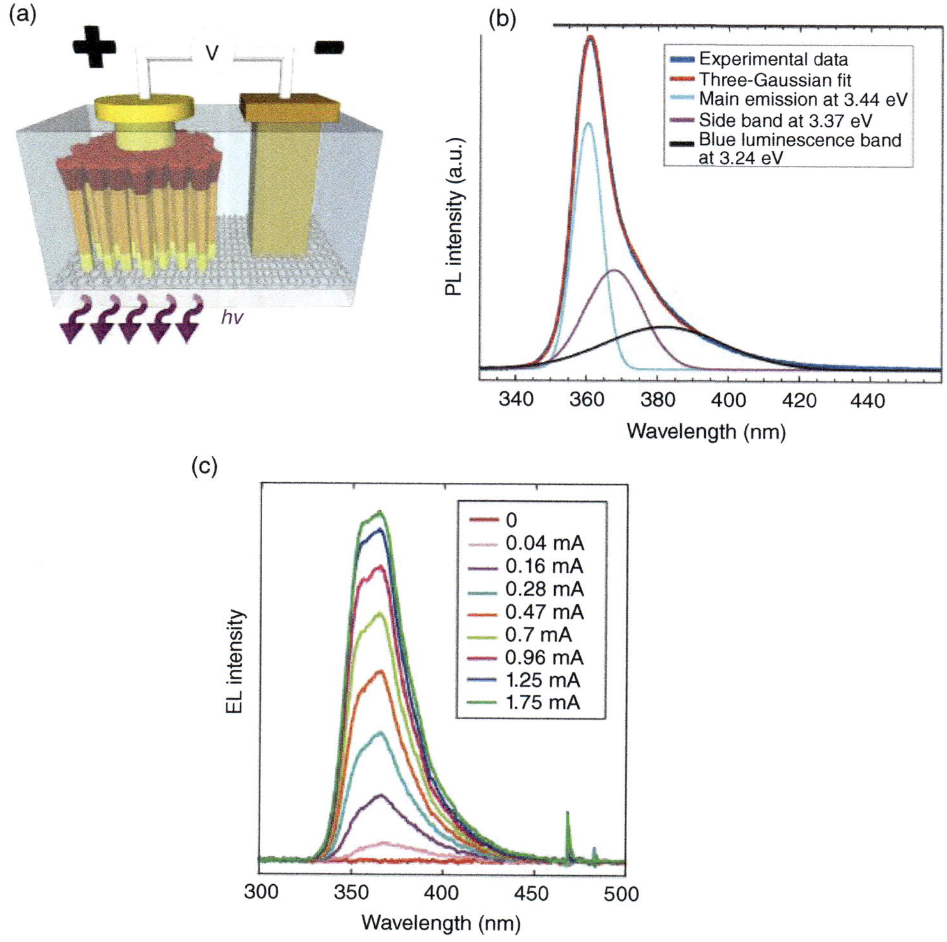

Figure 6.11 GaN/AlGaN nanowire UV LEDs using graphene as substrate and transparent electrode. (a) The LED device with light emitted through the graphene/silica glass substrate. (b) Low-temperature (10 K) photoluminescence spectrum of the LED. (c) EL spectra at different injection currents. *Source:* Høiaas et al. [25]/with permission of American Chemical Society.

platform). Figure 6.11b indicates a main emission peak at 360 nm (3.44 eV) in the spectra of PL measured at low temperature (10 K), which is attributed to Mg acceptor excitation. Figure 6.11c shows the increase of PL intensities with increasing injection currents, which are observed in p-GaN nanocolumn layer at a wavelength of 365 nm. The IQE was estimated to be ~46% while the EQE was measured to be below 0.015% for injection current of up to 1 mA [25]. Further studies need to be performed to optimize/improve the EQE for practical applications.

The challenges in the development of high-performance UV LEDs include minimizing surface defects of nanowires, fabricating metal contacts for nanowires, and extracting light from the nanowire LEDs. The solution to minimize defects has been proposed including passivation of nanowire surface area and selective growth of uniform nanowires [38]. The light extraction from LEDs requires transparent substrates and contacts, which suggest the development of new transparent LED platforms in future research.

6.2.3 ZnO Nanowire-Based LEDs

There has been some limitation in growth of GaN NWs array for LEDs applications. Therefore, it is desirable to have alternative platform for UV and near UV LEDs. ZnO NWs are considered as a potential candidate for growth of NWs based LEDs as it has a large bandgap (3.37 eV) and excellent luminescent properties. In addition, ZnO NWs can emit a wide range of visible wavelengths which are suitable to form white LEDs.

6.2.3.1 ZnO Nanowire LEDs Formed on Crystalline Substrates

It is important to select p-type crystalline substrates that have low lattice mismatch to ZnO to support the growth of n-type ZnO nanowires. ZnO NWs have been grown on p-type GaN and p-type SiC to fabricate white LEDs [41, 42]. For example, ZnO nanowires/p-4H-SiC heterostructures were demonstrated for white LEDs using EL measurements with three prominent peaks at violet (425 nm), green (527 nm), and red (683 nm) [41]. The bright white light was observed based on these three peaks which are attributed to different levels of defects in the ZnO NWs grown on SiC.

Zhang et al. demonstrated ZnO NWs grown on p-type GaN wafer as a LED platform for generating high brightness blue light [42]. Figure 6.12a shows the EL spectrum of the n-type ZnO NWs on p-type GaN film LED device under different forward bias voltages (e.g. 10, 15,

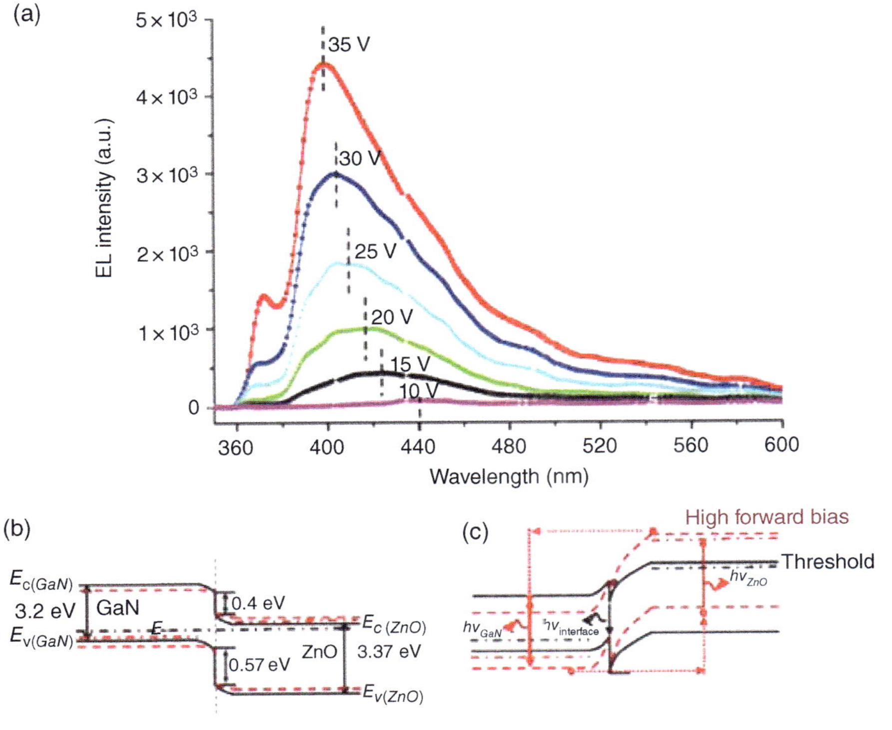

Figure 6.12 ZnO NWs on GaN LED device. (a) Electroluminescence spectrum of (n-ZnO NWs)/(p-GaN film) LED under various forward bias voltages. (b) Band diagram of GaN–ZnO heterostructure at zero (solid line) or smaller (dash line) forward bias. (c) Band diagram of GaN–ZnO heterostructure around threshold forward bias (solid line) and at forward bias higher than threshold (dash line). *Source:* Zhang et al. [42]/with permission of John Wiley & Sons.

20, 25, 30, 35 V). The results indicate broad emission peaks from UV to blue, and blueshift from 440 to 400 nm with increasing bias voltages. Figure 6.12b,c illustrates the band-structure diagrams of the LED for explaining the observed EL blueshift. The applied electric potential drives holes in p-GaN toward ZnO NWs in the valence band while electrons move in the opposite direction in the conduction band. Phonons with small energy are generated from recombination of electrons and holes, leading to the emission of light. When the applied voltage is large, holes and electrons can drive across the interface of GaN and ZnO without recombination, resulting in band-edge emission (i.e. 387 nm for GaN and 368 nm for ZnO). Both edge-emission and recombination made the broaden peaks, while high applied voltage causes an increase in the intensity of ZnO emission toward lower wavelengths or higher energy (see Figure 6.12a).

6.2.3.2 ZnO Nanowire LEDs Formed on Flexible Substrates

Owing to low temperature growth of ZnO NWs, flexible substrates including paper, plastic, and textile structures have been utilized to grown ZnO nanowire LEDs. The use of flexible substrates offers LEDs devices with low cost, simple fabrication, large device area, and low power consumption. It has been possible to develop p–n junction of ZnO NWs on p-type polymeric substrates. For example, ZnO NWs on SnO_2-coated glass with p-type poly(3,4-ethylenedioxythiophene) polystyrene sulfonate (PEDOT:PSS) polymeric contact were demonstrated for LEDs in an early work [43]. ZnO NWs grown on paper and flexible substrates were also attracted interest [44–46].

The application of ZnO nanowires for LEDs has been hindered by the difficulty in growth of high quality and reproducible p-type ZnO. Therefore, great effort has been paid to integrate ZnO nanowires into other p-type materials such as p-GaN. However, the development of high-performance ZnO nanowire LEDs has still faced great challenges due to the existence of defects at the interfaces caused by lattice mismatch between ZnO and other optical materials. The capability of growth of ZnO nanowires at low temperature (below 100 °C) will enable the development of ZnO nanowire LEDs on other material platforms, including flexible and stretchable substrate materials (e.g. plastic).

References

1 Agarwal, R. and Lieber, C. (2006). Semiconductor nanowires: optics and optoelectronics. *Applied Physics A* 85 (3): 209–215.

2 Liang, C., Meng, G., Zhang, L. et al. (2000). Large-scale synthesis of β-SiC nanowires by using mesoporous silica embedded with Fe nanoparticles. *Chemical Physics Letters* 329 (3, 4): 323–328.

3 Konstantinov, A., Henry, A., Harris, C., and Janzen, E. (1995). Photoluminescence studies of porous silicon carbide. *Applied Physics Letters* 66 (17): 2250–2252.

4 Wu, X., Fan, J., Qiu, T. et al. (2005). Experimental evidence for the quantum confinement effect in 3C-SiC nanocrystallites. *Physical Review Letters* 94 (2): 026102.

5 Wu, R., Li, B., Gao, M. et al. (2008). Tuning the morphologies of SiC nanowires via the control of growth temperature, and their photoluminescence properties. *Nanotechnology* 19 (33): 335602.

6 Zhong, Z., Qian, F., Wang, D., and Lieber, C.M. (2003). Synthesis of p-type gallium nitride nanowires for electronic and photonic nanodevices. *Nano Letters* 3 (3): 343–346.

7 Ha, B., Seo, S.H., Cho, J.H. et al. (2005). Optical and field emission properties of thin single-crystalline GaN nanowires. *The Journal of Physical Chemistry B* 109 (22): 11095–11099.

8 Li, K.-Z., Wei, J., Li, H.-J. et al. (2007). Photoluminescence of hexagonal-shaped SiC nanowires prepared by sol–gel process. *Materials Science and Engineering A* 460: 233–237.

9 Schlager, J.B., Bertness, K.A., Blanchard, P.T. et al. (2008). Steady-state and time-resolved photoluminescence from relaxed and strained GaN nanowires grown by catalyst-free molecular-beam epitaxy. *Journal of Applied Physics* 103 (12): 124309.

10 Hu, P., Dong, S., Zhang, X. et al. (2017). Synthesis and characterization of ultralong SiC nanowires with unique optical properties, excellent thermal stability and flexible nanomechanical properties. *Scientific Reports* 7 (1): 1–10.

11 Chin, A.H., Ahn, T.S., Li, H. et al. (2007). Photoluminescence of GaN nanowires of different crystallographic orientations. *Nano Letters* 7 (3): 626–631.

12 Qian, F., Gradecak, S., Li, Y. et al. (2005). Core/multishell nanowire heterostructures as multicolor, high-efficiency light-emitting diodes. *Nano Letters* 5 (11): 2287–2291.

13 Holmes, M.J., Choi, K., Kako, S. et al. (2014). Room-temperature triggered single photon emission from a III-nitride site-controlled nanowire quantum dot. *Nano Letters* 14 (2): 982–986.

14 Babinec, T.M., Hausmann, B.J., Khan, M. et al. (2010). A diamond nanowire single-photon source. *Nature Nanotechnology* 5 (3): 195–199.

15 Nguyen, H.P.T., Cui, K., Zhang, S. et al. (2011). Full-color InGaN/GaN dot-in-a-wire light emitting diodes on silicon. *Nanotechnology* 22 (44): 445202.

16 Zhao, C., Ng, T.K., ElAfandy, R.T. et al. (2016). Droop-free, reliable, and high-power InGaN/GaN nanowire light-emitting diodes for monolithic metal-optoelectronics. *Nano Letters* 16 (7): 4616–4623.

17 Kibria, M., Zhao, S., Chowdhury, F. et al. (2014). Tuning the surface Fermi level on p-type gallium nitride nanowires for efficient overall water splitting. *Nature Communications* 5 (1): 1–6.

18 Zhao, S., Nguyen, H.P., Kibria, M.G., and Mi, Z. (2015). III-Nitride nanowire optoelectronics. *Progress in Quantum Electronics* 44: 14–68.

19 Wu, J. and Walukiewicz, W. (2003). Band gaps of InN and group III nitride alloys. *Superlattices and Microstructures* 34 (1, 2): 63–75.

20 Gačević, Ž., Das, A., Teubert, J. et al. (2011). Internal quantum efficiency of III-nitride quantum dot superlattices grown by plasma-assisted molecular-beam epitaxy. *Journal of Applied Physics* 109 (10): 103501.

21 Ostapenko, I.A., Hönig, G., Rodt, S. et al. (2012). Exciton acoustic-phonon coupling in single GaN/AlN quantum dots. *Physical Review B: Condensed Matter* 85 (8): 081303.

22 Lounis, B. and Moerner, W.E. (2000). Single photons on demand from a single molecule at room temperature. *Nature* 407 (6803): 491–493.

23 Högele, A., Galland, C., Winger, M., and Imamoğlu, A. (2008). Photon antibunching in the photoluminescence spectra of a single carbon nanotube. *Physical Review Letters* 100 (21): 217401.

24 Bassani, F. (2005). *Encyclopedia of Condensed Matter Physics*. Elsevier Academic Press.

25 Høiaas, I.M., Liudi Mulyo, A., Vullum, P.E. et al. (2019). GaN/AlGaN nanocolumn ultraviolet light-emitting diode using double-layer graphene as substrate and transparent electrode. *Nano Letters* 19 (3): 1649–1658.

26 Wang, X., Summers, C.J., and Wang, Z.L. (2004). Large-scale hexagonal-patterned growth of aligned ZnO nanorods for nano-optoelectronics and nanosensor arrays. *Nano Letters* 4 (3): 423–426.

27 Koester, R., Sager, D., Quitsch, W.-A. et al. (2015). High-speed GaN/GaInN nanowire array light-emitting diode on silicon(111). *Nano Letters* 15 (4): 2318–2323.

28 Nguyen, H.P.T., Zhang, S., Cui, K. et al. (2011). p-Type modulation doped InGaN/GaN dot-in-a-wire white-light-emitting diodes monolithically grown on Si(111). *Nano Letters* 11 (5): 1919–1924.

29 Guo, W., Zhang, M., Banerjee, A., and Bhattacharya, P. (2010). Catalyst-free InGaN/GaN nanowire light emitting diodes grown on (001) silicon by molecular beam epitaxy. *Nano Letters* 10 (9): 3355–3359.

30 Kikuchi, A., Kawai, M., Tada, M., and Kishino, K. (2004). InGaN/GaN multiple quantum disk nanocolumn light-emitting diodes grown on (111) Si substrate. *Japanese Journal of Applied Physics* 43 (12A): L1524.

31 Sekiguchi, H., Kato, K., Tanaka, J. et al. (2008). Ultraviolet GaN-based nanocolumn light-emitting diodes grown on n-(111) Si substrates by rf-plasma-assisted molecular beam epitaxy. *Physica Status Solidi A Applications and Material Science* 205 (5): 1067–1069.

32 Bryan, Z., Bryan, I., Xie, J. et al. (2015). High internal quantum efficiency in AlGaN multiple quantum wells grown on bulk AlN substrates. *Applied Physics Letters* 106 (14): 142107.

33 Choi, I., Lee, K., Lee, C.-R. et al. (2019). Application of hexagonal boron nitride to a heat-transfer medium of an InGaN/GaN quantum-well green LED. *ACS Applied Materials & Interfaces* 11 (20): 18876–18884.

34 Luo, X., Song, W., Gao, F. et al. (2021). Strain-modulated light emission properties in a single InGaN/GaN multiple-quantum-well microwire-based flexible light-emitting diode. *Advanced Engineering Materials* 23: 2001430.

35 Zhao, S., Sadaf, S., Vanka, S. et al. (2016). Sub-milliwatt AlGaN nanowire tunnel junction deep ultraviolet light emitting diodes on silicon operating at 242 nm. *Applied Physics Letters* 109 (20): 201106.

36 Sadaf, S., Zhao, S., Wu, Y. et al. (2017). An AlGaN core–shell tunnel junction nanowire light-emitting diode operating in the ultraviolet-C band. *Nano Letters* 17 (2): 1212–1218.

37 Hai, X., Rashid, R., Sadaf, S. et al. (2019). Effect of low hole mobility on the efficiency droop of AlGaN nanowire deep ultraviolet light emitting diodes. *Applied Physics Letters* 114 (10): 101104.

38 Zhao, S., Lu, J., Hai, X., and Yin, X. (2020). AlGaN nanowires for ultraviolet light-emitting: recent progress, challenges, and prospects. *Micromachines* 11 (2): 125.

39 Iveland, J., Martinelli, L., Peretti, J. et al. (2013). Direct measurement of Auger electrons emitted from a semiconductor light-emitting diode under electrical injection: identification of the dominant mechanism for efficiency droop. *Physical Review Letters* 110 (17): 177406.

40 Sarwar, A.G., Carnevale, S.D., Yang, F. et al. (2015). Semiconductor nanowire light-emitting diodes grown on metal: a direction toward large-scale fabrication of nanowire devices. *Small* 11 (40): 5402–5408.

41 Willander, M., Nur, O., Bano, N., and Sultana, K. (2009). Zinc oxide nanorod-based heterostructures on solid and soft substrates for white-light-emitting diode applications. *New Journal of Physics* 11 (12): 125020.

42 Zhang, X.M., Lu, M.Y., Zhang, Y. et al. (2009). Fabrication of a high-brightness blue-light-emitting diode using a ZnO-nanowire array grown on p-GaN thin film. *Advanced Materials* 21 (27): 2767–2770.

43 Könenkamp, R., Word, R.C., and Schlegel, C. (2004). Vertical nanowire light-emitting diode. *Applied Physics Letters* 85 (24): 6004–6006.

44 Zainelabdin, A., Zaman, S., Amin, G. et al. (2010). Stable white light electroluminescence from highly flexible polymer/ZnO nanorods hybrid heterojunction grown at 50 °C. *Nanoscale Research Letters* 5 (9): 1442–1448.

45 Amin, G., Zaman, S., Zainelabdin, A. et al. (2011). ZnO nanorods–polymer hybrid white light emitting diode grown on a disposable paper substrate. *Physica Status Solidi RRL: Rapid Research Letters* 5 (2): 71–73.

46 Amin, G., Sandberg, M., Zainelabdin, A. et al. (2012). Scale-up synthesis of ZnO nanorods for printing inexpensive ZnO/polymer white light-emitting diode. *Journal of Materials Science* 47 (11): 4726–4731.

7

Thermal Properties of Wide Bandgap Nanowires

7.1 Thermal Conductivity

7.1.1 Fundamental of Thermal Transport and Thermal Conductivity

Thermal properties and transport of photon in nanowires (NWs) are affected by their nanostructures. The common structure of nanowires include core–shell and porous configurations. One of the most important thermal properties of nanowires is the thermal conductivity k (W/(m K)), which can be determined from the Fourier law:

$$J = -k\nabla T \tag{7.1}$$

where J and ∇T are the heat current and temperature gradient, respectively [1]. Thermal conductivity is determined by the lattice vibration, electron activity, and phonon radiation. In semiconductor materials, the thermal conductivity is dominated by the lattice vibration and electron scattering. Phonon transport plays the critical role in heat transfer of semiconductor nanowires. Dimensions of nanowires have been found as the dominant factor affecting the thermal conductivity. Selecting a material with a suitable thermal conductivity enables manipulation/control of heat flows in NWs. However, this approach limits the performance of nanowire-based thermal devices. Major research has conducted to understand the underlying physics of thermal transport properties of NWs, which is translated to high performance thermal and electronic devices. In a wide range of applications, a high thermal conductivity is required to enhance thermal conduction for cooling as high working temperature will degrade the performance of such nanowire devices [2]. The sources of heat include but are not limited to Joule heating that becomes significant when the size of devices is shrunk down to nanoscale. In application of thermoelectric nanowires, a low thermal conductivity is desirable to minimize heat losses and enhance their performance [3].

7.1.2 Measurement of Thermal Conductivity

Lee et al. utilized the four-point probe method to characterize the thermal conductivity of β-SiC nanowires [4]. Figure 7.1a shows the schematic sketch of the 3ω method for measurement of the thermal conductivity. Based on the measured voltage $V_{3\omega,rms,vac}$ from the 3ω method, the thermal conductivity is determined as [4–6]:

Wide Bandgap Nanowires: Synthesis, Properties, and Applications, First Edition. Tuan Anh Pham, Toan Dinh, Nam-Trung Nguyen, and Hoang-Phuong Phan.

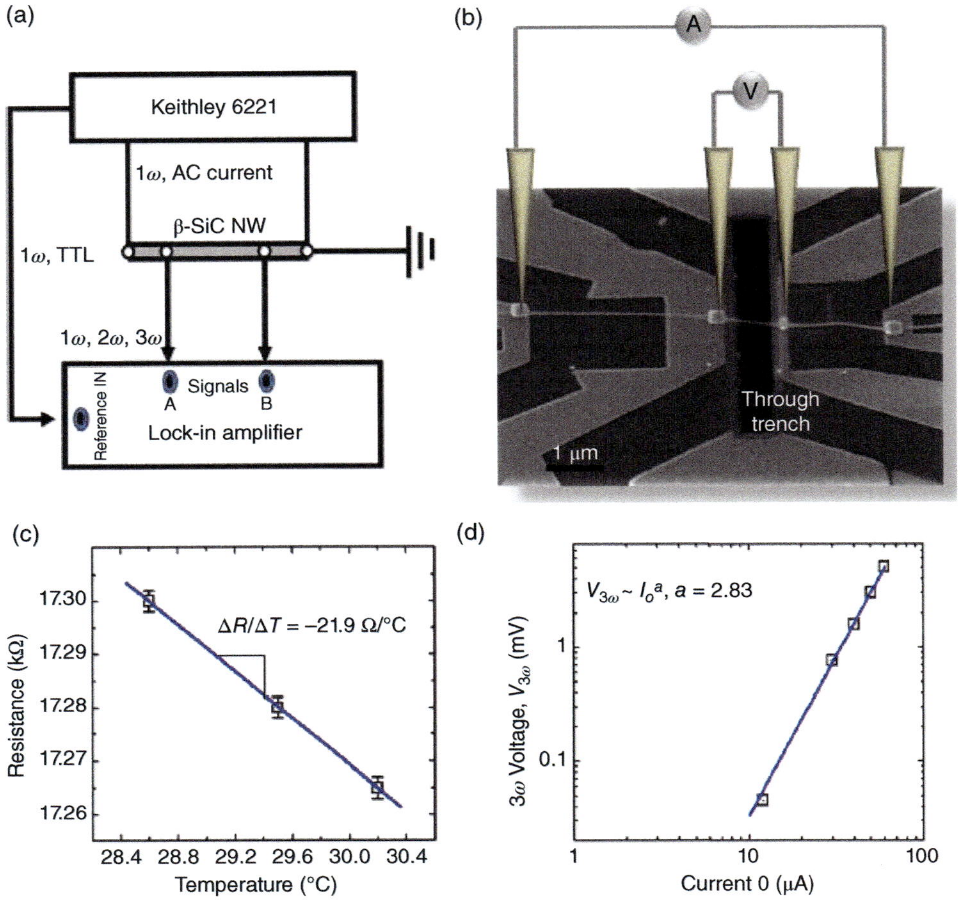

Figure 7.1 Measurement of thermal conductivity of SiC nanowire using 3ω method. (a) Four-point probe method. (b) SEM image of the device with four-point measurement method. (c) Dependence of resistance on temperature. (d) 3ω voltage vs current. Fitting shows the thermal conductivity of SiC nanowires 86.5 ± 3.5 W/(m K). *Source:* [4], Lee K-M, et. al., (2012)/with permission of Springer Nature.

$$k = \frac{\sqrt{2} I_o^{\,3} R R' L}{\pi^4 S V_{3\omega,rms,vac}} \tag{7.2}$$

where R is the resistance and $R' = dR/dT$ is the gradient of resistance; L and S are the length and cross-sectional area, respectively. I_o is the current amplitude. Figure 7.1b shows the SEM image of a β-SiC nanowire with four probe measurement. Figure 7.1c represents the temperature dependence of resistance, showing the gradient of resistance $R' = -21.9\ \Omega/°C$ for the range of temperature from 28.6 to 30.2 °C. Figure 7.1d shows the dependence of measured voltage $V_{3\omega,rms,vac}$ on the current I_o with the exponent of 2.83. The thermal conductivity was measured to be 86.5 ± 3.5 W/(m K), which is approximately four times smaller than that of bulk β-SiC materials (approximately 350 W/(m K)) [4, 7]. The thermal conductivity of double SiC NWs was measured to be 73 ± 5 W/(m K) using the same four point measurement method [7]. With a lower thermal conductivity and a good electrical conductivity, SiC NWs hold potential for enhancement of thermoelectric effect that can be used for energy harvesting applications.

7.1.3 Effect of Diameters on Thermal Properties

Thermal conductivity of nanowires depends on the boundary surfaces and phonon–phonon scattering. Theoretical study of phonon thermal conductivity of GaN nanowires/nanotubes has indicated that the thermal conductivity in nanostructures is much smaller than that of bulk materials and structures [8]. The decrease of diameter will result in the reduction of cross section in NWs, leading to the reduction of thermal conductivity due to the surface scattering effect [9]. In general, the decrease in the thermal conductivity of nanowires and nanotubes is expected with increasing surface roughness or decreasing thickness. Figure 7.2 shows the calculated thermal conductivity of GaN nanotubes with different dislocation densities and diameters, indicating the decreases of thermal conductivity with decreasing diameters [10]. For example, the thermal conductivity decreases from approximately 58 to 25 W/(m K) at 300 K when diameter decreases from 160 to 1000 nm. The dislocation density plays the important role in thermal transport properties of GaN NWs. As such, GaN NWs with high dislocation densities show a significant low thermal conductivity compared with GaN NWs with low dislocation densities (see Figure 7.2). In another simulation work, the thermal conductivity of 8 nm-diameter GaN nanotube is approximately 10 W/(m K), which is at least an order of magnitude smaller than that of bulk GaN materials [1, 8]. This is resulted from the increase in phonon scattering at boundaries and the variation of relaxation rate [8].

Notably, both experimental and theoretical studies have suggested that the thermal conductivity of SiC nanowires is much lower compared with that of the bulk SiC materials due to the size confinement effect [11, 12]. When the nanowire diameter decreases to a critical value, the thermal conductivity decreases as a result of boundary scattering limiting mean free paths of phonon. Takahashi et al. studied the thermal properties of SiC nanowires, thin films, and bulk materials [13]. The results of this study indicate that SiC nanowires have a higher thermal conductivity compared with that of the thin film but

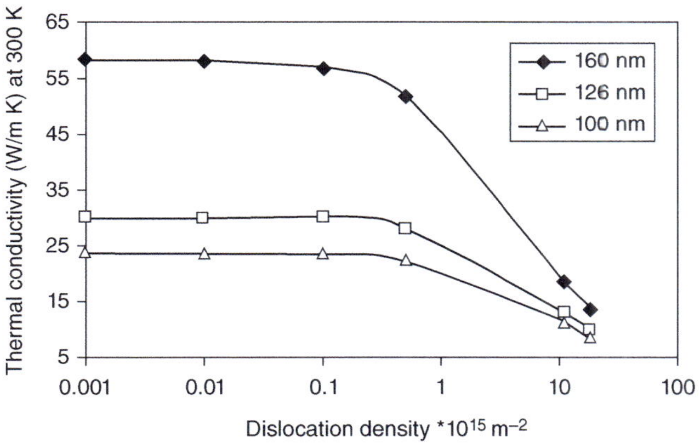

Figure 7.2 Calculated lattice thermal conductivity as a function of dislocation density at 300 K, of the GaN NWs for the diameters; 160, 126, and 100 nm. *Source:* Mamand et al. [10]/with permission of Elsevier.

smaller compared with that of bulk samples. In another work, Papanikolaou et al. used molecular dynamics (MD) simulation to study the thermal transport in SiC nanowire and bulk β-SiC [11]. The thermal conductivity of nanowire was simulated to be changed by 20% depending on surfaces and cross section of nanowires. In general, the small diameter nanowires have small thermal conductivity compared with the bulk materials, which is attributed to the presence of detects on the surfaces [11]. In another word, a larger cross section of the SiC nanowires shows a higher thermal conductivity. In addition, SiC nanowires with Si terminals have a lower thermal conductivity than those with C terminal. This is attributed to the stronger scattering on a more disordered Si surface compared with that of C surface [11].

The accuracy measurement of thermal conductivity has faced challenges due to the defects on the surfaces and the impact of contact resistance between NWs and electrodes. For example, 3-ω method has been utilized as a powerful measurement platform to reduce the contact resistance, lower heat loss, and measurement errors [14]. In addition, the existence of oxide layers on NWs or surface contamination/defect can significantly affect the thermal conductivity and thermal properties of nanowires [15]. Large energy gap nanowires such as ZnO are free from surface contamination or oxidation, which can be utilized for studying intrinsic thermal properties of NWs. Bui et al. studied the thermal transport properties of ZnO NWs with a range of diameters from 50 to 210 nm [15]. The reduction of the thermal conductivity of ZnO NWs is at least an order of magnitude compared to that of bulk materials, Figure 7.3a. Thermal conductivity increases with increasing temperature from 50 to 150 K, and then shows a reduction at higher temperature ranges. The reduction of thermal conductivity with decreasing diameter is attributed to enhanced phonon-boundary scattering when the dimension is shrunk down to nanoscale. Figure 7.3b shows the dependence of thermal conductivity k on temperature as $k{\sim}T^{-\alpha}$ where $\alpha = 1.42$–1.49, demonstrating a higher temperature dependence compared with silicon nanowires (SiNWs) with similar diameters [15–17]. This suggests that multiple scattering mechanisms are involved in the temperature dependence of thermal conductivity of ZnO NWs. The change of thermal conductivity of ZnO NWs is correlated to the cross-sectional area as $k{\sim}d^2$, with the rapid increase at low temperatures (e.g. 80 K) compared with high temperature (e.g. 300 K), Figure 7.3c. This is attributed to the influence of the phonon mean free path at low temperatures. In addition, the thermal transport properties of ZnO NWs were reported to be insensitive to amorphous carbon (a-C) contaminant coated on the surface of the ZnO NWs [15]; the thermal conductivity is greatly affected by the ion irradiation even at low doses. This suggests the importance of controlling the unwanted effect from contamination and defects in NWs.

7.1.4 Effect of Orientation on Thermal Properties

The dependence of thermal conductivity on crystal orientations has been studied for a wide range of nanowire materials, including Si [18], InAs [19], and GaN [20]. In silicon, atomic simulations proved that ⟨110⟩ has a thermal conductivity of up to 70% higher than that of ⟨111⟩ and ⟨100⟩ due to the higher group velocity of phonons existing in ⟨110⟩ direction [18]. In nanowire material with lower energy gap such as InAs, ⟨110⟩ has a thermal conductivity of three times higher than ⟨111⟩ and ⟨100⟩. While the thermal conductivity of ⟨110⟩ InAs decreases significantly with increasing temperature, its thermal conductivities in ⟨111⟩ and

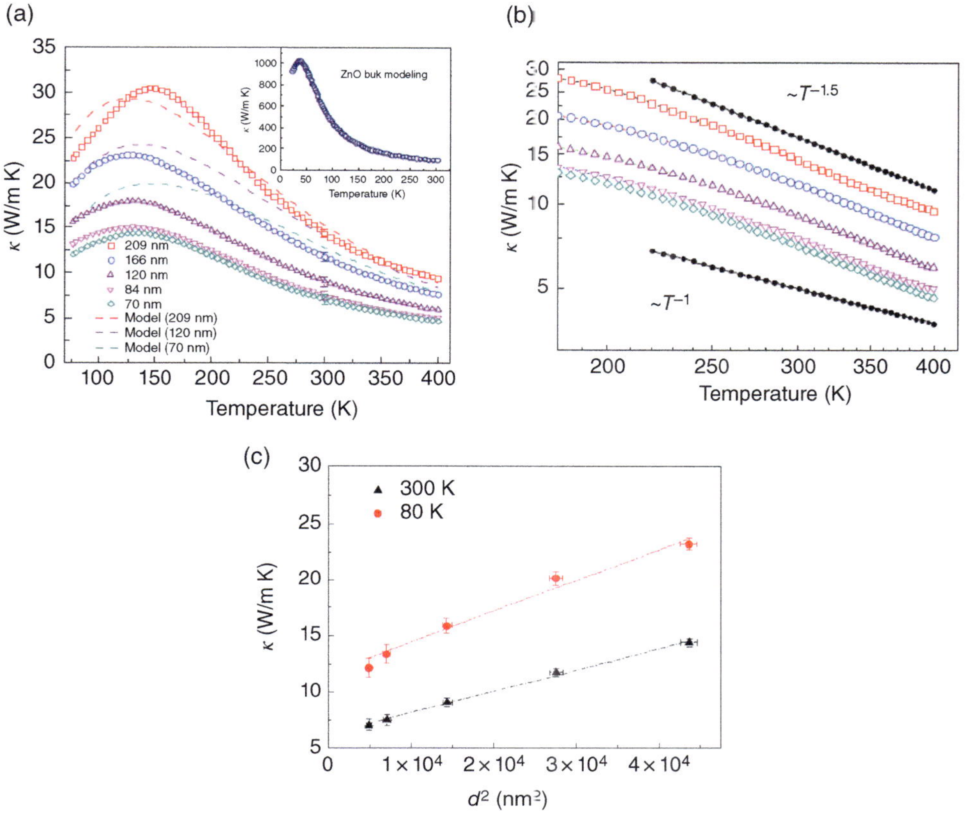

Figure 7.3 Thermal conductivity of ZnO nanowires. (a) Temperature dependence of thermal conductivity of ZnO nanowires with different diameters (b) Log–log plot of the dependence of the thermal conductivity on temperature and diameters. (c) Diameter dependence of thermal conductivity of ZnO NWs, showing the linear increase of thermal conductivity with squared cross-sectional area. *Source:* Bui et al. [15]/John Wiley and Sons.

⟨100⟩ directions are insensitive to temperature change, which are commonly observed for nanowires with zinc blende structures [19].

Wang et al. utilized atomistic simulation of thermal conductivity of GaN nanowires with different sizes and orientation [20]. GaN nanowires have a lower thermal conductivity compared with that of bulk crystal and the thermal conductivity increases with increasing of the diameter. Thermal conductivity in [001]-orientation is higher than that in $[1\bar{1}0]$ and [110] orientations. It is suggested that the dependence of the thermal conductivity on size and orientation is attributed to the surface scattering of phonons, which increases the surface-to-volume ratio with increasing GaN nanowire diameter. Davoody et al. described a computational work on the thermal properties of GaN nanowires with different thicknesses (3–9 nm) and for a wide range of temperatures from 300 to 1000 K [21]. The results suggested that thermal conductivity of GaN nanowires increase with increasing diameter, which is in solid agreement with other studies [1, 8]. However, the thermal conductivity exhibits a profound decrease at elevated temperatures [21].

7.1.5 Tenability of Thermal Properties

Recent theoretical studies suggested that the thermal properties of the thin-film wide bandgap (WBG) materials can be manipulated using some advanced methods, including application of strain [22–24] and engineering using buffer layer [25, 26]. For instance, Tang et al. theoretically studied the acoustic phonon and thermal properties of AlN/GaN/AlN nanofilms with the consideration of quantum confinement and pre-stress to tailor the thermal performance of nanostructures [22]. As such, pre-stress affects the dispersion and phonon group velocities, resulting in the alternation of acoustic phonon properties [24]. In addition, the variation of phonon characteristics is proven to be dependent on the thickness of core layers in heterostructures. The phonon thermal conductivity of different phonon modes depends on the level of pre-stress applied to the nanofilms/nanowires [22]. Strain or pre-stress could alter temperature dependence of the thermal conductivity of GaN heterostructures. The strain effect could be considered in design of high-performance thermal devices and effective thermal management of micro-/nanodevices.

In another work, Zhu and Ruan study the dependence of phonon thermal conductivity of GaN nanowires and nanostructures on the pre-stress fields [24]. Thermal conductivity increases with increasing pre-stress from −20 to 20 GPa. The change of thermal conductivity due to negative pre-stress is more significant compared with that caused by positive pre-stress. The increase of thermal conductivity under positive pre-stress is insignificant and reaches the saturation value. The change of thermal conductivity under application of negative pre-stress is much lower compared with that under the application of positive pre-stress. In another word, positive pre-stress leads to a strong dependence of thermal conductivity on nanowire size. In addition, nanowires with large diameters have their thermal conductivity strongly depending on the pre-stress field. The thermal conductivity of GaN nanowires is also dependent on temperature [24]. In both free standing and stressed/strained configurations, GaN nanowires exhibit the decrease in the thermal conductivity with increasing temperature [24]. The results are in solid agreement with other works reported in literature [27].

The phonon thermal conductivity of other GaN nanostructures is strongly dependent on pre-stress and surface/interface stress, including GaN nanofilms. Zhu and Luo demonstrated that negative stress, including pre-stress can increase the thermal conductivity of GaN nanofilms, while the positive stress causes a decrease of the thermal conductivity [28]. Figure 7.4 shows the effect on pre-stress and surface stress on the thermal conductivity of GaN nanofilms. The pre-stress and surface stress have more notably effect on the thermal conductivity of thicker GaN nanofilms [28].

Yalamarthy et al. reported on the modulation of thermal and electrical properties of two-dimensional electron gas (2DEG) at the AlGaN/GaN interface using buffer layer engineering [25]. For example, the GaN buffer layer can be utilized to separately control electrical transport and thermal conductance of the 2DEG. Figure 7.5a shows the microscope image of AlGaN/GaN heterostructure, which was utilized to measure the thermal and electrical properties of GaN 2DEG. It was proved that the electrical conductance of thin GaN sample exhibited a similar value to that measured for the bulk GaN sample. However, the thermal conductivity showed contrast properties. As such, thermal conductivity was measured to be approximately 115 W/(m K) for the bulk GaN sample, while it decreased to approximately 45 W/(m K) for the thin GaN sample. The difference in the thermal

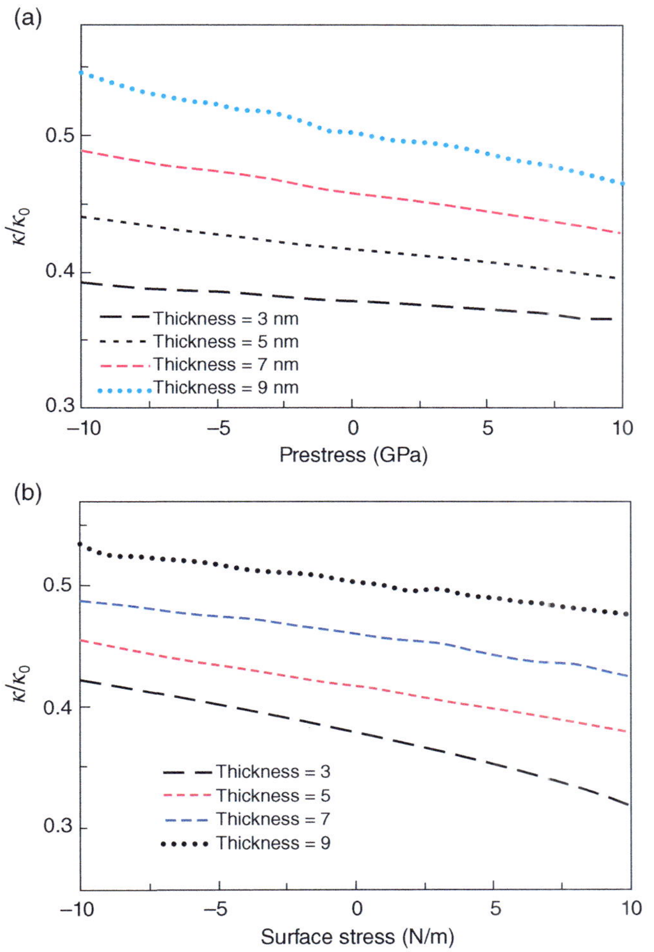

Figure 7.4 Photon thermal conductivity of GaN nanofilms. (a) Thermal conductivity vs pre-stress. (b) Thermal conductivity vs surface stress. *Source:* Zhu and Luo [28]/with permission of Elsevier.

conductivity of thin and bulk samples is attributed to the phonon boundary scattering effect. Figure 7.5b shows the dependence of thermal conductivity of 2DEG on a wide range of temperature. Thin GaN exhibited a decrease of thermal conductivity with increasing temperature $k \sim T^{-0.88}$ while bulk GaN showed a stronger temperature dependence of thermal conductivity as $k \sim T^{-1.18}$. The weaker dependence of the thermal conductivity in thin film GaN samples are attributed to the accumulation of electrons at the heterostructured interface [25]. Figure 7.5c shows the modeled in-plane thermal conductivities of the layers as a function of layer thickness. Figure 7.5d–g presents the energy bands of bulk and thin GaN heterostructures and the volumetric charge densities, indicating the visibility of 2DEG sheet density. The thickness of 2DEG region can be extracted to be from 4.4 (for thin GaN) to 6.6 nm for bulk heterostructures [25].

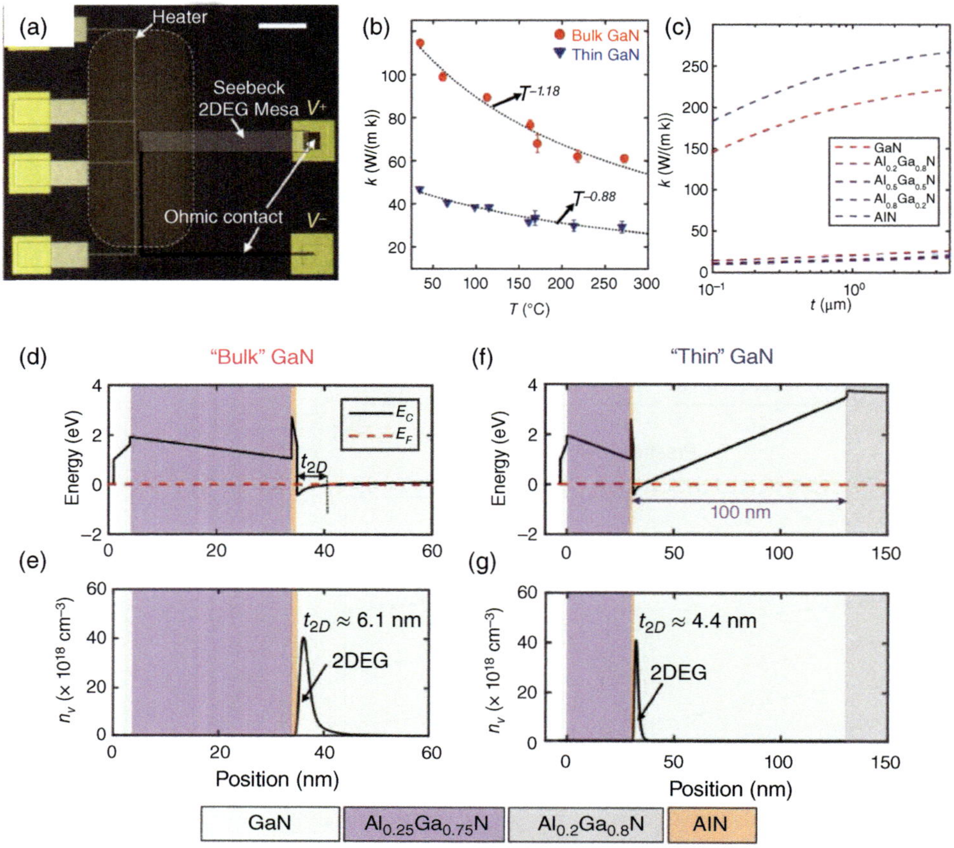

Figure 7.5 Thermal properties of AlGaN/GaN heterostructure. (a) Microscope image of AlGaN/GaN heterostructure. (b) Temperature-dependent thermal conductivity measurements for the bulk and thin GaN. (c) Modeled in-plane thermal conductivities of the layers. (d) Energy band diagram for the bulk GaN structure. (e) Charge density for bulk GaN. (f) Energy band diagram for the thin GaN structure. (g) Charge density for thin GaN. *Source:* [25], Yalamarthy AS, et. al., (2018)/with permission of John Wiley & Sons, Inc.

7.2 Thermoelectric Properties

7.2.1 Fundamental Thermoelectric Properties

WBG semiconductor nanowires possesses excellent thermal properties for fabricating electronic devices in a wide range of applications including thermal management and energy harvesting. The capability of generating electric power of WBG semiconductor nanowires can be described using the thermoelectric figure of merit (ZT) as [29]:

$$ZT = S^2\sigma T / k \tag{7.3}$$

where S and σ are the Seebeck coefficient and electrical conductivity; $k = k_e + k_l$ is the thermal conductivity of the nanowire, where k_e and k_l are the lattice thermal conductivity and electronic thermal conductivity, respectively. To evaluate the efficiency of thermoelectric properties, it is important to measure the Seebeck coefficient, electrical and thermal conductivities of nanowires. The thermoelectric efficiency depends upon the crystalline structure, impurities/dislocations, surface boundary, and scattering effect in WBG nanowires. These factors contribute to the change of thermal conductivity and electrical conductivity of WBG nanowires, including SiC nanostructures [7, 13, 30]. The Seebeck coefficient for semiconductor nanowires can be determined by:

$$S = \frac{8\pi^2 k_B^2}{3eh^2} m^* T \left(\frac{\pi}{3n} \right)^{2/3} \tag{7.4}$$

where e and m^* are the carrier charge and effective mass of charge carriers, respectively; k_B and h are the Boltzmann constant and Planck's constant, respectively. To enhance the performance of thermoelectric nanowires, the thermoelectric figure of merit (ZT) can be maximized by increasing the power factor ($S^2\sigma$), and decreasing the thermal conductivity k. However, the effort in maximizing the thermoelectric figure of merit (ZT) has faced great challenges due to the complex relationship between electrical conductivity and thermal conductivity. Currently, the value of thermoelectric figure of merit (ZT) is typically found to be approximately two for many thermoelectric materials.

7.2.2 Thermoelectric Properties of ZnO and GaN NWs

In early works, WBG GaN and ZnO NWs have been proven to show great potential for optoelectronics and high-temperature electronic applications [31]. Figure 7.6a shows the schematic sketch of the measurement approach for thermoelectric power of WBG NWs, which includes one microfabricated heater and two thermometers [31]. The thermometers function as electrodes for electrical measurements. A current is applied to the heater to generate a temperature gradient of ΔT due to Joule heating. The temperature gradient is measured by the resistance-based thermometers. Based on experimental measurements, the thermoelectric power is calculated by the thermoelectric voltage ΔV and the temperature gradient as $S = -\Delta V/\Delta T$. The linear characteristics of current–voltage measured at different temperatures are typically utilized observed to prove the good Ohmic behavior of ZnO NWs [31]. The thermoelectric power of ZnO nanowires is measured to be approximately $-400\,\mu V/K$ at room temperature. The negative value of thermoelectric power indicates that the carrier transport in ZnO nanowires is controlled by electron diffusion. The thermoelectric figure of merit (ZT) of n-type ZnO nanowires was investigated to be up to 0.6 at a temperature of 1000 °C [33].

ZnO NWs exhibit an increase of the thermoelectric power with increasing temperature. The slope of approximately $-1.36\,\mu V/K^2$ presents for the increase of the thermoelectric power with increasing temperature [31]. In a theoretical study, Wang et al. revealed the thermoelectric properties of ZnO NWs depending on the carrier concentrations and temperature. Figure 7.6b shows the increase of thermal conductivity of ZnO nanowires from 100 to 400 K, and the decrease trend for higher temperature ranges. The increase of thermal conductivity at low temperature is attributed to the boundary scattering; the impurity and phonon–phonon scattering is dominant at high temperatures. Figure 7.6c indicates the decrease of the Seebeck coefficient with increasing carrier concentration. Figure 7.6d illustrates the increase of the power factors with increasing carrier concentration. The maximum value of figure of merit ZT of 0.147 is achieved for Ga doped ZnO nanowires with the carrier concentration of $3.62 \times 10^{19}\ cm^{-3}$ [32].

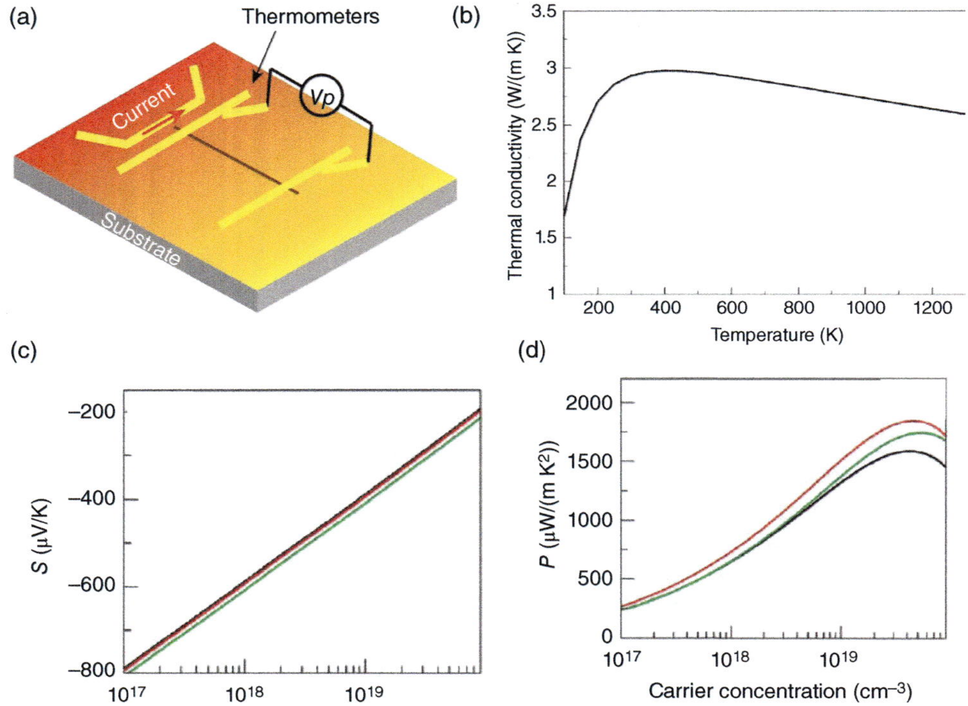

Figure 7.6 Measurement of thermoelectric power of ZnO NWs. (a) Schematic sketch of thermoelectric device measurements [31]. (b) Temperature dependence of the thermal conductivity for the ZnO nanowires. (c) Seebeck coefficient of Al, Ga, and Sb doped ZnO nanowires as a function of carrier concentration. (d) Power factor of ZnO nanowires. *Source:* Wang et al. [32]/Royal Society of Chemistry.

The thermoelectric power of 40 μV/K is measured for GaN nanowires, which is 1 order of magnitude lower than that of ZnO nanowires [31]. The decrease of thermoelectric power of GaN nanwires is about $-0.127\,\mu V/K^2$. Overall, WBG nanowires show a linear increase of thermoelectric power with increasing temperature. Guthy et al. studied the thermal conductivity of GaN for a range of diameter from 97 to 181 nm. The average thermal conductivity of GaN NWs is 17 W/(m K), which is about 1 order of magnitude lower than that of bulk GaN material [34]. The theoretical studies indicated that the thermoelectric figure of merit of GaN NWs is from 0.2 to 0.8 from room temperature to 1000 K [21]. The thermal conductivity decreases with decreasing diameter, while the Seebeck coefficient of GaN NWs showed opposite trend. These all factors would contribute to overall thermoelectric figure of merit of GaN NWs.

7.2.3 Thermoelectric Properties of SiC NWs

Within various WBG semiconductor nanowires, SiC NWs have been proven to show potential for thermoelectric devices working in harsh environments owing to excellent mechanical and electrical properties of SiC materials such as high resistance to oxidation and corrosion as well as large thermoelectric power generation [35, 36]. At nanoscale, the thermoelectric transport properties of SiC NWs are different to those reported for thin films and bulk forms [35]. The increase of surface boundary phonon scattering would potentially improve the thermoelectric power. Valentín et al. reported on the thermoelectric and transport properties of β-SiC nanowires (diameters of NW1, NW2, and NW3 are 90, 70, and 80 nm, respectively) in a temperature range of 190–370 K [30]. The thermal conductivity was found to be between 4 and 12 W/(m K), which is an order of magnitude lower than that of bulk SiC materials. To form the SiC NW thermoelectric device, a SiC nanowire is constructed over a suspended resistance thermodevice with electrodes formed by focused ion beam deposition of Pt [30]. The Seebeck coefficient of SiC NWs is measured to be dependent on temperature. For example, the Seebeck coefficient ranges from −18 to −68 μV/K for whole testing temperature range of 190 to 370 K [30], and it is from −22 to −56 μV/K measured at room temperature, which agrees with those reported for SiC nanofilms [37]. SiC NWs with higher Seebeck coefficients showed a higher value of the thermoelectric figure of merit (*ZT*). The value of *ZT* is dependent on thermal conductivity κ and electrical conductivity σ. SiC-NWs show a low thermal conductivity due to the quantum confinement at nanoscale and boundary scattering [30]. As NWs are scaled down to nanoscale, the thermal conductivity of nanowires would decrease significantly compared with that of bulk materials due to the difference in phonon transport and surface scattering. In addition, there is a variation in the impurities and dislocations existing in the SiC NWs, which leads to the difference in the electrical conductivity. All factors have significant effects on the thermoelectric figure of merit (*ZT*) of WBG nanowires.

SiC NWs have also been used to develop thermoelectric hybrids, including nanocomposites [38, 39]. Wang et al. demonstrated enhanced thermoelectric performance by integrating the organic and inorganic materials into a nanocomposite structure. The combination of defects in organic materials (low Seebeck coefficient) and inorganic materials (high thermal conductivity) has been demonstrated as an effective method to enhance thermal and electric transport in nanowire hybrids [39]. Figure 7.7a shows the dependence of Seebeck coefficient of PEDOT:PSS/SiC-NWs hybrid thermoelectric thin films on the content of bulk SiC nanowires (b-SiC-NWs) and high purity SiC nanowires (p-SiC-NWs). For example, the Seebeck coefficient increases to 20.9 μV/K with SiC NWs content increasing to approximately 10 wt% [39]. The enhancement of Seebeck coefficient is attributed to the energy filtering effect at the interface between SiC NWs and PEDOT:PSS. Figure 7.7b indicates the maximum thermoelectric power of 67.7 μW/(m K^2) achieved for the SiC NWs content approximately 10 wt% due to the large Seebeck coefficient of SiC NWs.

7.2.4 Optimization of the Thermoelectric Properties

In nanowires, the charge transport mechanisms and nobilities of charge carriers depend on the surface characteristics, including defect densities. The conductivities of nanowires are tailored by surface defects due to the trapping of charge carriers that prefer to become electrically inactive in nature. These factors need to be considered in design of high-performance thermoelectric devices based on WBG nanowires. In order to tune the thermoelectric properties of WBG nanowire-based devices, it is crucial to control the surface properties of the nanowires. In addition, the crystalline quality, which is associated growth direction of the nanowire, has significant effect on the carrier mobility and subsequent thermal and electronic transport properties of the nanowire. To express the dependence of the thermoelectric figure of merit on the thermal conductivity of nanowire, *ZT* is rewritten as follows:

$$ZT = \frac{S^2}{L}\left(1 + \frac{k_l}{k_e}\right)^{-1} \tag{7.5}$$

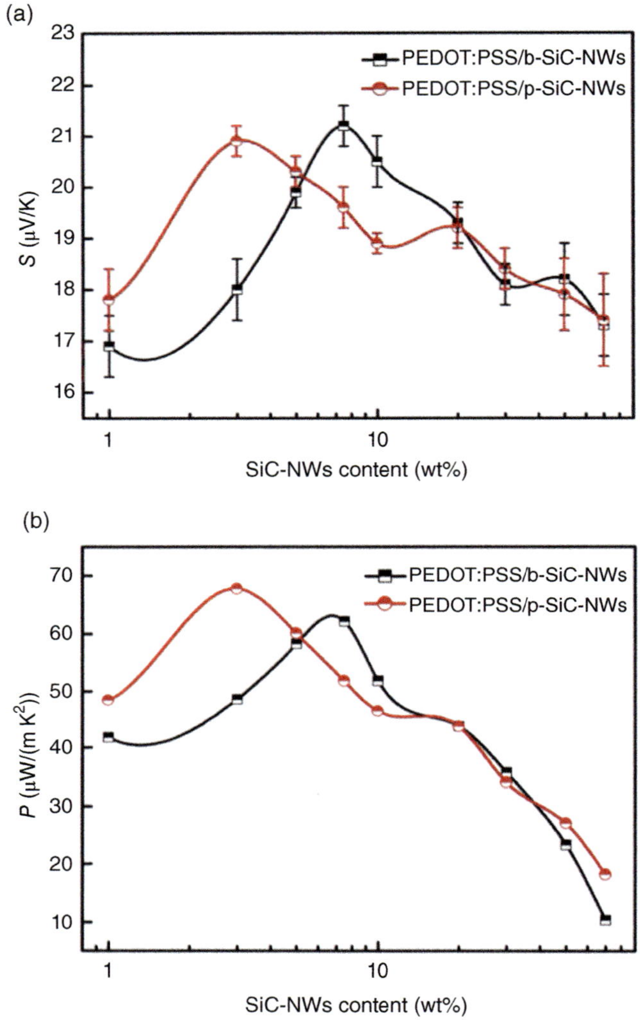

Figure 7.7 PEDOT:PSS/SiC-NWs hybrid thermoelectric. (a) Seebeck coefficient. (b) Power factor of PEDOT:PSS/SiC-NWs hybrid thin films depending on the content of SiC-NWs. *Source:* Wang et al. [39]/with permission of Elsevier.

where L is the Lorenz factor described in the Wiedemann–Franz law as $k_e = L\sigma T$. In order to increase the value of ZT, reducing the value of k_l is an effective method. As such, increase in phonon scattering at the interfaces and boundaries will lead to the reduction of k_l. This can be done by reducing the diameter of nanowires and increasing the surface-to-volume ratio. The decrease of the mean free path when scaling down the size of structures will result in the reduction of the thermal conductivity. In addition, roughness plays an important role in electric transport of carriers and phonon scattering on the surface [40]. For example, rough surfaces of nanowires cause the incoherent scattering of phonon, and subsequent decrease of thermal conductivity [41]. In addition, the high acoustic mismatch at the surfaces due to the high roughness causes the reduction of thermal conductivity. Therefore, control of both diameter and roughness of nanowires could reduce the thermal conductivity and significantly enhance the thermoelectric figure of merit. Post-synthesis methods including etching of NWs could be implemented to control the roughness and diameter of nanowires. However, higher roughness will couple with electrostatic mismatch at the nanowire interface, lowering the electrical conductivity and subsequent thermoelectric figure of merit.

References

1 Liu, W., Balandin, A.A., Lee, C., and Lee, H.Y. (2005). Increased thermal conductivity of free-standing low-dislocation-density GaN films. *Physica Status Solidi A Applications and Material Science* 202 (12): R135–R137.

2 Dinh, T., Phan, H.P., Kashaninejad, N. et al. (2018). An on-chip SiC MEMS device with integrated heating, sensing, and microfluidic cooling systems. *Advanced Materials Interfaces* 5 (20): 1800764.

3 Zhao, D. and Tan, G. (2014). A review of thermoelectric cooling: materials, modeling and applications. *Applied Thermal Engineering* 66 (1, 2): 15–24.

4 Lee, K.-M., Lee, S.-K., and Choi, T.-Y. (2012). Highly enhanced thermoelectric figure of merit of a β-SiC nanowire with a nanoelectromechanical measurement approach. *Applied Physics A* 106 (4): 955–960.

5 Yi, W., Lu, L., Dian-Lin, Z. et al. (1999). Linear specific heat of carbon nanotubes. *Physical Review B* 59 (14): R9015.

6 Choi, T.Y., Poulikakos, D., Tharian, J., and Sennhauser, U. (2005). Measurement of thermal conductivity of individual multiwalled carbon nanotubes by the 3-ω method. *Applied Physics Letters* 87 (1): 013108.

7 Lee, K., Choi, T., Lee, S., and Poulikakos, D. (2010). Focused ion beam-assisted manipulation of single and double β-SiC nanowires and their thermal conductivity measurements by the four-point-probe 3-ω method. *Nanotechnology* 21 (12): 125301.

8 Zhou, G. and Li, L. (2012). Phonon thermal conductivity of GaN nanotubes. *Journal of Applied Physics* 112 (1): 014317.

9 Zhang, Z. and Chen, J. (2018). Thermal conductivity of nanowires. *Chinese Physics B* 27 (3): 035101.

10 Mamand, S., Omar, M., and Muhammad, A. (2012). Nanoscale size dependence parameters on lattice thermal conductivity of Wurtzite GaN nanowires. *Materials Research Bulletin* 47 (5): 1264–1272.

11 Papanikolaou, N. (2008). Lattice thermal conductivity of SiC nanowires. *Journal of Physics: Condensed Matter 20* (13): 135201.

12 Termentzidis, K., Barreteau, T., Ni, Y. et al. (2013). Modulated SiC nanowires: molecular dynamics study of their thermal properties. *Physical Review B* 87 (12): 125410.

13 Takahashi, K., Ito, Y., Ikuta, T. et al. (2008). Thermal conductivity of SiC nanowire formed by combustion synthesis. *High Temperatures-High Pressures* 37 (2): 119–125.

14 Lee, S.-Y., Lee, M.-R., Park, N.-W. et al. (2013). Temperature-dependent thermal conductivities of 1D semiconducting nanowires via four-point-probe 3-ω method. *Nanotechnology* 24 (49): 495202.

15 Bui, C.T., Xie, R., Zheng, M. et al. (2012). Diameter-dependent thermal transport in individual ZnO nanowires and its correlation with surface coating and defects. *Small* 8 (5): 738–745.

16 Hochbaum, A.I., Chen, R., Delgado, R.D. et al. (2008). Enhanced thermoelectric performance of rough silicon nanowires. *Nature* 451 (7175): 163–167.

17 Chen, R., Hochbaum, A.I., Murphy, P. et al. (2008). Thermal conductance of thin silicon nanowires. *Physical Review Letters* 101 (10): 105501.

18 Holmes, J.D., Johnston, K.P., Doty, R.C., and Korgel, B.A. (2000). Control of thickness and orientation of solution-grown silicon nanowires. *Science* 287 (5457): 1471–1473.

19 Zhou, W.-X., Chen, K.-Q., Tang, L.-M., and Yao, L.-J. (2013). Phonon thermal transport in InAs nanowires with different size and growth directions. *Physics Letters A* 377 (43): 3144–3147.

20 Wang, Z., Zu, X., Gao, F. et al. (2007). Atomistic simulation of the size and orientation dependences of thermal conductivity in GaN nanowires. *Applied Physics Letters* 90 (16): 161923.

21 Davoody, A., Ramayya, E., Maurer, L., and Knezevic, I. (2014). Ultrathin GaN nanowires: electronic, thermal, and thermoelectric properties. *Physical Review B* 89 (11): 115313.

22 Tang, X., Wang, J., Zhu, L., and Yin, W. (2018). Simulating stress-tunable phonon and thermal properties in heterostructured AlN/GaN/AlN-nanofilms. *Materials Research Express* 6 (1): 015018.

23 Zhang, S., Tang, X., Ruan, H., and Zhu, L. (2019). Effects of surface/interface stress on phonon properties and thermal conductivity in AlN/GaN/AlN heterostructural nanofilms. *Applied Physics A* 125 (10): 1–14.

24 Zhu, L. and Ruan, H. (2014). Influence of prestress fields on the phonon thermal conductivity of GaN nanostructures. *Journal of Heat Transfer* 136 (10): 102402-1–102402-7.

25 Yalamarthy, A.S., So, H., Muñoz Rojo, M. et al. (2018). Tuning electrical and thermal transport in AlGaN/GaN heterostructures via buffer layer engineering. *Advanced Functional Materials* 28 (22): 1705823.

26 Power, M., Pomeroy, J.W., Otoki, Y. et al. (2015). Measuring the thermal conductivity of the GaN buffer layer in AlGaN/GaN HEMTs. *Physica Status Solidi A Applications and Material Science* 212 (8): 1742–1745.

27 Jung, K., Cho, M., and Zhou, M. (2011). Strain dependence of thermal conductivity of [0001]-oriented GaN nanowires. *Applied Physics Letters* 98 (4): 041909.

28 Zhu, L. and Luo, H. (2016). Phonon properties and thermal conductivity of GaN nanofilm under prestress and surface/interface stress. *Journal of Alloys and Compounds* 685: 619–625.

29 Ali, A., Chen, Y., Vasiraju, V., and Vaddiraju, S. (2017). Nanowire-based thermoelectrics. *Nanotechnology* 28 (28): 282001.

30 Valentín, L., Betancourt, J., Fonseca, L. et al. (2013). A comprehensive study of thermoelectric and transport properties of β-silicon carbide nanowires. *Journal of Applied Physics* 114 (18): 184301.

31 Lee, C.-H., Yi, G.-C., Zuev, Y.M., and Kim, P. (2009). Thermoelectric power measurements of wide band gap semiconducting nanowires. *Applied Physics Letters* 94 (2): 022106.

32 Wang, C., Wang, Y., Zhang, G. et al. (2014). Theoretical investigation of the effects of doping on the electronic structure and thermoelectric properties of ZnO nanowires. *Physical Chemistry Chemical Physics* 16 (8): 3771–3776.

33 Brockway, L., Vasiraju, V., Sunkara, M.K., and Vaddiraju, S. (2014). Engineering efficient thermoelectrics from large-scale assemblies of doped ZnO nanowires: nanoscale effects and resonant-level scattering. *ACS Applied Materials & Interfaces* 6 (17): 14923–14930.

34 Guthy, C., Nam, C.-Y., and Fischer, J.E. (2008). Unusually low thermal conductivity of gallium nitride nanowires. *Journal of Applied Physics* 103 (6): 064319.

35 Feng, Z., Mascarenhas, A., Choyke, W., and Powell, J. (1988). Raman scattering studies of chemical-vapor-deposited cubic SiC films of (100) Si. *Journal of Applied Physics* 64 (6): 3176–3186.

36 Kim, J.G., Choi, Y.Y., Choi, D.J. et al. (2009). A study on the thermoelectric properties of chemical vapor deposited SiC films with temperature and diluent gases variation. *Journal of the Ceramic Society of Japan* 117 (1365): 574–577.

37 Abu-Ageel, N., Aslam, M., Ager, R., and Rimai, L. (2000). The Seebeck coefficient of monocrystalline-SiC and polycrystalline-SiC measured at 300–533 K. *Semiconductor Science and Technology* 15 (1): 32.

38 Yang, W., Araki, H., Tang, C. et al. (2005). Single-crystal SiC nanowires with a thin carbon coating for stronger and tougher ceramic composites. *Advanced Materials* 17 (12): 1519–1523.

39 Wang, X., Meng, F., Wang, T. et al. (2018). High performance of PEDOT:PSS/SiC-NWs hybrid thermoelectric thin film for energy harvesting. *Journal of Alloys and Compounds* 734: 121–129.

40 Toberer, E.S., Baranowski, L.L., and Dames, C. (2012). Advances in thermal conductivity. *Annual Review of Materials Research* 42: 179–209.

41 Martin, P., Aksamija, Z., Pop, E., and Ravaioli, U. (2009). Impact of phonon-surface roughness scattering on thermal conductivity of thin Si nanowires. *Physical Review Letters* 102 (12): 125503.

8

Ultraviolet Sensors

8.1 Introduction

Ultraviolet (UV) sensors are one of the most attractive sensing devices used in system and environmental monitoring, civil engineering, and military applications [1, 2]. In this regard, wide bandgap (WBG) semiconductors are highly suitable for detecting short wavelengths from near UV (green and blue) to deep UV [3–5]. Several WBG materials exhibit a direct energy bandgap that offers a high quantum efficiency, allowing for excellent UV sensitivity. The capability of converting photon energy into an output voltage through the photovoltaic effect enables the development of self-powered photodetectors.

It is noteworthy to mention that the 2014 Nobel Prize in Physics was awarded for the invention of blue light emitting diodes (LEDs) using GaN materials. Along with LED that are imperative for numerous display applications, WBG semiconductors also play an important role in the sustainable growth of the UV sensors market, with the market size being expected to exceed US$2.7 billion in 2021 (https://www.technavio.com/report/global-uv-sensor-market). For instance, commercialized applications can be found in flame detectors, air/water quality assessment systems, and portable health care device (e.g. dosimeter). These evidences demonstrate the importance of WBG materials for highly effective light emitting and detecting applications.

Compared with the bulk platform, nanowires are expected to further enhance the performance of UV sensors due to its large surface areas, active electron layer exposed on their surfaces, and high optical absorption [6–8]. These unique features in nanowires can significantly enhance sensitivity, stability, responsivity, response speed, and quantum efficiency. In addition, the advancement in the synthesis and fabrication technologies also enables the seamless integration of WBG nanowires into a wide range of device architectures to meet specific requirements.

This chapter provides an overview on the working principle, fabrication process, materials of choices for specific wavelength, device performance, and examples of applications of UV sensors in the real world.

8.2 Sensing Mechanism

Understanding the advantage, drawback, and remaining challenges of each type of UV sensors is critically important to meet the requirement from niche, practical applications. To evaluate and standardize sensors performance, common physical parameters such as

Wide Bandgap Nanowires: Synthesis, Properties, and Applications, First Edition. Tuan Anh Pham, Toan Dinh, Nam-Trung Nguyen, and Hoang-Phuong Phan.

the cutoff wavelength, the quantum efficiency, and time response are widely employed. The cutoff wavelength refers to the longest wafer length that a UV sensor can detect, which is defined by the energy gap of materials ($\lambda = hc/E$). The cutoff wavelength can be measured from the absorption spectra of the used materials. In several cases that require a high selectivity, materials that are only sensitive to a narrow broadband are desired. The sensitivity of UV sensors or optical sensors in general is defined by the responsivity:

$$R_e = \frac{I_p - I_d}{hv} \tag{8.1}$$

Here I_p and I_d are the photo and dark currents, respectively. v is the wavelength of the incident light. The sensitivity can also be represented through quantum efficiency, which is the ratio of the responsivity to the incident photon (hc/λ):

$$\eta(\lambda) = R_\lambda \frac{hc}{q\lambda} \tag{8.2}$$

Similar to the bulk platform, semiconductor nanowire (NW) sensors employ four main architectures: (i) metal–semiconductor–metal (i.e. photoconductors), (ii) Schottky barrier, (iii) p–n junction, and (iv) p–i–n junction.

8.2.1 Photoconductor Architectures

In this architecture, two Ohmic contacts are formed between metal electrodes and the two terminals of a nanowire [9]. The formation of the Ohmic contact can be achieved using a relevant metal that has the Fermi level aligning with or being close to the conduction band (for n-type) or the valance band (for p-type) of semiconductor nanowires. Thermal annealing or laser irradiation can also be employed to form an interfacing layer between metal and semiconductors to reduce the built-in potential, which in turn enhances the Ohmic characteristics. Under UV light illumination, semiconductor nanowires will absorb photon energy ($E = hv$ where h is the Planck constant, and v is the optical frequency), generating electron–hole pairs that directly result in an increase in the charge carriers. The increase in the charge carriers leads to a significant enhancement in the electrical conductance, which can be detected using a voltage or current meter. As the responsivity is proportional to the difference between the photocurrent and the dark current, a small dark current is preferred for a highly sensitive UV sensor. However, due to the nature of the Ohmic contact, the photoconductor architecture exhibits a large dark current, which diminishes its photo sensitivity. In several cases, this drawback can be overcome by utilizing the large surface area of nanowires that can enhance the efficiency of photon absorption. Furthermore, due to the simplicity of fabrication process (e.g. compared with the p–n junctions), the metal–semiconductor–metal architecture enables large-scale productions, reduces the device cost, and offers excellent performance uniformity.

Figure 8.1 shows an example of photoconductive UV sensors using Si doped GaN nanowires [10]. The GaN nanowires, with a length of 1.2 μm and diameter of approximately 50–80 nm, consist of two highly Si-doped edges and an undoped middle section. The current–voltage measurement shows a symmetrical and linear characteristic in a bias range from −4 to 4 V, indicating an Ohmic contact between metal (Ti/Al/Ti/Au) and Si-doped GaN. Under UV exposure, the conductivity of the GaN nanowire increased and then returned to the

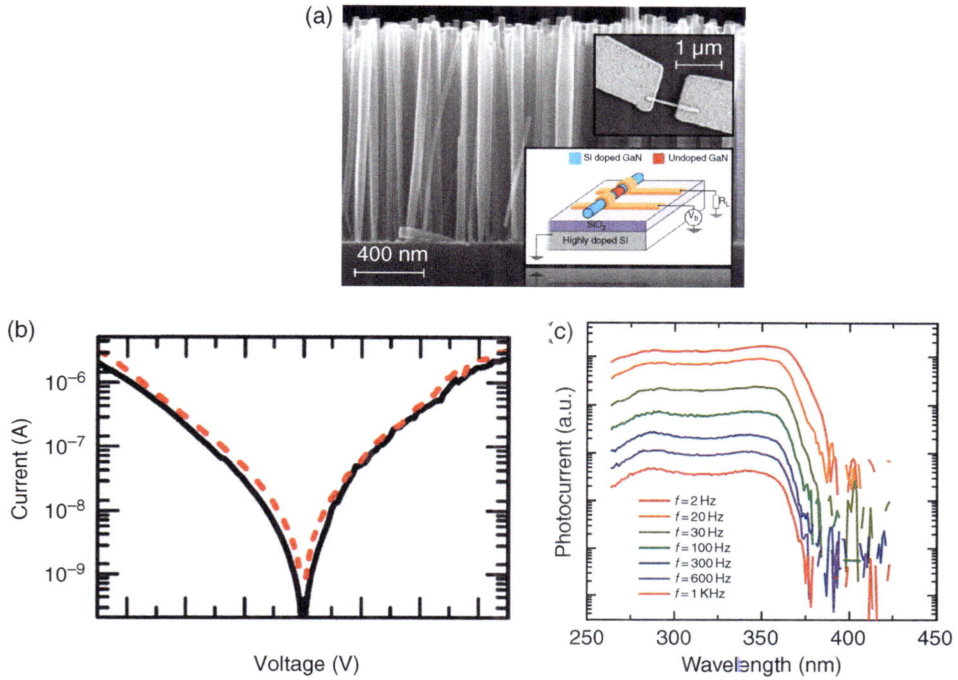

Figure 8.1 UV sensor using the photoconductive configuration. (a) A photoconductive sensor using a single GaN nanowire. (b) The current–voltage characteristic of the GaN nanowire. (c) The response of the UV sensors under wavelengths ranging from 250 to 450 nm at different measurement frequencies. *Source:* [10], González-Posada et al., (2012)/American Chemical Society.

initial value when UW was switched off, Figure 8.1b. Figure 8.1c shows the change in the photocurrent under different UV intensity and frequency, showing a high photocurrent gain in a range from 10^5 to 10^8. The sensor exhibits a large visible rejection of over 6 orders of magnitude, which is independent of the measurement frequency. This demonstrates the high selectivity in GaN based photoconductive types for UV sensing applications.

8.2.2 Schottky Diode Photo Sensors

UV sensors operating using a Schottky contact employ the current change between the metal–semiconductor junction under light illumination. In contradiction to the photoconductor architecture, Schottky contact based devices use metals that have significant potential barrier to the Fermi level of semiconductor nanowires [11, 12]. This barrier rectifies the current following through the metal–semiconductor junctions and is in fact applied in high-power modules that require large operational voltage of several hundred volts or thousand volts. The photon energy from a UV source can be absorbed in WBG nanowire that generates electron–hole pairs (similar to the case of the photoconductor devices). The modification in the charge concentration of nanowires will shift the Fermi level (e.g. toward the conduction band for n-type, or toward the valance band for p-type). The change in the Fermi level leads to a change in the built-in potential ($q\Phi_b$), that alter the turn-on voltage as well as the photon current [13]. Since Schottky diode sensors exhibits relatively small dark current, they typically offer much higher responsivity than the photoconduction type sensors.

Furthermore, compared with the p–n or o p–i–n junctions-based devices, the Schottky diodes generally have a faster response due to the elimination of the electron and hole recombination phenomenon that occurs in the semiconductor heterojunctions.

Zhao et al. demonstrated a Schottky-type UV sensor using ZnO nanowires [14]. The sensor consists of two Schottky junctions of Ag/ZnO and Au/ZnO with an asymmetric *I–V* characteristic, Figure 8.2a. By illuminating UV light locally at each junction, the author revealed that the responsivity of the sensors depends on the barrier high of the junction. In particular, the Ag/ZnO diode has a larger Schottky barrier height than the Au/ZnO diode, resulting in a larger current in positive bias compared with the negative bias (i.e. when the Au electrode was grounded and an external voltage was applied to the Ag contact), Figure 8.2b, left. The sensors showed repeatable response when the UV was locally applied at the Ag/ZnO junction (approximately 100 nA current change under 0 V bias), and the Au/ZnO junction (approximately 100 nA current change under 0 V bias), respectively, Figure 8.2b, right. Under a UV intensity of 0.68 mW/cm^2 the sensors possess a large photo-to-dark current ratio from 10^5 to 10^6, and a fast raising and decay time of less than 30 ms.

The main disadvantage of Schottky diode UV sensors is the stability and uniformity of metal contact along with the critical selection of metals that allows UV light to penetrate through and reach the metal–semiconductor junction underneath. As the junction current is the main conducting mechanism in the metal/semiconductor interface, the presence of impurity or defect at the junctions could significantly affect the characteristic of the potential barrier. In addition, as the sensing area is the metal and semiconductor interface, the metal must exhibit high UV transmittance, while at the same time provides relevant barrier to the semiconductor nanowires to form the Schottky behavior [15, 16]. Recent studies suggest that the use of ultrathin materials like graphene or two-dimensional electron gas (2DEG) enable the development of highly sensitive UV sensors. For instance, Xu and coworkers developed a UV sensor in which a ZnO nanowire was sandwiched between two graphene layers, Figure 8.2c [17]. Au electrodes were then patterned onto these two graphene layers to apply bias, while UV light with a wavelength of 325 nm from a He–Cd laser was illuminated on the graphene/ZnO Schottky junction. The current under the forward bias is mainly dominated by the thermionic emission of electrode, and is given by:

$$
\begin{aligned}
I &= I_{SAT}\left[\exp\left(\frac{qV}{nkT}\right)-1\right] \\
I_{SAT} &= AA^{*}T^{2}\exp\left(-\frac{q\Phi_{SB}}{nkT}\right)
\end{aligned}
\tag{8.3}
$$

where I_{SAT} is the saturation current, q is the electron charge, n is the ideality factor (varying from 1 to 2), and T is the temperature. These equations imply that the photocurrent is highly sensitive to the barrier height between graphene and ZnO. The *I–V* curve of the sensors clearly shows an exponential increase in the photocurrent when increase the light intensity. A high photo on–off ratio of 8×10^2, and short raising time of 0.7 seconds were obtained, which are attributed to the existence of the Schottky barrier between the highly transparent graphene electrode and the ZnO nanowire.

8.2.3 Semiconductor p–n Junction

The working principle of a p–n junction UV sensor is based on the modification of the potential barrier between the p-type nanowires and n-type nanowires under UV light

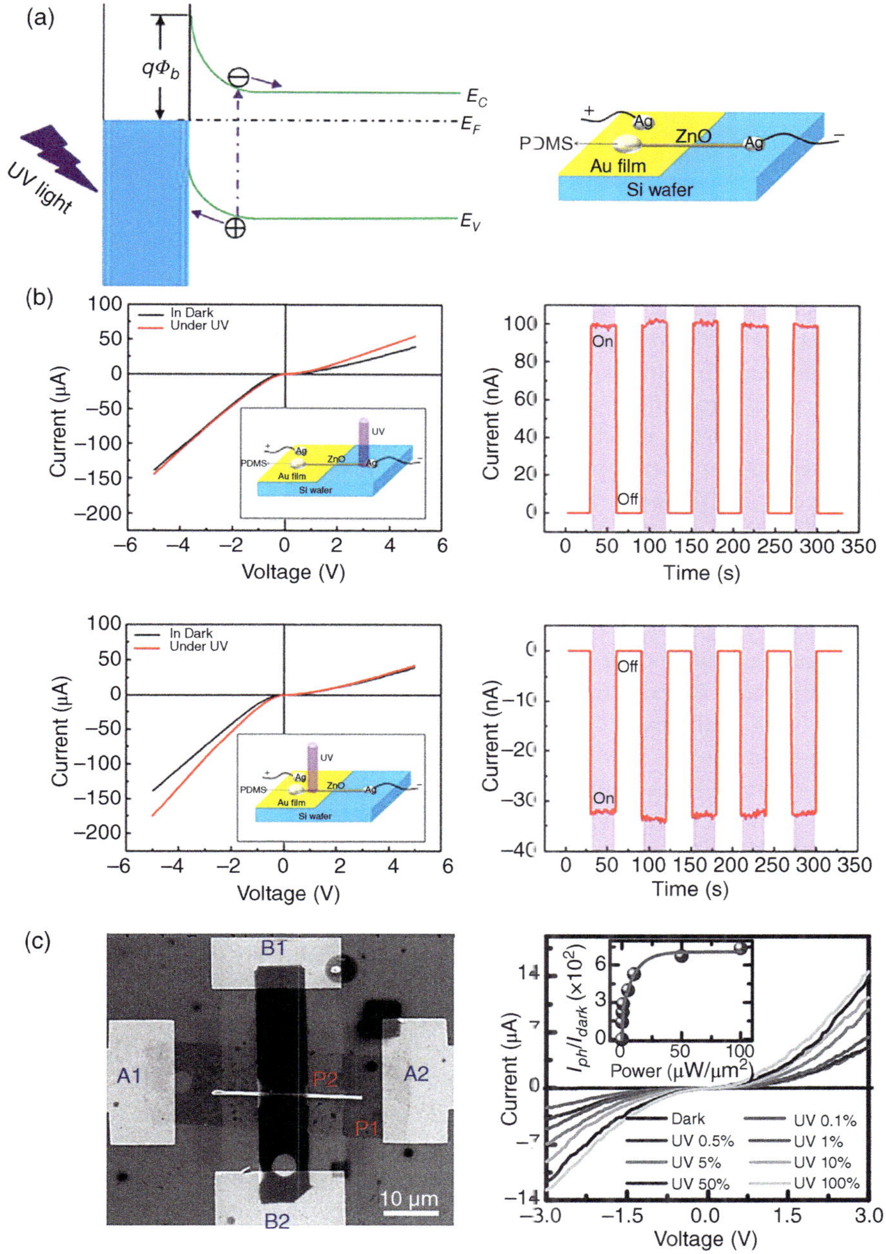

Figure 8.2 Principle of Schottky diode UV sensors. (a) Energy diagram of metal–nanowire Schottky junction. (b) Au/ZnO nanowire/silver UV sensors and its response to UV light. *Source:* Zhao et al. [14]/ Elsevier (c) Transparent electrode using graphene for ZnO UV sensors. *Source:* Fu et al. [17]/American Institute of Physics (AIP). [14], Zhao, et. al., (2015)/Elsevier.

illuminations. Most p–n junction UV sensors function under the reverse bias regime to minimize the dark current that allows for a high sensitivity [18–20]. The photon energy of a UV light source is absorbed by the p–n junction nanowire, increasing the holes in p-type and electron in the n-type, which alters the built-in potential and leads to a marked change in the reverse photo current. Different from the Schottky diode, the sensing element locates within the semiconductor nanowire rather than under metal/semiconductor interface. This property enables a more effective photon absorption and work well with several types of metals that do not exhibit good UV transmittance. The current following to a p–n junction is given by:

$$I(V) = I_s\left[\exp\left(\frac{eV}{nKT}\right) - 1\right] - eG \tag{8.4}$$

where I_d is the dark current, and $-eG$ correspond to the photocurrent (which is proportional to the generation rate G). The previous equation also suggests a shift in the photocurrent with respect to the dark current in the I–V curve. This means that, even at zero bias voltage, there is a current following through the p–n junction under UV illumination. This phenomenon is known as the photovoltaic effect, which can be employed for self-powered UV sensors (i.e. photodetector that can operate without external bias voltage).

The p–n junction in UV sensors can be formed in various configurations, Figure 8.3. Common configurations include core–shell p–n junction where the outer layer is p-type and the inner layer is n-doped or vice versa. The p–n junction can also be formed in the longitudinal direction of nanowires, in which the concentration of dopant can be *in situ* controlled. In some other cases, p–n junctions are formed by placing two nanowires with different semiconducting types overlapping each other. Simple approach grows n-type nanowires onto a p-type semiconducting substrate. In this concept, the sensing area (p–n junction) is defined at the interface between the nanowire and the bulk substrate. Leung et al. developed a UV sensor where the p–n junction was formed at the interface of a n-type ZnO nanowire and a p-type ZnO film formed by co-doping Al and N, Figure 8.3b [21]. Ohmic contacts were confirmed between the metal and semiconductors (Au/p-AZNO:N, and indium tin oxide [ITO]/n-ZnO nanowires). The I–V curve measured by sweeping bias voltages from −5 to 5 V shows a typical p–n characteristic, with a small reverse current of 5 μA, and a rectification ratio of 150. The excellent transparency of the glass substrate and the ITO electrodes enable UV light to pass through and to be effectively absorbed in the ZnO forest. The sensors exhibit a peak responsivity at 384 nm and a good UV–visible selectivity of approximately 70 at −3 V bias, Figure 8.3b. A small responsivity was observed in the red light regime, which was considered to be caused by the presence of deep-level defects in the material.

Compared with Schottky diode sensors, the p–n junction type sensors can offer enhanced sensitivity and is suitable for several type of metal contacts. However, similar to standard p–n diode used in power electronics, p–n junctions generally exhibit lower time response due to the recombination of electron–hole pairs. Furthermore, as this architecture requires both n-type and p-type semiconductors within a single nanowire, additional doping techniques are imperative to form the junction.

8.2.4 Field Effect Transistor-Based UV Sensors

Recent studies suggest a new approach to boost the sensitivity of nanowire UV sensors using field effect transistor (FET) configuration. As the current in the conducting channel can be controlled using a gate bias voltage, FETs can provide a small dark current. This ability typically leads to a high photo-to-dark current ratio, making FET WBG nanowires highly sensitive to UV light [22, 23].

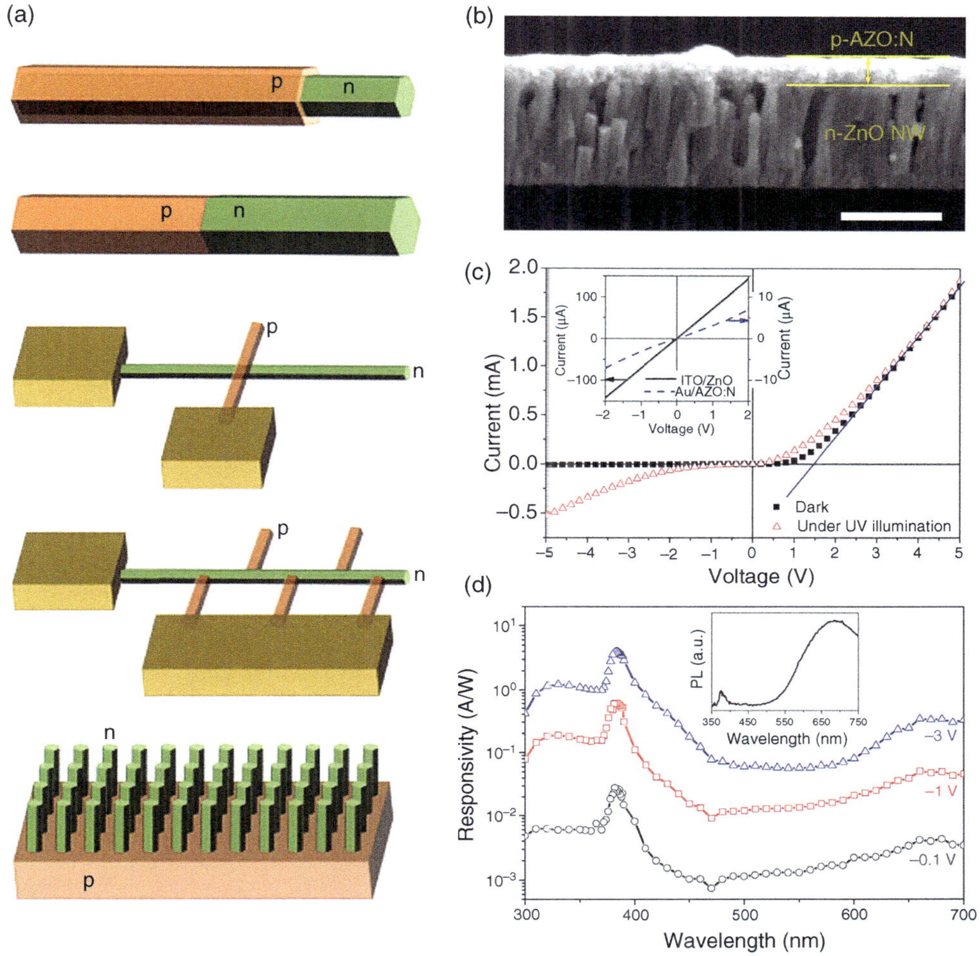

Figure 8.3 UV sensors using p−n junction. (a) Common p−n configuration. (b) Scanning electron microscope (SEM) image of vertical ZnO nanowire with the p-doped Al,N-codoped ZnO (AZO:N) layer on top and n-doped layer at the bottom. (c) The *I*−*V* characteristic of ZnO p−n junction. Inset: Ohmic contact of ITO/ZnO and Au/AZO:N. (d) The responsivity of the sensor. *Source:* [21], Leung, et. al., (2010)/AIP Publishing LLC.

Kim and Chu demonstrated a ZnO FET UV sensors using p^{++} Si as the gate and SiO_2 as the dielectric layer, Figure 8.4a [24]. The I_{ds}–V_{gs} characteristics of the ZnO FET show that the n-type ZnO nanowires can be depleted using a negative gate bias with small dark current down to 10^{-12} A. Under UV light illumination (365 nm and power density of 0.47 mW/cm^2), the FET changed from semiconducting state to metal-like state behavior with little gate dependency, Figure 8.4b. The photocurrent was measured in a range of 10^{-6} A with gate bias ranging from −5 to 5 V. As the dark current can be tuned by the gate voltage, the photo gain of the ZnO sensors can be controlled using gate bias and can reach to a high value of 10^6 in the negative gate regimes, Figure 8.4(c). The study also found that the photocurrent decayed exponentially over time ($\Delta I = I_{ph} - I_{dark} = A \times exp(-t/\tau)$, where τ is the exponential time constant). The time decay constants were found to be dependent on gate voltage, and were relatively slow (185, 107, and

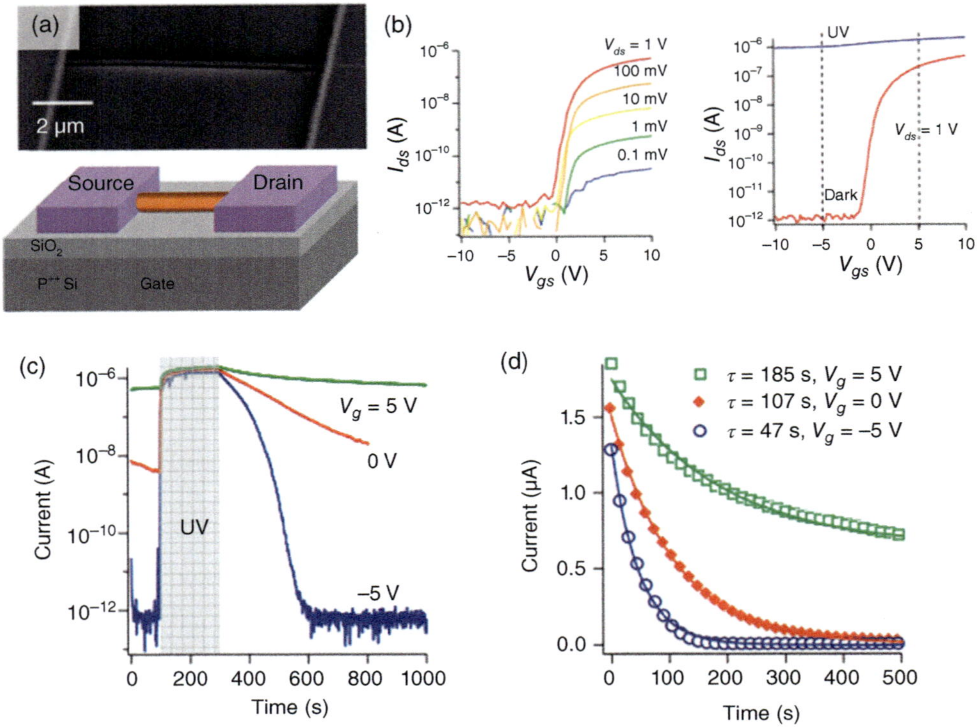

Figure 8.4 UV sensors employing field effect transistors. (a) SEM image of a ZnO FET. (b) The $I_{ds}-V_{gs}$ characteristic of the FET. (c) The time response and (d) current decay of the sensors under different gate voltages. *Source:* [24], Kim, et. al., (2009)/John Wiley & Sons, Inc.

47 seconds for gate voltages of 5, 0, and −5, respectively), Figure 8.4(d). This result indicates that the decay process in the ZnO FET is limited due to oxygen absorption rather than due to the recombination of the electron–hole pairs.

8.3 Device Development Technologies

8.3.1 The Choice of Wide Bandgap Materials for UV Sensing

The cutoff wavelength, responsivity, time response (or recovery time), and operational temperatures of UV sensors heavily depend on the selection of semiconductors. The WBG material family covers a wide variety of physical properties and chemical characteristics that can be utilized to develop UV sensors for specific operating conditions. Based on their unique properties, the WBG semiconductors can be grouped into three main categories of (i) metal oxide semiconductors, (ii) III-nitride group, and (iii) silicon carbide materials. Table 8.1 lists the specification of common WBG semiconductors for UV detectors.

8.3.1.1 Metal Oxide Semiconductors

WBG metal oxide semiconductors are formed by the covalent bonding between metal and oxygen atoms in their crystal [38, 39]. Common materials from this group include ZnO,

Table 8.1 Performance of various photodetectors using WBG semiconductor nanowires and their heterostructures.

Materials	Wavelength (nm)	Photocurrent (A)	Dark current (A)	Photocurrent (A)	Responsivity (A/W)	Response time (s)	References
3C-SiC NWs	—	4.3×10^{-6}	4.3×10^{-8}	4.3×10^{-6}	—	—	[25]
2H-SiC NWs	375	1.8×10^{-6}–2.7×10^{-6}	—	1.8×10^{-6}–2.7×10^{-6}	—	<0.003	[26]
AlN NWs	193	2.4×10^{-8}	1×10^{-14}	2.4×10^{-8}	0.39	<0.1	[27]
AlN NWs	325	—	—	—	2.7×10^{6}	0.001	[28]
GaN NWs	320–400	—	—	—	1.74×10^{7}	0.144	[29]
GaN NWs	325	10^{-8}	—	10^{-8}	2.2×10^{4}	<0.026	[30]
GaN/AlN NWs	300	—	5.2×10^{-14}	—	2×10^{3}	—	[31]
GaN NWs/Pt	380	—	—	—	6.39×10^{4}	1.1	[32]
GaN NWs/graphene	357	1.28×10^{-4}	—	1.28×10^{-4}	25	—	[33]
AlN/GaN/AlN NWs	325	10^{-9}–10^{-12}	—	10^{-9}–10^{-12}	200–700	—	[34]
GaN/AlN NWs	280–330	0.5×10^{-9}–3.4×10^{-9}	—	0.5×10^{-9}–3.4×10^{-9}	3.1×10^{6}–12×10^{6}	—	[35]
Diamond NWs	300	3×10^{-6}	7×10^{-8}	3×10^{-6}	338	0.02	[36]

Source: Adapted from Pham et al. [37].

TiO_2, SnO_2, and Ga_2O_3. Among these semiconductors, ZnO is the most widely used material for UV sensing applications owing to its worldwide availability and relatively well-established and simple synthesis process [40]. The biocompatibility and biodegradability along with capability for photocatalysis and conversion of mechanical energy make ZnO the material of choice for multifunctional micro-/nano-electromechanical systems (MEMS/NEMS) applications. As a direct bandgap material ($E_g = 3.365\,eV$), ZnO provide a high photon efficiency at wavelengths around 365 nm, highly suitable for UV sensors. ZnO nanowires can be synthesized using different processes such as vapor phase synthesis, physical vapor deposition, chemical vapor deposition (CVD), and laser assisted growth with excellent uniformity and large scale. Besides its significant surface to volume ratio compared with the bulk or thin ZnO platforms, nanowires offer much higher sensitivity due to the nanowire carrier transport channel and enhanced exciton bonding energy. Examples of UV sensors using ZnO nanowires include the work reported by Mallampati et al. in which the nanowires were synthesized using a vapor phase transport growth [41]. The UV sensors employing the metal–semiconductor–metal configuration offer excellent responsivity of 10^5 A/W.

Besides ZnO, other metal oxide semiconductors with different bandgaps such as TiO_2 ($E_g = 3.2\,eV$), SnO_2 ($E_g = 3.6\,eV$), and Ga_2O_3 ($E_g = 4.8\,eV$) have also been employed to detect UV light at specific wavelength of approximately 330, 320, and 250 nm, respectively [42–44]. For instance, Ga_2O_3 with its very large energy gap is the material of choice for deep UV sensing applications, with the cutoff wavelength around 270 nm. Another interesting property in Ga_2O_3 that has gained significant attention is the capability to form a 2DEG layer out of this material, opening exciting opportunities to integrate sensing elements with regulating electronics circuits (e.g. high electron mobility transistors [HEMTs]) [45]. However, as the investigation into Ga_2O_3 2DEG has just been promoted recently, further studies are required to compare its performance with other counterparts such as III-nitride, whose theoretical frameworks and fabrication technologies are more advanced.

A UV sensor using a network of Ga_2O_3 nanowires was developed by Du et al., Figure 8.5 [46]. The sensor showed a high responsivity of 377 A/W to wavelength in the range of 200–250 nm

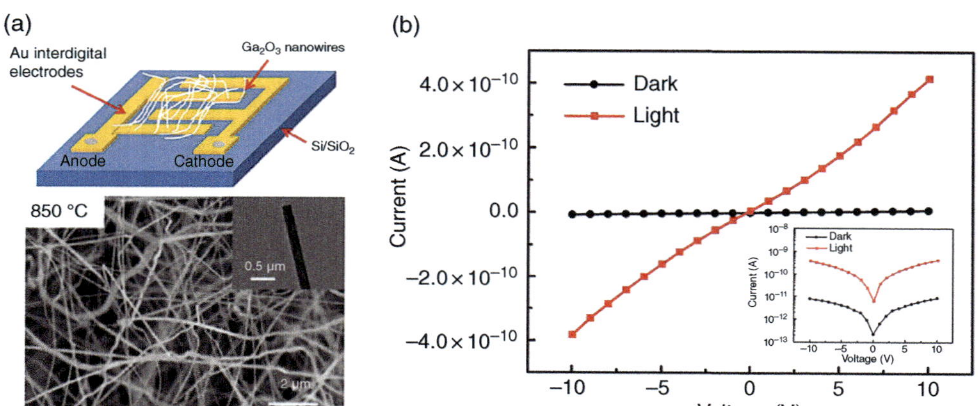

Figure 8.5 Ga_2O_3 nanowire UV sensors. (a) Schematic of a UV sensor using high density of Ga_2O_3 nanowire network. (b) *I–V* curves of the Ga_2O_3 nanowires sensors under 230 nm UV exposure. *Source:* [46], Du et al., (2016)/IOP Publishing.

with a corresponding quantum efficiency of $2 \times 10^5\%$. The solar-blind rejection ratio was found to be over 10^3, indicating the excellent selectivity of Ga_2O_3 for deep UV sensing applications.

8.3.1.2 III-Nitride Materials

Common III-nitride materials used for UV sensors are GaN and AlN [47, 48]. With their bandgap of 3.4 eV for GaN and 6.2 eV for AlN, UV sensors employing these semiconductors are suitable for wavelengths around 200–400 nm. The energy bandgap of III-nitride group can be engineered within this range by changing the mixing compound of $Al_xGa_{1-x}N$, making this family of WBG materials an excellent candidate for deep UV, UV, and blue light sensing applications. Compared with ZnO, the synthesis of III-nitride is less matured, and these materials still present significant density of crystal defects and dislocation in the nanowire platform, that hinders its full potential in optoelectronic devices. The sophisticated growth process along with material costs are some other challenges in III-nitride nanowire-based UV sensors. However, III nitride offer several advantages over ZnO based devices, particularly for applications used in extreme environments. GaN and AlN exhibit excellent chemical inertness, making them suitable for sensors in corrosion conditions along with high temperatures [49]. Another unique feature in III-nitride semiconductor is the formation of a 2DEG layer between AlGaN and GaN due to the built-in stress between the two layers that create a quantum well at the AlGaN/GaN interface. This quantum well drives electrodes toward the interface between AlGaN/GaN junction, forming 2DEG with exceptionally high mobility [50]. This 2DEG layer is considered the key toward HEMTs with much faster switching frequency compared with conventional Si-based FETs. GaN based HEMT have been demonstrated to function at temperature above 400 °C [51]. This property leads to new functions such as embedding III-nitride sensors into the same devices (or same chip) with HEMT components for a fully integrated system, capable of operating in extreme environments. In addition, due to its extremely low resistance, 2DEG can function as the UV transparent electrodes for GaN, enabling the Schottky diode photodetector with excellent quantum efficiency. For instance, the Senesky group has demonstrated high responsivity and low dark current UV sensors using 2DEG interdigitated electrodes, Figure 8.6 [52]. The GaN sensors exhibit ultra-high photo to dark current ratio of 6 × 10 [13] with excellent responsivity of 7800 A/W. The sensors also show significantly enhanced time response, with the raise/fall time being reduced from 20/60 seconds using metal contacts to 32/76 ms, thanks to the use of 2DEG electrode. Although this concept has only been demonstrated in GaN thin films, a similar approach can be applied in nanowires to enhance overall device performance.

8.3.1.3 Silicon Carbide

Different from metal oxide semiconductors (e.g. ZnO, TiO_2, or Ga_2O_3) and the III-nitride family, silicon carbide is an indirect semiconductor [25, 53]. This indirect bandgap in SiC leads to a longer absorption length and longer minority carrier lifetime, which is undesirable for UV sensors that require sharp absorption edge and high quantum efficiency. Nevertheless, among several WBG materials, perhaps SiC is the most chemically robust material. The extremely low oxidation rate allows SiC based devices to operate at high temperatures. For instance, SiC electronics have been demonstrated to work at temperatures of above 800 °C [54]. Therefore, despite of its indirect bandgap nature, the radiation hardness, chemical inert, and mechanical stability make SiC UV sensors suitable for harsh conditions.

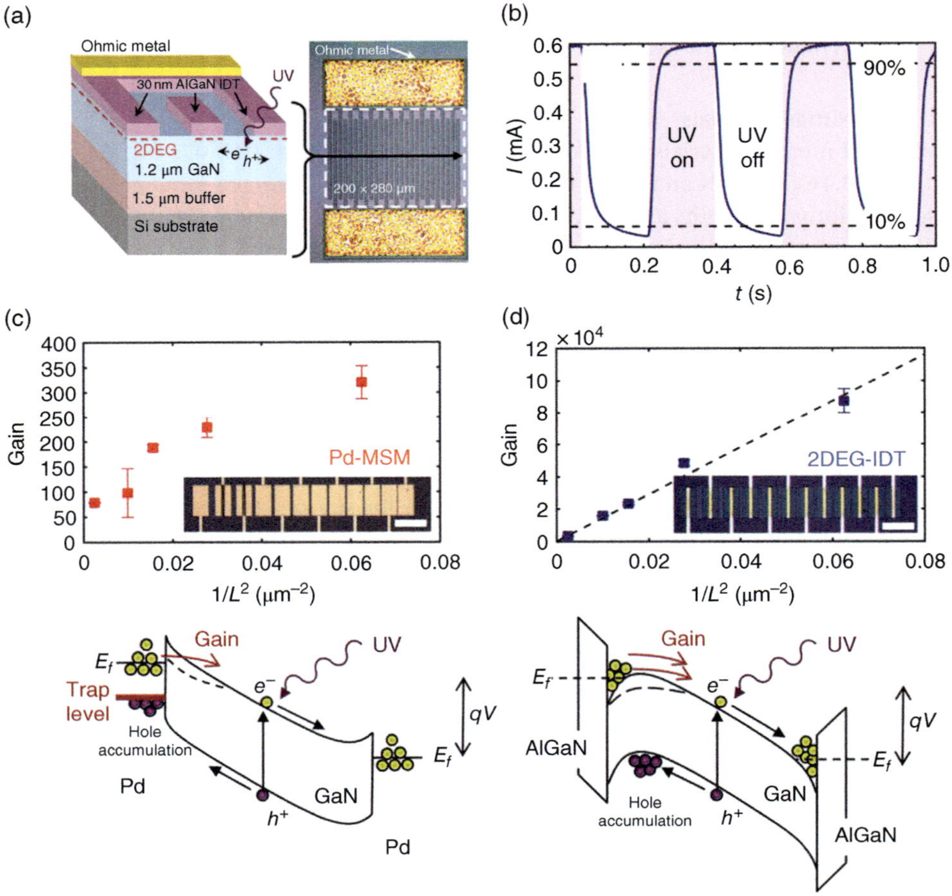

Figure 8.6 (a) Schematic sketch and optical microscope image of the AlGaN/GaN UV sensors. (b) The fast time-response of the sensors to 365 nm wavelength illumination. (c) Comparison of internal gain vs length for Pd-metal semiconductor metal (MSM) and (d) 2DEG-intedigited (IDT) photodetectors. *Source:* [52], Satterthwaite, et. al., (2018)/American Chemical Society.

For instance, Yang et al. developed a high-performance UV sensor using single crystalline β-doped 3C-SiC nanobelts [55]. The nanoribbons with a thickness of 120 nm and a width of 4 μm were synthesized via catalyst-free pyrolysis of polymeric precursors of polysilazane (Figure 8.7a,b). The SiC nanobelts exhibit a high responsivity of 6.37×10^5 A/W and a quantum efficiency of 2×10^8% under 405 nm light. Due to its excellent oxidation tolerance and high temperature operation capability, the SiC nanoribbon sensor showed consistent and repeatable photo current against 300 °C over a long period of time of 180 days, Figure 8.7c,d. This perhaps is one of the longest tests so far performed in nanoscaled UV sensors at elevated temperatures. This demonstration indicates the promising possibility of using SiC for highly sensitive and long-term stable UV sensors in extreme conditions.

8.3.2 Top-Down Fabrication of Wide Bandgap Nanowire UV Sensors

Nanowires can be synthesized or fabricated using either top-down or bottom-up process. Compared with the bottom-up approach, the top-down process has numerous advantages

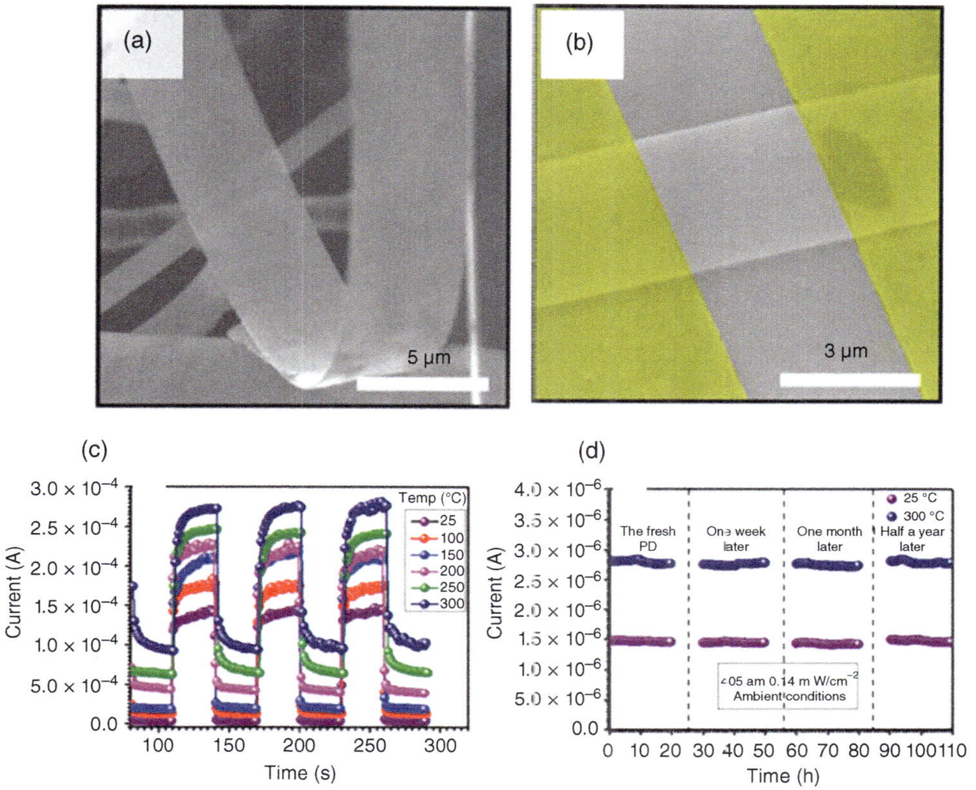

Figure 8.7 (a) SEM image of SiC nanobelt. (b) Metal electrodes to form SiC photoconductance sensors. (c) Photocurrent of the SiC nanobelt photo diodes (PDs) as a function of temperatures ranging from 25 to 300 °C under 405 nm light illumination. (d) The long-term stability of the SiC UV sensor at 300 °C up to 180 days. *Source:* [55], Yang, et. al., (2019)/John Wiley & Sons, Inc.

such as uniformity and the ease for integration with other complementary metal-oxide-semiconductor (CMOS) devices [56]. A common configuration of UV sensors fabricated using top-down fabrication is an array of parallelly aligned nanowires on an insulating substrate with two metal contact pads formed at the two edges of the nanowire array. Typically, the nanowires are formed from a thin film of WBG material deposited on a seeding wafer. Nanolithography such as electron-beam, focused ion beam, or X-ray lithography is then applied to transfer nanowire structures from a mask-design on to the as-growth substrate. Subsequently, the WBG nanowires are formed using dry-etching (e.g. inductive coupling plasma [ICP] or reactive ion etching [RIE]). Finally, electrodes are formed through standard metal deposition followed by metal patterning.

An example of top-down nanowire UV sensors is the work reported by Zhou et al. using ultra-nanocrystalline diamond, Figure 8.8 [57]. The authors implemented hot filament CVD to form ultra-nanocrystalline diamond thin films on a standard and low-cost Si substrate. Boron atoms were *in situ* doped into the diamond nanowires to form n-type semiconductors. Following the growth of diamond films, electron beam lithography (EBL) and deep reactive ion etching (DRIE) were applied to fabricate high density of parallel nanowires with a width of approximately 70 nm. Metallic layer of 10 nm-Ti/100 nm-Au was then deposited and etched to form contact areas. The diamond sensors exhibit high photoresponsivity of

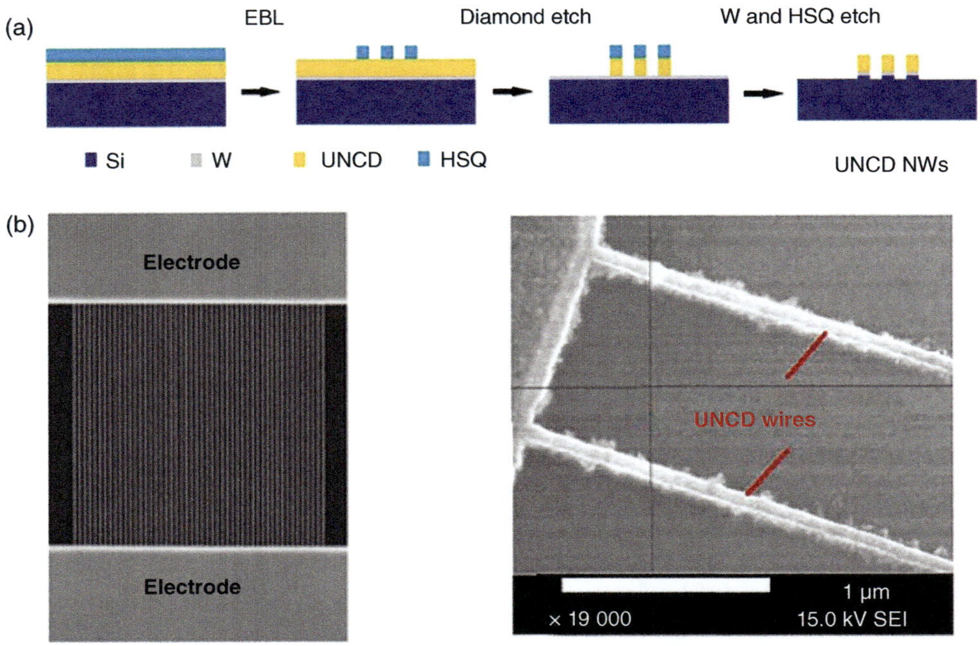

Figure 8.8 Top-down fabrication of diamond nanowires using EBL and RIE. (a) Schematic sketch of the fabrication process. (b) SEM images of an array of diamond nanowires. *Source:* [57], Zhou, et. al., (2019)/American Chemical Society.

388 A/W at 300 nm wavelength along with a fast response time of 20 ms. The sensors also possess good UV-to-visible rejection ratio of approximately 10^5 under zero bias condition.

8.3.3 UV Sensors Using Standing Nanowire Sandwiched Between Electrodes

Compared with top-down nanowires, bottom-up process has been more widely employed to construct UV sensors [58, 59]. The main reason for the preferable choice in bottom-up nanowires is the capability to obtain high density of nanowire networks, high quality (fewer surface defects induced from the dry etching step), and a smaller diameter that enables the quantum effect. In many cases, UV sensors can be directly constructed on as-synthesis nanowires without the requirement of transferring process. To enable this, nanowires are typically grown on a conductive substrate such as doped Si or ITO that can subsequently be used as the bottom electrode [60]. Another metal layer is required to form the top contact.

Many WBG nanowires are synthesized using nanodots of metal as the catalyst. These nanodots of metal generally remain at the tip of vertical nanowires after the growth process. Therefore, they can be utilized as the top-contacting electrode, Figure 8.9a [61]. A transparent, highly conductive film of ITO on a glass wafer is generally employed to form a metal connection between these residual metal nanodot to external wiring. The transparency of ITO allows UV light to pass through and then be absorbed in the nanowire forest underneath. This technique is simple and highly relevant for simple prototyping to demonstrate the UV sensing function as well as basic electrical characteristics of nanowires. However,

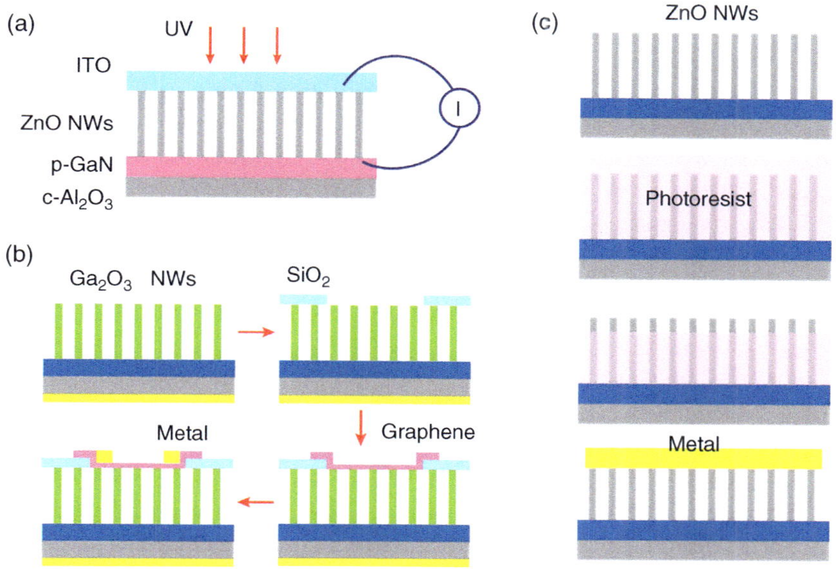

Figure 8.9 Sensors using vertical nanowire forest sandwiched between two electrodes. (a) ZnO nanowires array with p-GaN electrode at the bottom and transparent ITO electrodes on top. (b) Fabrication of top metal contacts using sacrificial polymer. (c) Fabrication method of vertical Ga_2O_3 nanowires with transparent graphene electrodes. *Source:* Liu et al. [61]/Optical Publishing Group.

miniaturization, fabrication scalability, and maintaining long-term stable contacts are the main challenges that hinder this approach from practical applications.

Another approach uses graphene as the top electrode for vertical nanowire UV sensors, Figure 8.9b [62]. In this process, a few layers of graphene are transferred from the seeding substrate using exfoliation, and then placed on to the top surface of the nanowire array. A metal layer is then deposited onto the graphene film and the patterned to create a better contact for external wiring. The benefit of using graphene in this case is that it functions as the Schottky contact for the UV sensors (e.g. graphene/Ga_2O_3), and at the same time, as a physical barrier preventing the deposited metal to directly contact to the conductive substrate (i.e. the short circuit issue). Although this is promising approach to maintain stable metal/nanowire contact, its main disadvantages are the complexity of the graphene transfer process and suitability for mass production.

A more scalable process was developed by Park et al. using polymer as a sacrificial layer during metal deposition process [63]. Figure 8.9c represents schematic sketch of the fabrication process applied on a vertically grown ZnO nanowire array. In the first step, ZnO nanowires were synthesized on the top surface of a ZnO nanothin film, pre-deposited onto a Pt layer. This ZnO/Pt substrate functions at the bottom electrode for the sensors. After the growth process, a photoresist layer was spin-coated and thermally baked on the ZnO nanowires. The photoresist layer is expected to fully cover the nanowire network. A plasma etching back method was then applied to selectively remove a small portion of the photoresist on the top surface, leaving the ZnO nanowire tips to expose. The majority of the nanowires at the bottom were still fully covered by the resist layer. Metal films (e.g. Ni/Au) were then deposited through a metal mask to form the top electrodes, while the remaining photoresist was completely removed by dipping the devices into acetone.

8.3.4 Transfer Process for Nanowires

For several practical applications, integration of UV sensors with standard planar electronic devices (e.g. conditioning circuits) is desired. In this regard, transferring vertically grown nanowires onto two-dimensional surface is critically important to develop UV sensors.

One of the most common methods is using microfluidic to assembly nanowires into a functional network. This technique was developed by the Lieber group for a range of nano-architecture including nanowires and nanotubes with different dimensions and growth techniques. In this method, the as-synthesis nanowires are suspended in ethanol solution. The suspension is then passed through a polydimethylsiloxane (PDMS) channel that previously formed on a glass substrate, Figure 8.10a,b [64]. The nanowires in the suspension are aligned in the flow direction by the shear flow and then immobilized on the substrate. Experimental results show that nanowires well align with the direction of flow (i.e. the longitudinal direction of the PDMS channel), Figure 8.10. The degree of the alignment and nanowire separation can also be controlled by varying the flow rate as such increasing the flow velocity results in a better alignment. For PDMS channels with widths ranging from 50 to 500 μm and lengths from 6 to 200 mm, over 80% of nanowires are aligned within ±5° of the flow direction at the maximum flow rate of 10 mm/s. Furthermore, by utilizing cross micro-channel configurations, it is possible to align different groups of nanowires in several planar directions. This technique has been employed to several nanowire based electronic devices including p–n junction UV sensors with a pair of p-type and n-type nanowires assembled onto each other. Upon the alignment process using liquid flow, metal deposition and micropatterning are applied to form the electrical contact for functional devices.

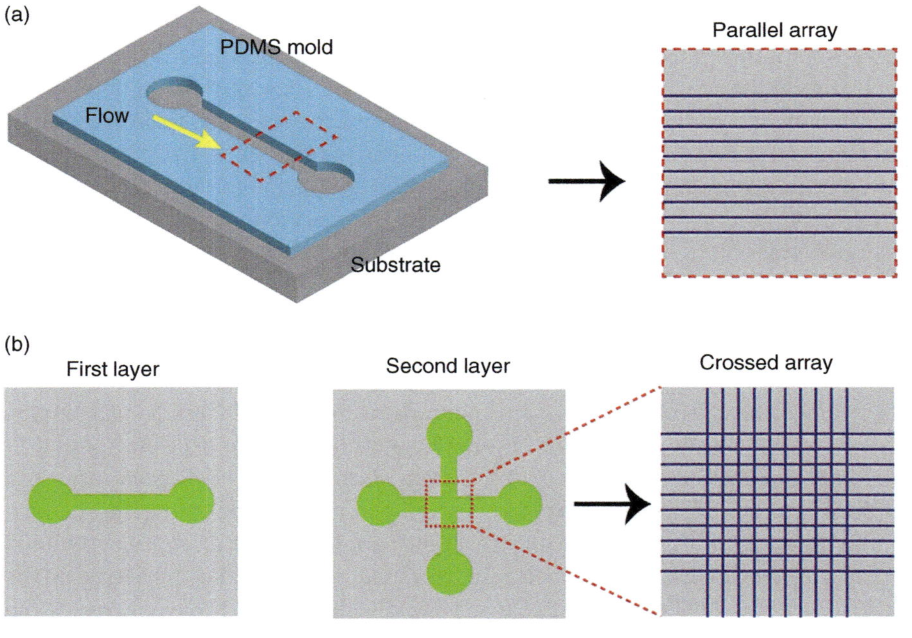

Figure 8.10 Assembly of nanowires into functional networks using fluidic flows. (a) Schematic sketch of the fabrication method. (b) SEM images showing nanowires aligning with the flow direction. *Source:* Huang et al. [64]/American Association for the Advancement of Science – AAAS.

As well as employing shear flow to assemble nanowires for UV sensors, the Langmuir–Blodgett (LB) method is also widely applied to control the orientation of nanowires [65]. Specifically, in the LB method, bottom-up grown nanowires are dispersed in a volatile solvent, which is immiscible in water. The solvent that consists randomly distributed nanowire is then spread onto the water surface and a mobile barrier is applied onto the top surface of the solution. By controlling the uniaxial compression from the mobile barrier, nanowires are aligned along the moving direction of the barrier, Figure 8.11. The greatest advantage of the LB method is the elimination of specialized equipment or pre-patterned microstructures (e.g. PDMS channel for the fluidic flow method), enabling a rapid prototype of UV sensors using bench-top fabrication. The LB method can be applied to several types of nanoarchitectures including nanowire, nanotubes, and nanodots, making it a versatile technique for the development of nanoelectronics and sensing devices.

Besides the assembly method using liquid, solid mechanic approach that uses polymer stamping was also developed to transfer nanowires onto different platforms. Figure 8.12 presents an example of ZnO nanowires transferred onto glass and polymer substrates using the physical stamping technique [66]. The fabrication process starts with CVD growth of vertically aligned ZnO nanowires on a silicon wafer. A thin PDMS film was spin coated on a glass substrate and then pressed again the as-grown ZnO nanowires. The glass substrate with the thin PDMS layer was slightly shifted in a lateral direction, which transferred the nanowire onto PDMS surface with orientation well aligning with the shifting direction. The UV sensors were then formed by pressing a glass wafer with prepatterned interdigitated Ag electrodes. This fabrication method can be applied for both rigid and soft substrates, thereby suitable for the development of standard solid-state UV sensors or flexible sensing devices.

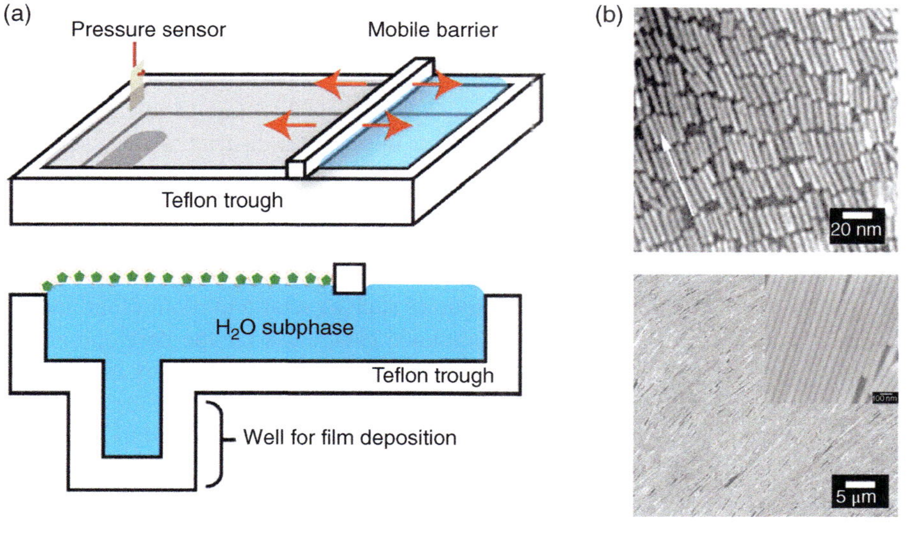

Figure 8.11 The Langmuir–Blodgett method to align bottom-up nanowires onto a planar surface. (a) Schematic sketch of the LB method. (b) SEM images of nanowires aligned along the moving direction of the mobile barrier. *Source:* [65], Tao, et. al., (2008)/American Chemical Society.

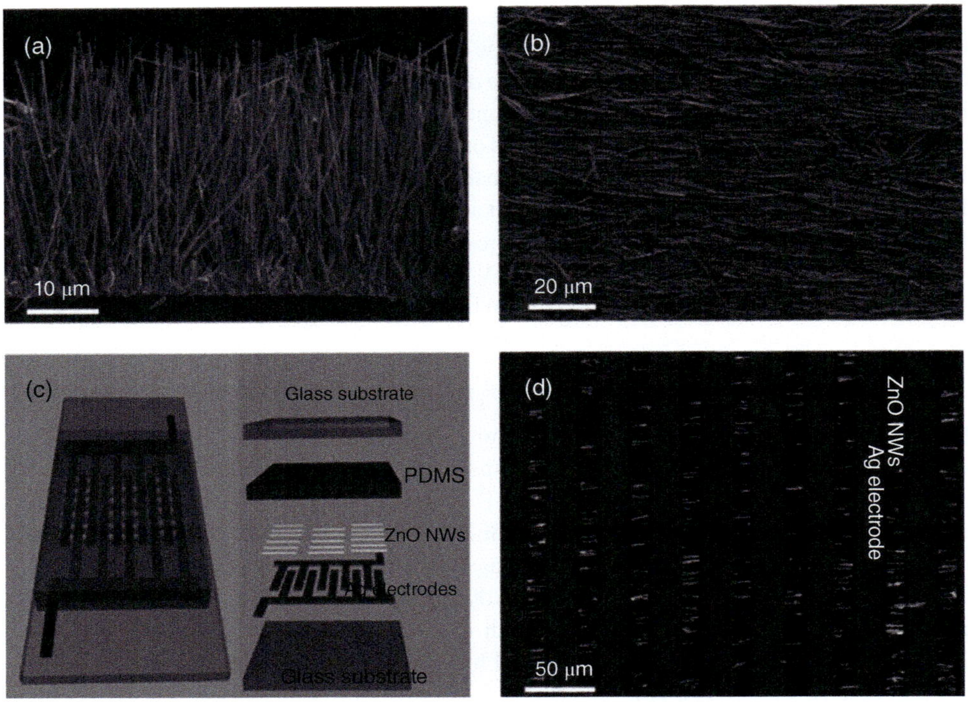

Figure 8.12 PDMS stamping method for UV sensors. (a) A SEM image of vertical ZnO nanowires. (b) Nanowires stamped on PDMS. (c) Schematic sketch of the fabrication. (d) ZnO nanowires under interdigitated electrodes. *Source:* Bai et al. [66]/John Wiley & Sons, Inc. Bai, et. al., (2011)/John Wiley & Sons, Inc.

8.4 Applications of Nanowire UV Sensors

8.4.1 Flame Sensors

Flame detectors play an important part in detecting fire and activate alarm in emergency case for subsequent evacuation and corresponding actions in several industrial systems [67, 68]. As UV light is emitted from flame, real time, and continues measurement of UV helps detecting the event of flame in a timely manner with minimized error induced from the visible range (e.g. from solar light). Typical flame sensors are required the capability of detecting light with wavelengths ranging from 250 to 280 nm. In addition, to avoid fall detection, flame sensors must also be insensitive to other light sources such as solar and room illumination. WBG materials, typically blind to solar light, are good candidates for the development of flame sensors. Examples of flame sensors using WBG materials include the work reported by Oshima et al., which employed Ga_2O_3 as the UV sensing element [69] (Figure 8.13). Using the Schottky configuration, the sensors exhibit a cutoff wavelength of 300 nm, a responsivity of 37 mA/W, and rejection ratio (to solar light) of 1.5×10^4. The detector exhibits a fast raise and fall time of 9 ms, sufficiently good for a rapid response flame detector.

Besides safety purpose, detection of flame is also of significant interest for the optimization of combustion engines. UV sensors are utilized to detect the flame ignite time and temperatures, which are used as feedback signals to enhance the efficiency of fuel and reduces the level of pollutant emission. For instance, the Senesky group developed a GaN based UV for use in flame detection as well as in combustion monitoring of a hybrid rocket motor igniter

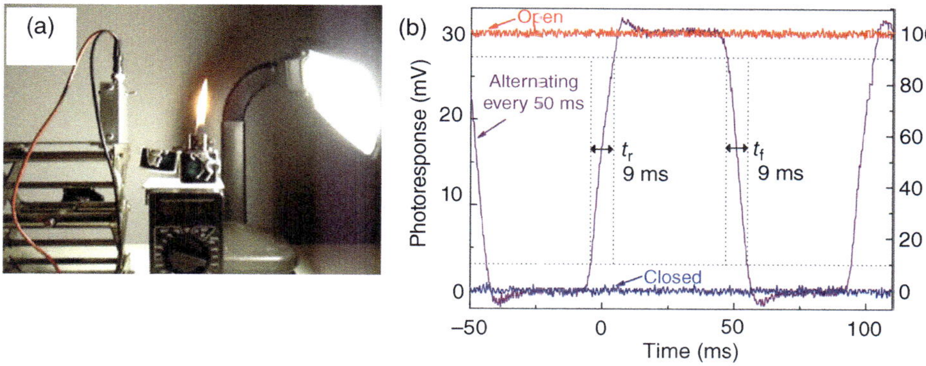

Figure 8.13 Flame sensor using wide bandgap materials. (a) An example of experimental setup to detect flame. (b) Photoresponse of an Al_2O_3 UV sensor. *Source:* [69], Oshima, et. al., (2009)/IOP Publishing.

plume [70]. The sensors were developed using a 2DEG with a width of 5 μm as the electrodes, which are spaced by an intrinsic GaN substrate. The UV sensors possessed high responsivity of 7800 A/W at 362 nm and a UV-to-visible rejection ratio of 4×10^6 which is highly suitable for UV detection in flame. The AlGaN/GaN sensors are well functioned at high temperature up to 250 °C with the peak of responsivity shifting from 362 nm at room temperature to 375 nm at high temperatures. The sensors were also demonstrated their capability of detecting flame generated from an igniter plume with varying oxidizer-to-fuel ratios (e.g. O_2 to CH_4). Specifically, the photocurrent of the UV sensors increased with increasing the emission intensity from the plume (i.e. increasing fuel concentration). The relationship between the optical power and the measured current was obtained using a calibrated data of responsivity. By treating the plume as a black body, the authors estimated the plume temperature in a range of 850–900 K in all tested configurations. These experimental studies demonstrate the promising possibility of WBG material-based UV sensors for flame monitoring in harsh environments.

8.4.2 Environmental Monitoring

In the era of internet of things, environmental monitoring is increasingly important to minimize pollution and natural disasters [71, 72]. In these monitoring systems, UV sensors are typically used with temperatures, gas, pressure, and humidity sensing devices to obtain multiple parameters from environments. These sensors are connected to a cloud through wireless communication, Figure 8.14. Data obtained from the sensors can be employed for machine learning to predict climate change and/or detect unusual phenomena. The use of UV index can help estimating the condition of ozone layers through the intensity of UV at different wavelengths. UV sensors can also be applied to monitor pollution. For instance, hydrocarbons generated from fuel spills can impair the quality of water and cause environmental pollution [73]. Hydrocarbons are harmful to human and are the cause of many health issues. Therefore, the development of portable sensing devices that can detect hydrocarbon content in water pipes is critically important. Several studies suggested that electron transition in the molecules of hydrocarbons and petroleum products lead to a significant intense light absorption in the UV range. Therefore, the use of WBG UV sensors can precisely detect the concentration of hydrocarbon in flowing liquid. Commercial products of UV sensors for the measurement of pollution content include silicon carbide devices that can detect wavelength ranging from 200 to 290 nm.

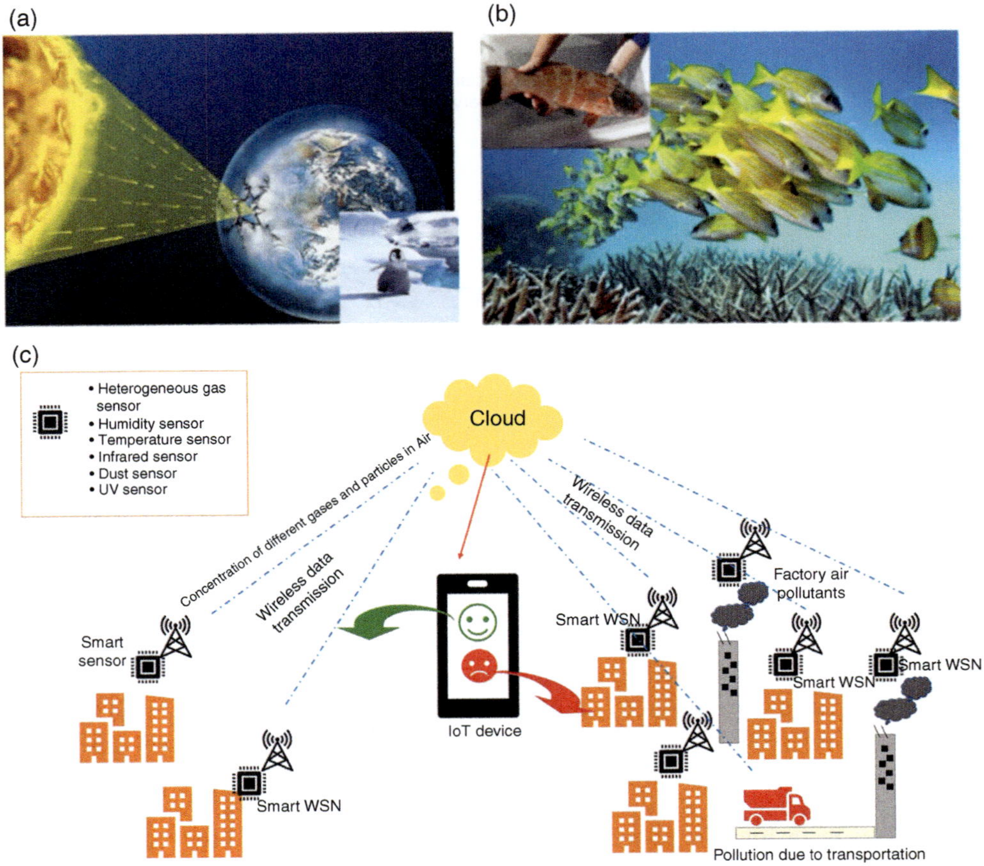

Figure 8.14 Ozone irradiation and its impact to the environment. (a) Illustration of the ozone hole over Antarctica, implying the important of continuous ozone index monitoring. (b) Photo of a group of blue-stripe snappers. Inset: *Plectropomus leopardus* with skin cancer caused by water pollution. *Source:* [71], Chen, H., et. al., (2015)/Elsevier/CC BY-NC-ND 4.0. (c) UV sensors employed in a sensor network for internet of things and big data analysis. *Source:* Ullo and Sinha [72]/MDPI/CC BY 4.0.

8.4.3 Biological Sensors and Health Care Applications

Sunlight is an abundant natural source of vitamin D, which is necessary for calcium homeostasis in the human body. However, high intensity UV exposure can cause several health problems, including sunburn, inflammation, and DNA damage [74, 75]. Furthermore, long-term effects of excessive UV exposure can lead to severe health issues including skin cancer. As such, strong UV rays can damage the DNA and mutations cell-cycle regulatory genes that results in skin cancer. According to the World Health Organization (WHO), skin cancer is one of the most common cancer diseases globally, with over 1.2 million cases worldwide in 2020. To minimize these risks, monitor human skin exposure to UV light is critically important. In this regard, WBG materials that are blind to the visible wavelength raise as excellent candidates personalized UV detection [76, 77].

There are several wearable UV sensors using WBG materials that have been successfully commercialized in the medical technology market. For instance, the Rogers group utilized AlGaN-based material for ultraviolet B (UVB) (with wavelength ranging from 280 to 315 nm)

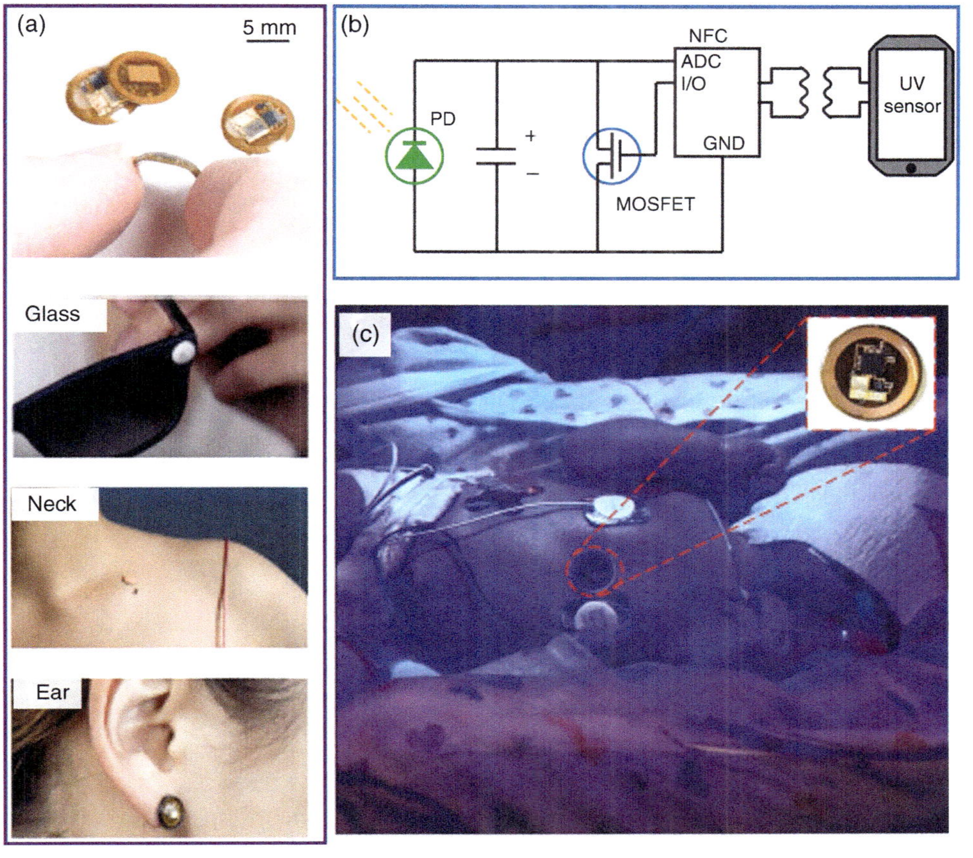

Figure 8.15 Miniaturized, battery-free, and wireless ultraviolet A (UVA) radiation sensors. (a) Wearable UVA detectors integrated into flexible circuits. (b) The circuit diagram for wireless sensing with data being transmitted to smart phones. (c) Application of the wireless wearable UV sensors in therapy. *Source:* [78], Heo, S.Y., et. al., (2018)/American Association for the Advancement of Science – AAAS.

sensing and integrated the sensor into a wireless module for flexible dosimeters [78] (Figure 8.15). The devices exhibit excellent mechanical robustness with miniaturized footprint that eliminates the requirement for an external battery. By integrating all functioning components of antenna, integrated circuit (IC)-chip, and UV sensors into a soft copper-on-polyimide circuit, the sensors can be comfortably attached onto several locations on skin surface including fingernails, earrings, and sunglasses. The output of the sensors can be quickly read using smartphone through a near field communication protocol.

References

1 Espid, E. and Taghipour, F. (2017). *Critical Reviews in Solid State and Materials Sciences* 42 (5): 416–432.

2 Sang, L., Liao, M., and Sumiya, M. (2013). *Sensors* 13 (8): 10482–10518.

3 Zou, Y., Zhang, Y., Hu, Y., and Gu, H. (2018). *Sensors* 18 (7): 2072.

4 BenMoussa, A., Soltani, A., Schühle, U. et al. (2009). *Diamond and Related Materials* 18 (5–8): 860–864.
5 Foisal, A.R.M., Dinh, T., Nguyen, V.-T. et al. (2019). *IEEE Transactions on Electron Devices* 66 (4): 1804–1809.
6 Rai, S.C., Wang, K., Ding, Y. et al. (2015). *ACS Nano* 9 (6): 6419–6427.
7 Yan, C., Singh, N., and Lee, P.S. (2010). *Applied Physics Letters* 96 (5): 053108.
8 Nguyen, N.K., Nguyen, T., Nguyen, T.-K. et al. (2021). *ACS Applied Electronic Materials* 19 (3): 1959.
9 Gimenez, A.J., Yáñez-Limón, J.M., and Seminario, J.M. (2011). *Journal of Physical Chemistry C* 115 (1): 282–287.
10 González-Posada, F., Songmuang, R., Den Hertog, M., and Monroy, E. (2012). *Nano Letters* 12 (1): 172–176.
11 Pikhtin, A.N., Tarasov, S.A., and Kloth, B. (2003). *IEEE Transactions on Electron Devices* 50 (1): 215–217.
12 Hu, Y., Zhou, J., Yeh, P.H. et al. (2010). *Advanced Materials* 22 (30): 3327–3332.
13 Zahedi, F., Dariani, R.S., and Rozati, S.M. (2013). *Sensors and Actuators A: Physical* 199: 123–128.
14 Zhao, Y., Qi, J., Biswas, C. et al. (2015). *Materials Chemistry and Physics* 166: 116–121.
15 Li, C., Zhang, Z., Chen, H. et al. (2013). *Thin Solid Films* 548: 456–459.
16 Ikenoue, T., Kameyama, N., and Fujita, S. (2011). *Physica Status Solidi C: Current Topics in Solid State Physics* 8 (2): 613–615.
17 Fu, X.-W., Liao, Z.-M., Zhou, Y.-B. et al. (2012). *Applied Physics Letters* 100: 223114.
18 Liu, W.W., Yao, B., Li, B.H. et al. (2010). *Solid State Sciences* 12: 1567–1569.
19 Shi, L., Wang, F., Li, B. et al. (2014). *Journal of Materials Chemistry C* 2: 5005–5010.
20 Bie, Y.-Q., Liao, Z.-M., Zhang, H.-Z. et al. (2011). *Advanced Materials* 23: 649–653.
21 Leung, Y.H., He, Z.B., Luo, L.B. et al. (2010). *Applied Physics Letters* 96: 053102.
22 Heo, Y.W., Tien, L.C., Kwon, Y. et al. (2004). *Applied Physics Letters* 85 (12): 2274–2276.
23 Yoshikawa, A., Yamamoto, Y., Murase, T. et al. (2016). *Japanese Journal of Applied Physics* 55 (5S): 05FJ04.
24 Kim, W. and Chu, K.S. (2009). *Physica Status Solidi A: Applications and Materials Science* 206 (1): 179–182.
25 Peng, G., Zhou, Y., He, Y. et al. (2015). *Journal of Physics D: Applied Physics* 48: 055102.
26 Sun, B., Sun, Y., and Wang, C. (2018). *Small* 14: 1703391.
27 Zheng, W., Huang, F., Zheng, R., and Wu, H. (2015). *Advanced Materials* 27: 3921.
28 Liu, F., Li, L., Guo, T. et al. (2012). *Nanoscale Research Letters* 7: 454.
29 Zhang, X., Liu, B., Liu, Q. et al. (2017). *ACS Applied Materials & Interfaces* 9: 2669.
30 Wang, X., Zhang, Y., Chen, X. et al. (2014). 12009. *Nanoscale* 6.
31 Rigutti, L., Techrnycheva, M., De Luna Bugallo, A. et al. (2010). *Nano Letters* 10: 2939.
32 Zhang, X., Liu, Q., Liu, B. et al. (2017). *Journal of Materials Chemistry C* 5: 4319.
33 Babichev, A.V., Zhang, H., Lavenus, P. et al. (2013). *Applied Physics Letters* 103: 201103.
34 Spies, M., Polaczynski, J., Ajay, A. et al. (2018). *Nanotechnology* 29: 255204.
35 Spies, M., den Hertog, M.I., Hille, P. et al. (2017). *Nano Letters* 17: 4231.
36 Zhou, A.F., Velazquez, R., Wang, X., and Feng, P.X. (2019). 38068. *ACS Applied Materials & Interfaces* 11.
37 Pham, T.-A., Qamar, A., Dinh, T. et al. (2020). *Advanced Science* 7 (21): 2001294.
38 Nunes, D., Pimentel, A., Gonçalves, A. et al. (2019). *Semiconductor Science and Technology* 34 (4): 043001.

39 Prades, J.D., Hernandez-Ramirez, F., Jimenez-Diaz, R. et al. (2008). *Nanotechnology* 19 (46): 465501.

40 Galdámez-Martinez, A., Santana, G., Güell, F. et al. (2020). *Nanomaterials* 10 (5): 857.

41 Mallampati, B., Nair, S.V., Ruda, H.E., and Philipose, U. (2015). *Journal of Nanoparticle Research* 17: 176–186.

42 Shao, D., Sun, H., Xin, G. et al. (2014). *Applied Surface Science* 314: 872–876.

43 Kim, D., Shin, G., Yoon, J. et al. (2013). *Nanotechnology* 24 (31): 315502.

44 Oshima, T., Okuno, T., Arai, N. et al. (2008). *Applied Physics Express* 1 (1): 011202.

45 Zhang, Y., Neal, A., Xia, Z. et al. (2018). *Applied Physics Letters* 112 (17): 173502.

46 Du, J., Xing, J., Ge, C. et al. (2016). *Journal of Physics D: Applied Physics* 49 (42): 425105.

47 Khan, M.A., Shatalov, M., Maruska, H.P. et al. (2005). *Japanese Journal of Applied Physics* 44 (10R): 7191.

48 Lee, C.J., Won, C.H., Lee, J.H. et al. (2019). *Sensors* 19 (5): 1051.

49 Chapin, C.A., Miller, R.A., Dowling, K.M. et al. (2017). *Sensors and Actuators A: Physical* 263: 216–223.

50 Chung, J.W., Hoke, W.E., Chumbes, E.M., and Palacios, T. (2010). *IEEE Electron Device Letters* 31 (3): 95–197.

51 Kargarrazi, S., Yalamarthy, A.S., Satterthwaite, P.F. et al. (2019). *IEEE Journal of the Electron Devices Society* 7: 931–935.

52 Satterthwaite, P.F., Yalamarthy, A.S., Scandrette, N.A. et al. (2018). *ACS Photonics* 5 (11): 4277–4282.

53 Foisal, A.R.M., Dinh, T., Tanner, P. et al. (2018). *IEEE Electron Device Letters* 39: 1219–1222.

54 Neudeck, P.G., Spry, D.J., Chen, L. et al. (2017). *IEEE Electron Device Letters* 38 (8): 1082–1085.

55 Yang, T., Chen, S., Li, X. et al. (2019). *Advanced Functional Materials* 29 (11): 1806250.

56 Hobbs, R.G., Petkov, N., and Holmes, J.D. (2012). *Chemistry of Materials* 24 (11): 1975–1991.

57 Zhou, A.F., Velázquez, R., Wang, X., and Feng, P.X. (2019). *ACS Applied Materials & Interfaces* 11 (41): 38068–38074.

58 Feng, W., Ma, J., and Yang, W. (2012). *CrystEngComm* 14: 1210.

59 Hsu, C.H. and Xu, J. (2012). *Nanoscale* 4: 5293.

60 Zhao, Q., Wang, D., Peng, L. et al. (2007). *Chemical Physics Letters* 434 (1–3): 96–100.

61 Liu, Y., Zhang, X., Su, J. et al. (2014). *Optics Express* 22 (24): 30148–30155.

62 He, T., Zhao, Y., Zhang, X. et al. (2018). *Nanophotonics* 7 (9): 1557–1562.

63 Park, J.Y., Song, D.E., and Kim, S.S. (2008). *Nanotechnology* 19 (10): 105503.

64 Huang, Y., Duan, X., Wei, Q., and Lieber, C.M. (2001). *Science* 291 (5504): 630–633.

65 Tao, A.R., Huang, J., and Yang, P. (2008). *Accounts of Chemical Research* 41 (12): 1662–1673.

66 Bai, S., Wu, W., Qin, Y. et al. (2011). *Advanced Functional Materials* 21 (23): 4464–4469.

67 Wu, Q., Gong, L.X., Li, Y. et al. (2018). *ACS Nano* 12 (1): 416–424.

68 Cheong, P., Chang, K.F., Lai, Y.H. et al. (2011). *IEEE Transactions on Industrial Electronics* 58 (11): 5271–5277.

69 Oshima, T., Okuno, T., Arai, N. et al. (2009). *Japanese Journal of Applied Physics* 48 (1R): 011605.

70 Alpert, H.S., Yalamarthy, A.S., Satterthwaite, P.F., et al. (2019). *IEEE Aerospace Conference, Big Sky, MT, USA, June 2019*, 1–8.

71 Chen, H., Liu, K., Hu, L. et al. (2015). *Materials Today* 18 (9): 493–502.

72 Ullo, S.L. and Sinha, G.R. (2020). *Sensors* 20 (11): 3113.

73 Shepelev, V.A., Altukhov, A.A., Gladchenkov, E.V. et al. (2017). *Russian Engineering Research* 37 (3): 273–275.

74 Heilingloh, C.S., Aufderhorst, U.W., Schipper, L. et al. (2020). *American Journal of Infection Control* 48 (10): 1273–1275.

75 Zou, W., Sastry, M., Gooding, J.J. et al. (2020). *Advanced Materials Technologies* 5 (4): 1901036.

76 Kim, S.J., Moon, D.I., Seol, M.L. et al. (2018). *ACS Applied Materials & Interfaces* 10 (46): 40198–40202.

77 Wu, F., Wu, T., and Yuce, M.R. (2019). *Sensors* 19 (1): 21.

78 Heo, S.Y., Kim, J., Gutruf, P. et al. (2018). *Science Translational Medicine* 10 (470): aau1643.

9

Mechanical Sensors

9.1 Introduction

Mechanical sensors that can measure several types of mechanical impacts such as stress (or strain), pressure, vibration, and acceleration have claimed a significant role in micro electromechanical systems (MEMS) [1, 2]. Several mechanical sensors have been translated from research to industrial applications, including the shock sensors used to manipulate air bag in automobile [3], strain sensors for measurement of impacts during taking-off and landing in aerospace [4], and pressure sensors to monitor conditions inside reaction chambers in chemical production industry [5].

Wide bandgap semiconductor nanowires (NWs) with their unique strain effects such as piezoresistive, piezoelectric, and piezotronics phenomena show promising potentials for mechanical sensing and mechanical-energy harvesting applications. Compared with standard materials such as Si nanowires, wide bandgap nanowires such as SiC, III-nitride, and diamond exhibit excellent mechanical sensitivity under hostile conditions (e.g. high temperature, high radiation, and increased corrosion). Therefore, some main applications of wide bandgap nanowires focus on mechanical sensing in extreme environments where Si material is no longer applicable [6]. The unique feature of quantum confinement effect in the low dimension of nanowires leads to unprecedented phenomenon of strain effect in semiconductor nanowires. For instance, with its exceptionally high piezotronics effect, ZnO and GaN nanowires offer ultra-high sensitivity along with the capability to harvest tiny mechanical energy.

This chapter summarizes the general concept of wide bandgap semiconductors nanowires, covering the functioning principles, transducing mechanism to sense different mechanical parameter, and applications.

9.2 Sensing Mechanisms and Corresponding Materials

9.2.1 The Piezoresistive Effect

The piezoresistive effect is defined as the change in electrical conductivity under mechanical load [7]. For semiconductor nanowires, the application of mechanical strain or stress changes their dimensions, and more importantly alters their crystalline lattice constant. The change in the distance between atoms in their crystal leads to a modification of energy band structures including the shift of energy bandgap The deformation in the energy bandgap

Wide Bandgap Nanowires: Synthesis, Properties, and Applications, First Edition. Tuan Anh Pham, Toan Dinh, Nam-Trung Nguyen, and Hoang-Phuong Phan leads to the redistribution of charge carriers following the Boltzmann distribution. This also results in a change in the charge carrier mobility and contributes to the change in the electrical conductivity.

The piezoresistive effect in semiconductors is quantified by a parameter named gauge factor, which is the ratio of relative resistance change over applied strain $\left(GF = \frac{\Delta R / R}{\varepsilon}\right)$. A larger gauge factor corresponds to higher sensitivity to strain.

The gauge factor of cubic SiC nanothin films has been reported to be approximately ± 30 for both p-type and n-type [1]. Although this gauge factor is smaller than that of Si materials, pressure sensors using silicon carbide material have been proven to function at high temperatures up to 800 °C [8]. This indicates the high feasibility of the piezoresistive effect in SiC for mechanical sensors used in extreme environments. The piezoresistive effect in silicon carbide nanowires has also been investigated to be compared with their bulk counterpart. Using focus ion beam machining as a top-down approach, our group developed silicon carbide nanowires with diameter of 300 nm from a SiC nanothin film grown on Si [9]. The gauge factor of these nanowires was found to be comparable to the thin films, indicating that the focused ion beam, the quantum confinement, and the surface state does not influence the piezoresistance of SiC at this scale (i.e. several hundred of nanometers). Using nanowires with the same dimension (300 nm in width and 10 μm in length), we demonstrated that the sensitivity of SiC mechanical sensors can be significantly enhanced using a strain concentration design [10]. In particular, by locally fabricating SiC nanowires on a free standing microbridge, a large stress can be concentrated on the nanowire areas, leading to a significant change in the electrical resistance, Figure 9.1.

One of the unique properties in nanowires for mechanical sensing is their excellent tolerance to large strains. As such, Shao et al. performed a SiC nanowire (diameter: 156 nm) stretching experiments in scanning electron microscopy (SEM) and observed an ultrahigh strain of 9.35% [11]. The gauge factor of the stretched SiC nanowires was found to be −6.9, much smaller than the thin films. However, owing to its tolerance to high strain regime, wide bandgap nanowires are suitable for extreme mechanical applications.

Another approach to investigate the piezoresistive effect employed atomic force microscopy (AFM) cantilever to induce stress into a localized area of 3C-SiC nanowires. The nanowires were placed on a metallic graphite substrate to form the bottom contact while the AFM tip was utilized as the top contact, creating a back-to-back diode structure, Figure 9.2. The transverse gauge factor of the 3C-SiC nanowires (diameter: 230 nm) varied from 4.5 to 46.2. Two main mechanisms were hypothesized to attribute to the measured gauge factor, which are (i) strain-induced changes in surface state and (ii) the band structure change under mechanical strain. Using a similar experimental setup to apply strain from an AFM tip, Gao et al. reported a relatively large piezoresistive coefficients ranging from 51.2 to $159.5 \times 10^{-11}\ \text{Pa}^{-1}$ in p-type 6H-SiC NWs [13]. These results are perhaps the largest piezoresistive coefficient so far reported in SiC nanowires. However, the authors also measured a small Young's modulus of approximately 50 GPa in 6H-SiC NWs, which is at least sixfold smaller than that of bulk SiC. Therefore, the gauge factor of the characterized 6H-SiC nanowires ranges from 25.6 to 79, which cannot be considered as a giant piezoresistive effect that reported in Si nanowire materials.

Most of the experimental studies on the piezoresistive effect in SiC nanowires were conducted on relatively large wires, with diameters of several hundred nanowires. Under this scale, the quantum confinement is considered negligible and the surface state (e.g. a thin layer of Si-O formed on the surface of SiC that creates a small built-in potential) may have

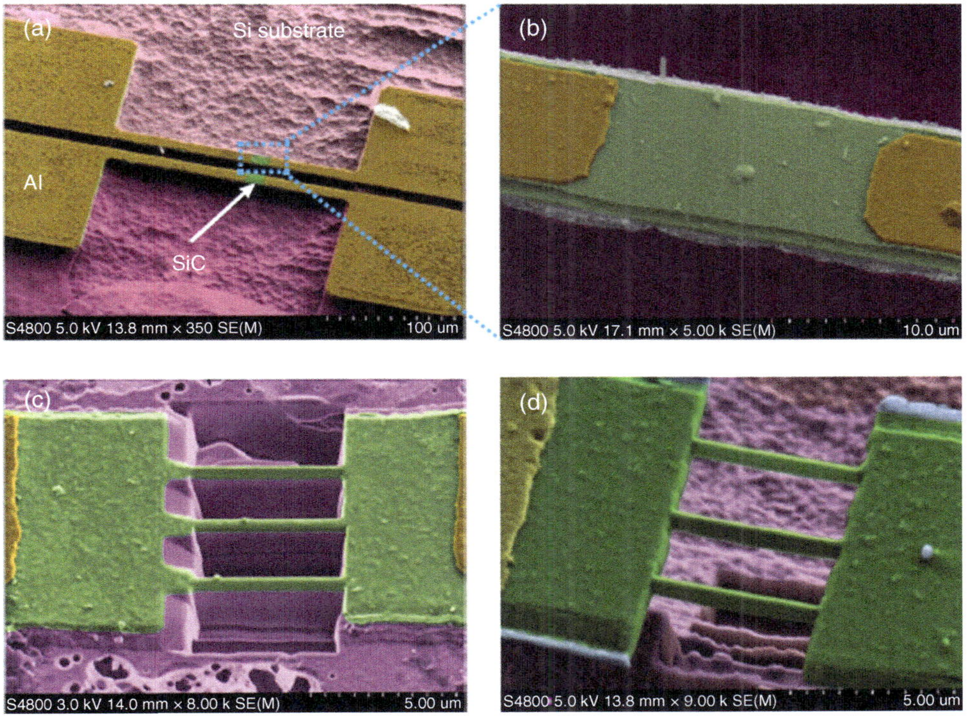

Figure 9.1 An example of strain sensors using SiC nanowires. (a, b) SiC micromembranes; (c, d) the membranes were then patterned into nanowires using focused ion beam. *Source:* [10], Phan, et. al., (2016)/AIP Publishing LLC.

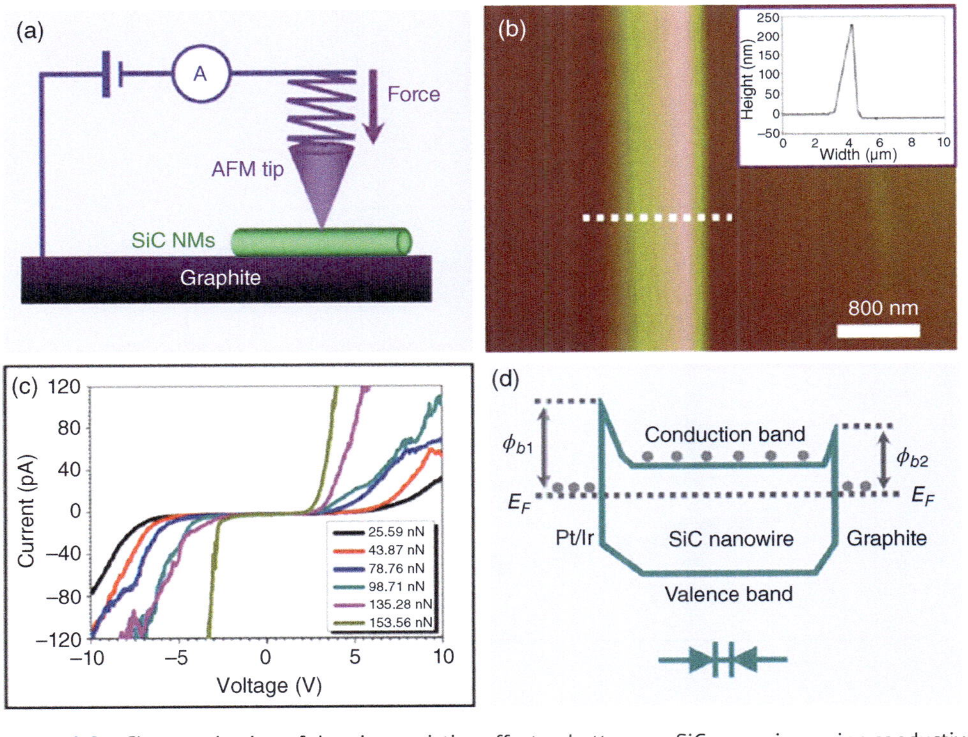

Figure 9.2 Characterization of the piezoresistive effect in bottom-up SiC nanowires using conductive AFM probe. (a) Concept of experimental setup to induce lateral strains. (b) AFM image of the nanowire. (c) *I–V* characteristic of the nanowire under applied force. (d) Energy diagram of the nanowire with two back-to-back Schottky contacts. *Source:* [12], Bi., et. al., (2013)/Royal Society of Chemistry.

insignificant influence on the electrical properties of nanowires. A number of theoretical studies were performed to simulate the gauge factor of SiC nanowires at atomic scale.

Nakamura et al. reported a simulation study on the piezoresistive effect of n-type SiC nanosheets with different polytypes of 2H-, 4H-, 6H-, and 3C-SiC [14]. The authors employed

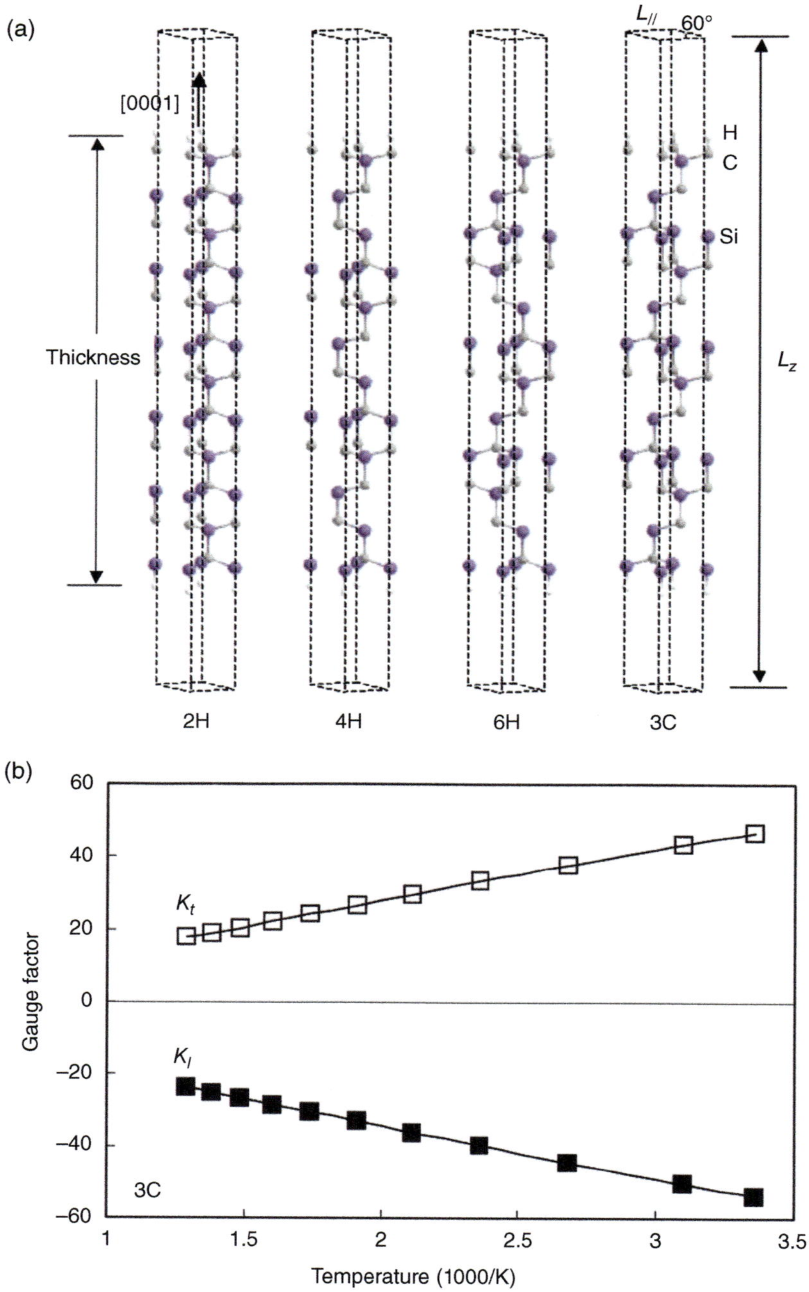

Figure 9.3 Theoretical study on the piezoresistive effect in nanoscaled SiC. (a) Periodic boundary models of 3 nm-thick 2H-, 4H-, 6H-, and 3C-SiC nanosheets with $Si_{12}C_{12}H_2$ unit supercells. (b) Numerical calculation of 3C-SiC nanosheet. *Source:* Nakamura et al. [14]/IOP.

the density functional theory (DFT) to estimate the energy band structures of nanosheets with thickness of 3 nm, at which the quantum confinement is considered effective, Figure 9.3. The results show a clear energy band deformation under 1% tensile strain. Based on the change in the energy band structure, the authors estimated the gauge factor in 3C-SiC nanosheet ranging from 30 to 60 at room temperature. The gauge factors decrease approximately 50% at high temperature of 500 °C, Figure 9.3. The principle calculation results suggest that the piezoresistive effect of SiC nanosheet is comparable with that of bulk SiC platform. No giant piezoresistance was theoretically predicted based on the quantum confinement effect. Note that, in this simulation model, the author did not consider the electrostatic depletion in nanowires and nanosheets when reducing the doping concentration, a phenomenon that was deployed to explain the giant piezoresistive effect in Si [15]. Therefore, further studies are required to validate the possibility of large piezoresistive effect in nanoscaled SiC devices.

As well as SiC, the piezoresistive in diamond has also attracted significant research attention. This is due to the advantage of diamond for sensing application in extreme environments along with its high mechanical stiffness. For instance, a B-doped diamond nanowire strain sensor was developed using a top-down process from a diamond thin film deposited on a SiO_2/Si substrate [16, 17]. The diamond nanowire was released from the substrate by under etching the SiO_2 box layer. Strains were induced using an SEM/scanning tunneling microscopy (STM) probe station, where the longitudinal strain was estimated based on the SEM image of the deflected nanowire, Figure 9.4a,b. Figure 9.4c shows a decrease in the current of diamond nanowire under tensile strains, indicating a positive gauge factor. The resistance of

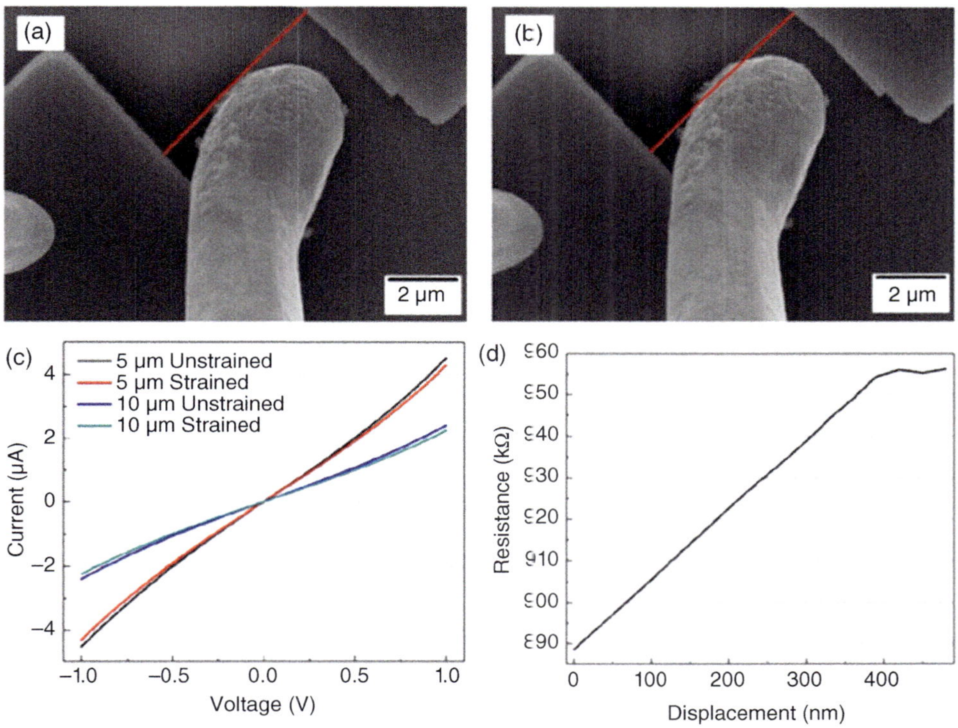

Figure 9.4 The piezoresistive effect in diamond nanowires. (a, b) SEM image of unstrained and strained nanowires. (c) *I*–*V* curve of a nanowire under different strains. (d) A linear relationship between resistance and applied displacement. *Source:* [16], Wang, et. al., (2012)/MDPI/CC BY 4.0.

the nanowire kept increased with increasing the magnitude of deflection, Figure 9.4d. The gauge factor was found to be 70, comparable with that of Si materials. Considering the potential of diamond for extreme environments, the large gauge factor indicates promising possibility for mechanical sensing applications where Si material cannot be used.

9.2.2 Piezotronics Effect in Nanowires

The piezotronics effect is in fact the piezoelectric effect coupled by the surface effect under low dimensions. Briefly, piezoelectric is a two-way phenomenon where an applied mechanical stimulation generates an electrical polarization (or voltage) in materials, and in turn, applying an electrical bias to piezoelectric materials generates a mechanical displacement [18]. External mechanical strains or stresses change the distance between the center of positive charges and the center of negative charges in the lattice of a piezoelectric materials. This deformation generates electric dipole moments, resulting in the formation of polarization charges and an electric field. The magnitude of the effect is quantified by a tensor of piezoelectric coefficient (d_{ijk}), which present the relationship between generated polarization E_k (here the indices k corresponds to x, y, z direction in the Cartesian coordinate) and strain tensors corresponding materials S_{ij} (e.g. S_{xy} is the shear strain in x and y directions): $d_{ijk} = \delta S_{ij}/\delta E_k$ [19]. For simplification, the tensor indices ij are generally converted to a single digit indices from 1 to 6 (e.g. 11 → 1; 22 → 2; 33 → 3, 23 or 32 → 4; 13 or 31 → 5; and 12 or 21 → 6). For wurtzite materials (i.e. basic hexagonal symmetry), common piezoelectric constants that are used in designing sensors and actuators are d_{31} (i.e. the force and polarization are perpendicular) and d_{33} (the force and polarization are in the same direction). The constant in bulk ZnO and AlN are $d_{31} = -5$, $d_{33} = 12$, and $d_{31} = -2$, $d_{33} = 5$, respectively. Generally, high piezoelectric coefficients are desired for sensing and energy harvesting applications as the material can generate a higher voltage (or displacement) at a certain applied strain (or bias). The piezoelectric effect in bulk materials has been widely used in several sensing and actuation applications. Common applications include speakers, ultrasound scanners, resonators, micro-/nano-manipulators [20].

For several decades, the enhancement of the sensitivity and the energy harvesting efficiency in piezoelectric material were mainly based on optimization of mechanical designs and selection of the piezoelectric orientation. In early 2000s, the Wang group discovered a giant piezoelectric effect in ZnO nanowires, which consequently shaped an innovative and highly active research direction in piezoelectric materials and devices [21]. The group coined the term piezotronics for nanoscaled semiconductors. In their breakthrough discovery, Wang and Song employed a Si AFM tip coated with a metal film (e.g. Pt) to deflect a vertically aligned ZnO nanowire. The output voltage generated from the piezotronics effect in each individual nanowire was measured using a large electrical load of 500 MΩ. The authors observed a significant voltage of 8 mV from a single nanowire which was caused by the polarization of the non-uniformly bent nanowires, Figure 9.5. In the experimental configuration, the use of high work function metal such as Pt and Ag create a back-to-back Schottky configuration with the ZnO nanowire. Under strain-free state, there is an insignificant current flowing through the back-to-back Schottky barrier. When pressing the AFM tip against the free end of the ZnO nanowire, it can cause a tensile and compressive strain on each side of the nanowires. These strains result in polarization that leads to the generation of charges in the nanowires. The electric field from negative charges increases the barrier height, while positive charges reduce the Schottky barrier. Due to the asymmetric barrier change, a positive voltage is generated under tensile strain, while compressive strain generated a negative output voltage. The piezoelectric coefficients in ZnO nanowires (d_{33}) can reach up to

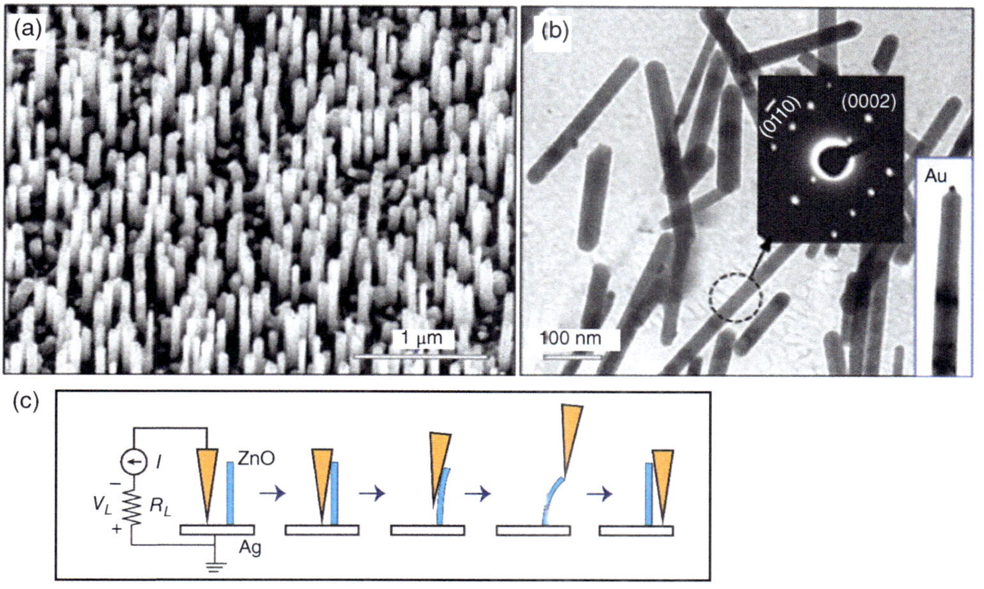

Figure 9.5 The piezotronics effect in ZnO nanowires. (a b) SEM images and selected area electron diffraction (SAED) data of ZnO nanowires. (c) Measurement of the piezotronics effect using a conductive AFM tip. *Source:* [21], Wang, Z.L., et. al., (2006)/American Association for the Advancement of Science – AAAS.

26.7 pm/V, much higher than that of the bulk material. More importantly, considering each individual nanowire can generate 8 mV, a high-density array of ZnO nanowires is expected to generate sufficient electrical energy that can power the functional device itself [22]. The relationship between the output voltage and applied pressure can also be utilized for several types of mechanical sensing applications. The discovery of Wang and coworkers has defined a new research field of nano-energy, which has been a highly active research direction in the last 15 years. Several sensing and electronic devices have been developed based on the piezotronics effects in nanowires, such as strain sensors, pressure sensors, tactile sensors, and self-powered electronics [23].

Inspired by the work on ZnO nanomaterial, the piezotronics effect in other piezoelectric wide bandgap nanowires has been extensively investigated. Among these, the piezotronics effect in III-nitride attracted great attention as this material exhibits large piezoelectricity in the bulk form. The Espinosa group studied the scale effect in GaN nanowires using a scanning probe microscopy (SPM) [24]. The nanowires were placed on a silicon substrate that functions as the bottom electrode, and a conductive AFM tip was employed as the top electrodes. A transverse electric field was induced into the GaN nanowires by applying a voltage between the AFM tip and the substrate. The displacement of the wire under external electric field was monitoring through the SPM system. Figure 9.6 shows the experimental results of the piezoresponse in the GaN nanowires, with good repeatability after several scanning cycles. The piezoelectric coefficient of GaN was found to be $d_{33} = 12.4$ pm/V for nanowires with a height of 156 nm. The experimental result was consistent with numerical calculation where d_{33} was estimated to be 12.8 pm/V using finite element modelling. The results indicate that GaN nanowires exhibit stronger piezoelectricity than the bulk platform. This enhancement in the piezoresistive effect is attributed to the increase in the polarization per unit

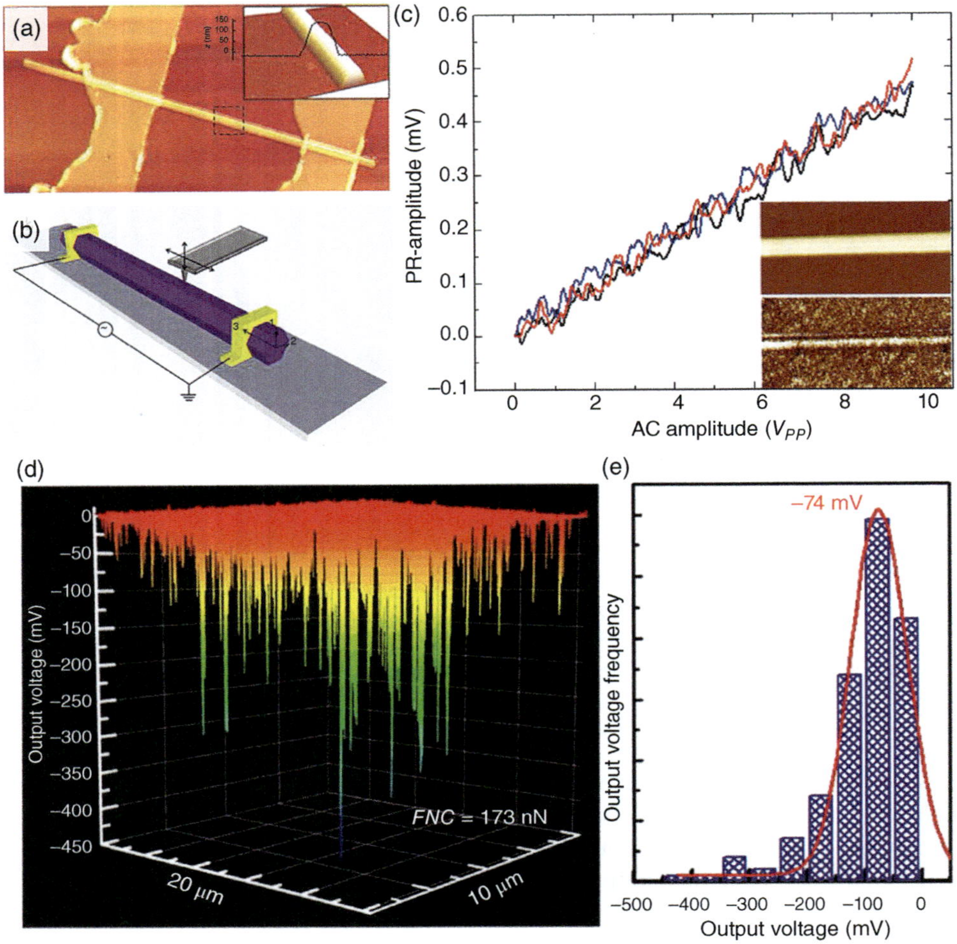

Figure 9.6 Piezoelectric in GaN nanowires. (a) SEM image of a GaN nanowire. (b) Setup for characterization of the piezoelectric in GaN using scanning AFM. (c) Piezoelectric response under different AC amplitude. (d) 3D voltage output collected by an AFM Resiscope on partially encapsulated GaN NWs at 173 nN. (e) Histogram distribution of the generated voltages at 173 nN. *Source:* Gogneau et al. [25]/John Wiley & Sons, Inc.

volume as the diameter of nanowires decreases. The large piezoelectric coefficient in GaN nanowires demonstrates the potential of wide bandgap semiconductor nanowires for mechanical sensing and energy harvesting applications. This potential was further confirmed by the work of Gogneau et al., which shows a high generated voltage of −443 mV under a constant applied force of 173 nN induced from an AFM tip to a vertical GaN nanowire [25]. The author estimated a maximum power density of 130 mW/cm^2, which can be obtained from a nanowires forest.

9.2.3 Capacitive Sensing

Another common type for mechanical sensor is based on the capacitive configuration. In this sensing approach, wide bandgap nanowires are mixed with polymer to form the dielectric layer for mechanical sensors. Two metal layers are formed on two side of the polymer for electrodes. Under external forces or pressures, the polymeric dielectric layer is deformed that changes the capacitance of devices. Mixing conductive nanowires with polymer could enhance the dielectric constant and ease the electrical measurement. More importantly, nanowires with the piezotronics effect can further boost the sensitivity of devices by coupling this phenomenon with the capacitive effect. Specifically, under applied forces, an electrical potential is generated in nanowires, which attract positive and negative charges toward opposite ends of nanowires. These generated charges significantly change the dielectric constant of the composite, Figure 9.7a. Therefore, the capacitance of the device changes as a result of both dielectric constant modifications along with the change in the geometry (e.g. the distance between the two electrodes). For instance, Chen et al. developed a ZnO-polymer capacitive sensor for pressure sensing applications [26]. The devices exhibit a linear relationship between capacitive change and the applied pressure. The sensitivity of the sensors was found to be 9.95×10^{-3} cm^2/gf, approximately 23 times higher than that of poly(methyl methacrylate) (PMMA) sensor (~0.43×10^{-3} cm^2/gf), Figure 9.7b. The sensor outputs show excellent repeatability over several applying/releasing pressure cycles, Figure 9.7c.

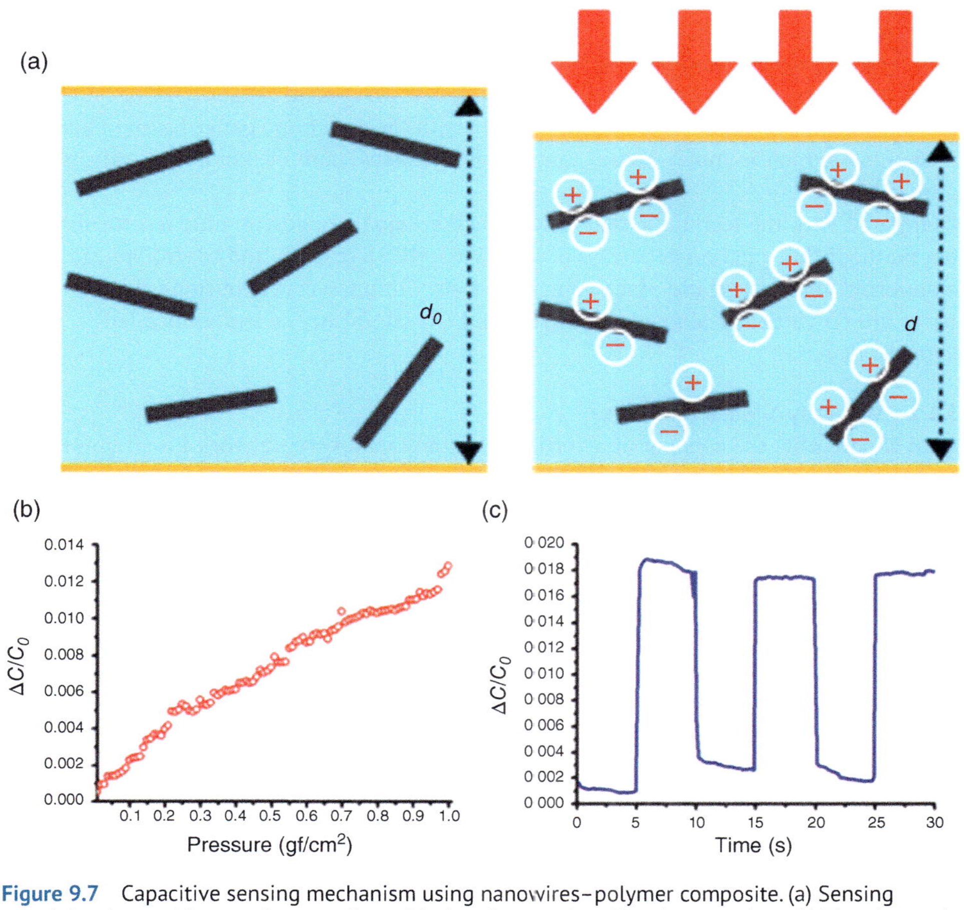

Figure 9.7 Capacitive sensing mechanism using nanowires–polymer composite. (a) Sensing principle of capacitive sensing coupling with the piezotronics effect in nanowires. (b) The change in capacitance under varying external pressure stress. (c) Good repeatability of capacitance change over several pressurizing cycles. *Source:* Chen et al. [26]/American Society of Chemistry.

9.3 Transducer Configurations and Fabrication Technologies

9.3.1 Strain Sensors

Strain sensors can be built using a single nanowire or an array of wide bandgap nanowires. In the concept of the single nanowire, sensing devices are typically mounted on a target surface using adhesion. The strain applied to the target surface is considered to equally induce into the nanowire mounted on its surface. A common method to readout the electrical signal of a piezoresistive strain sensor is the use of the Wheatstone bridge, which contains the nanowire piezoresistors and three other reference resistors. The change in the resistance of the nanowire is converted into an output voltage through the Wheatstone bridge: $V_o = \frac{V_{in}}{4} \times \frac{\Delta R}{R}$.

Zhou et al. reported a strain sensor using a single ZnO nanowire on a soft polystyrene substrate [27]. A ZnO nanowire network was grown using thermal evaporation with lengths ranging from a few hundred micrometers to several millimeters. A single nanowire was transferred on to a polymer substrate and metal contacts were formed using silver paste, which created a back-to-back Schottky barrier. The entire structure was encapsulated in a polydimethylsiloxane (PDMS) film for strain sensing, Figure 9.8a. The device was then fixed at one end while the other end was stretched using a linear stage, Figure 9.8b. Figure 9.8c shows the *I–V* characteristic of the sensor under 0–0.98% strain regime. The sensor exhibited an increase in the current with increasing the applied strain. The gauge factor of the sensor was found to be dependent on the applied strain, with the maximum gauge factor of approximately 1250, which is much higher that of piezoresistive carbon nanotube strain sensor, Figure 9.8d.

Compared with the technique that only transfers a single nanowire onto a soft substrate, other methods that can transfer arrays of nanowires with a scalable and controllable manner are preferred for low cost and practical applications. Common configurations of these strain sensors are (i) vertically aligned nanowires and (ii) lateral nanowires embedded in a soft polymeric material.

9.3.1.1 Vertically Aligned Nanowires

Figure 9.9 presents the concept of a vertically aligned nanowire network for strain sensors developed by Jenkins and Yang [28]. ZnO nanowires were chemical vapor deposition (CVD) grown on a Si substrate coated by a 100 nm thick ZnO film. The lengths of the nanowire were approximately 100 μm. Next, a PDMS layer was drop casted and then spin-coated to fully cover the as-synthesized nanowires. Once the PDMS film was cured, the PDMS film with ZnO nanowires embedded inside was mechanically detached from the Si substrate. The detached film was then covered by a polyethylene terephthalate (PET) polymer mask with an opening window of 2 mm × 10 mm. The top surface of the PDMS layer in the opening area was then removed using dry etch with SF_6 and O_2 to expose nanowire tips. A 5 nm Cr/100 nm Au film was deposited to form the top electrical contact with the exposed nanowire tips. The sensor was mounted onto a stepped shaft using silver paste, which served as the bottom electrode of the strain sensors. Strain induced into the shaft changed the Schottky barrier between the metal and ZnO nanowires, resulting in a change in the conductance of the sensors. Experimental results showed a large gauge factor of approximately 64 when the free end of the shaft was deflected by 2.5 mm. The sensors showed excellent repeatability over severable bending cycles.

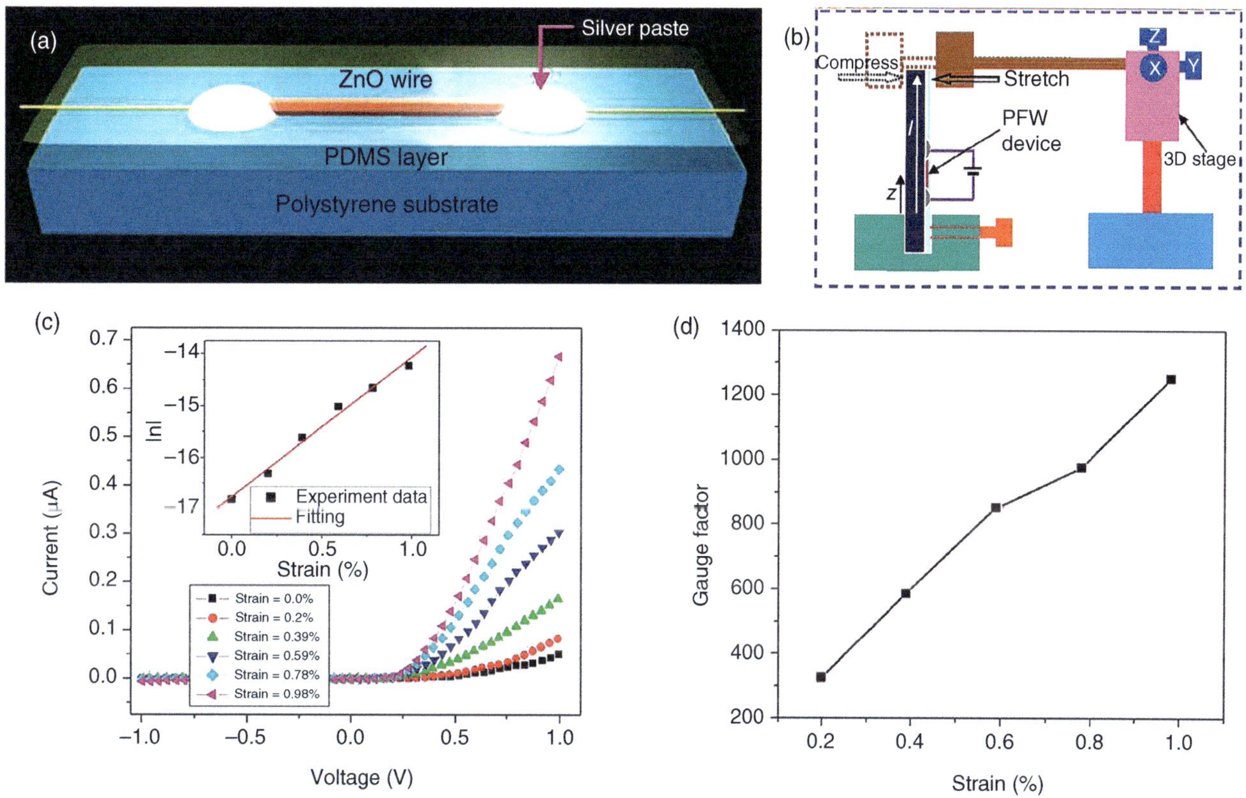

Figure 9.8 Strain sensor using a single nanowire configuration. (a) A ZnO nanowire on a PDMS substrate. (b) Experimental setup. (c) *I*–*V* characteristic of the nanowires. (d) Gauge factor vs applied strains. *Source:* [27], Zhou, et. al., (2008)/American Chemical Society.

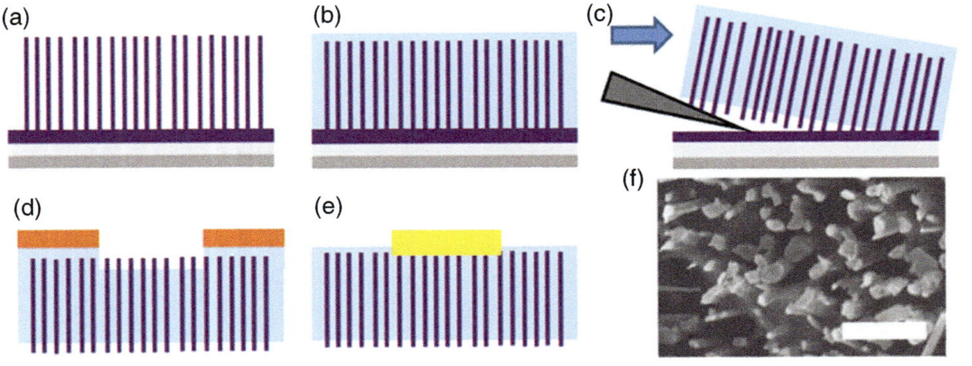

Figure 9.9 Vertical nanowires strain sensors. (a–e) Fabrication process. (f) SEM image of the vertical ZnO nanowires. *Source:* [28], Jenkins, et. al., (2017)/IOP Publishing.

9.3.1.2 Horizontal Nanowires Network (Nanowires and Polymer Composite)

Another configuration employs lateral nanowires mixed with polymer (e.g. composite). Xiao et al. developed a strain sensor using ZnO nanowires/polystyrene nanofiber (PSNF) on a flexible PDMS film, Figure 9.10 [29]. The PSNF fibers were electrospun on a PDMS film with dimensions of 2 cm × 1 cm × 1 mm. ZnO seeds were formed on the surface of PSNF by immersing these fibers into a solution containing ZnO nanoparticles. These ZnO seeds on PSNF enabled the growth of ZnO nanowires using a hydrothermal process. SEM image clearly shows that ZnO nanowires with diameters ranging from 50 to 100 nm and lengths of approximately 3 μm were densely and uniformly growth on the PSNFs network. Besides mechanical flexibility, the ZnO/PSNF hybrid structure exhibits a semi-transparent property with a transmittance of over 16% at 600 nm wavelength. As the ZnO/PSNFs were laterally distributed on a flat substrate (i.e. PDMS), electrodes (e.g. Ti/Ag) can be easily deposited to form metal contact for the sensors using a simple shadow mask. The ZnO/PSNFs were then fully encapsulated by another layer of PDMS. The sensors were characterized using a stretching setup, which show a monotonic increase in the resistance when increasing tensile strains from 0% to 50%. The sensors can tolerate high strain of up to 50% owing to the excellent stretchability of the PSNF network. The sensors show excellent repeatability after several stretching and releasing cycles. Based on the *I–V* curve under varying strains, the gauge factor of the sensors was found to be approximately 116. This high gauge factor was attributed from the change in the contact areas between nanowires under applied strain. Specifically, tensile strains lead to a reduction in the contact surface between adjacent nanowires, resulting in a decrease in the electrical conductance.

9.3.2 Pressure Sensors

Pressure sensors are one of the most important MEMS applications, widely used in several fields of industry [30–32]. Standard configurations for pressure sensors use a thick membrane to transform applied (air or liquid) pressure into a strain that is induced into piezoresistive or piezoelectric sensing elements. The applied pressure causes an out of plane deflection in the membranes, and the maximum induced strain occurs at the edge of the

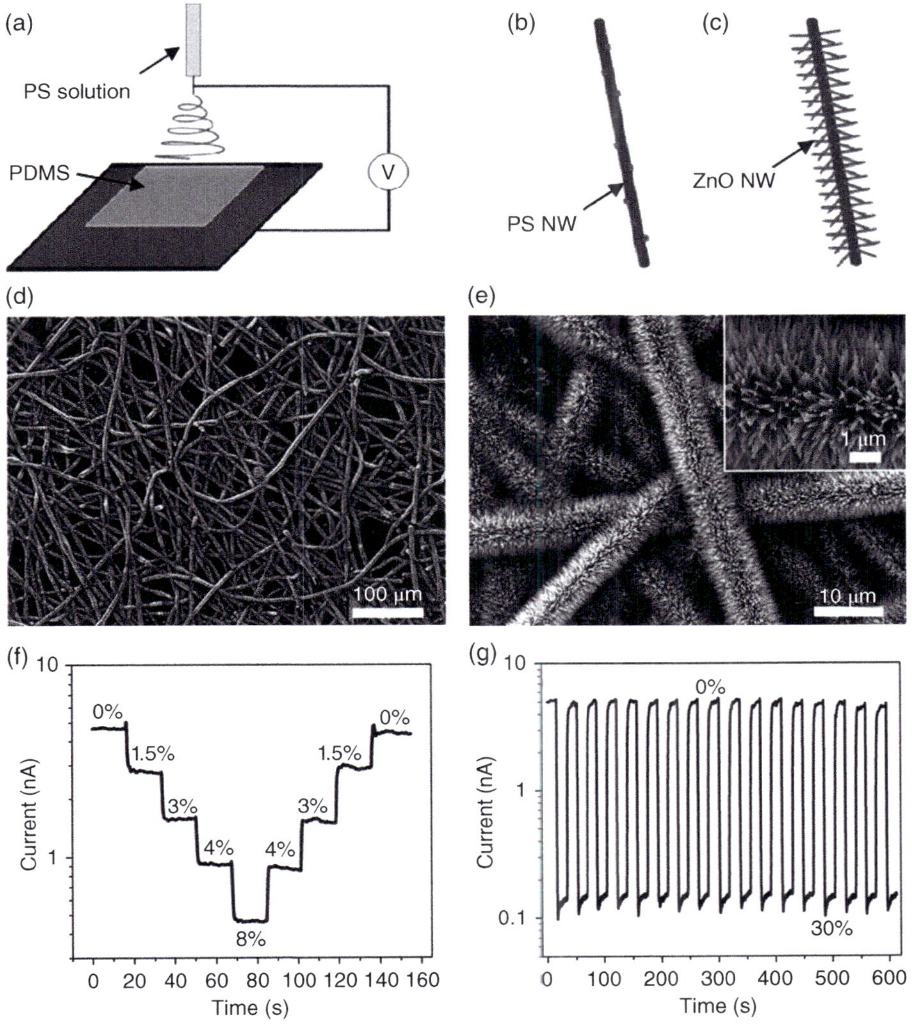

Figure 9.10 Nanowire composite sensors. (a) Polystyrene nanofiber was electronspun on PDMS. (b, c) The growth of ZnO nanowires; (d, e) SEM images of PSNF and ZnO nanowires; (f) current changes under applied strains; (g) repeatability of the sensor. *Source:* Suma et al. [34]/Springer.

membrane. For square membranes, the relationship between the maximum stress and applied pressure is given by [32]:

$$\sigma = \frac{0.38pa^2}{t^2}$$

where a and t are the width and thickness of the square membrane, respectively. The square configuration is a common design in MEMS pressure sensing, mainly due to its compatibility with most of anisotropic wet and dry etching processes on an open square-shaped window. Compared with square shape, circular pressure sensors (with a similar surface area) can withstand large pressure, and also provide a uniform strain along the edge of the membrane. However, since circular-shaped sensors have a symmetric structure, some anisotropic etchings (e.g. wet etching of the Si substrate using KOH or tetramethylammonium hydroxide [TMAH]) are not applicable. The strain induced into a sensing element located at the edge of circular pressure sensors is given by [33]

$$\sigma = \frac{3pr^2}{4t^2}$$

where r and t are the radius and thickness of the circular membrane, respectively.

Figure 9.11 presents an example of pressure sensors using wide bandgap material nanowires formed on a free-standing membrane [34]. The fabrication of this pressure sensor is relatively similar to the strain sensors using vertical ZnO nanowires described in Section 9.3.1. Briefly, ZnO nanowires were synthesized at a low temperature of 85 °C on a highly rugged, elastic substrate of Phynox. To improve robustness, the nanowires on Phynox substrate were coated with a 2 μm-thick parylene-C layer. A Kapton film was used to partly cover some areas on top and bottom, enabling electrical contacts to the nanowires using silver paste. The whole parylene film with embedded nanowires was mounted on a deflectable membrane with a diameter of 14 mm and a thickness of 380 μm. Pressure was applied using nitrogen gas, and the output of the sensor was observed using an oscilloscope. Experimental results revealed that when nitrogen was purged using a gas regulator, a shockwave was propagated to the membranes, causing a deflection in the ZnO nanowire array. The deflection of the nanowire array generated an output voltage as a result of the piezotronics effect. For instance, at a pressure of 17.22 bar, the ZnO sensor generated a voltage of 30.8 mV. Increasing applied pressure leads to an increase in the output voltage of the sensors. The sensitivity of the sensor was found to be approximately 1.65 mV/bar, with a fast rise time and discharge time of 1.38 and 1.79 ms, respectively. This good performance indicates the potential of ZnO nanowires for dynamic pressure sensing devices.

ZnO nanowires with their simple and well-established synthesis technologies, along with large piezotronics effects are preferred for ubiquitous and low-cost pressure sensing applications. On the other hand, GaN, SiC, and diamond are materials of choice for pressure sensors used in extreme environments. For III-nitride materials, the formation of two dimensional electron gas (2DEG) layer, a unique feature to create high electron mobility transistors (HEMTs), is attractive for the integration of sensing devices and electronic circuits on the same chip. The stable electrical properties in III-nitride also enable its pressure sensing devices to operate at high temperatures. Chapin et al. developed a pressure sensor using the 2DEG layer created at the interface of InAlN/GaN [33]. The sensors were fabricated from an InAlN/GaN on a Si wafer with a 2 nm GaN capping layer, a 15 nm InAlN barrier layer, a 1 nm AlN interlayer, a 1 μm GaN layer, and an AlGaN buffer layer that were deposited on a (111) Si substrate using metal organic chemical vapor deposition (MOCVD). A ring-shaped InAlGaN/GaN HEMT used as the pressure sensing element, was fabricated near the edge of the circular diagram, Figure 9.12a. The diagram (1 μm thick and 500 μm wide) was created by removing the Si substrate using deep reactive-ion etching (DRIE) etching. Figure 9.12b shows an SEM image of the pressure sensors, where an Ohmic contact (Ti/Al/Pt/Au) was used for drain and source, while a Schottky contact (Ni/Au) was formed at the gate of the HEMT device. The I_{ds}–V_{ds} was characterized at room temperature up to 300 °C, Figure 9.12c. Both released and non-released devices show a typical HEMT characteristic with a slightly decreased current in the released sensor, which is attributed to a reduction in the film stress

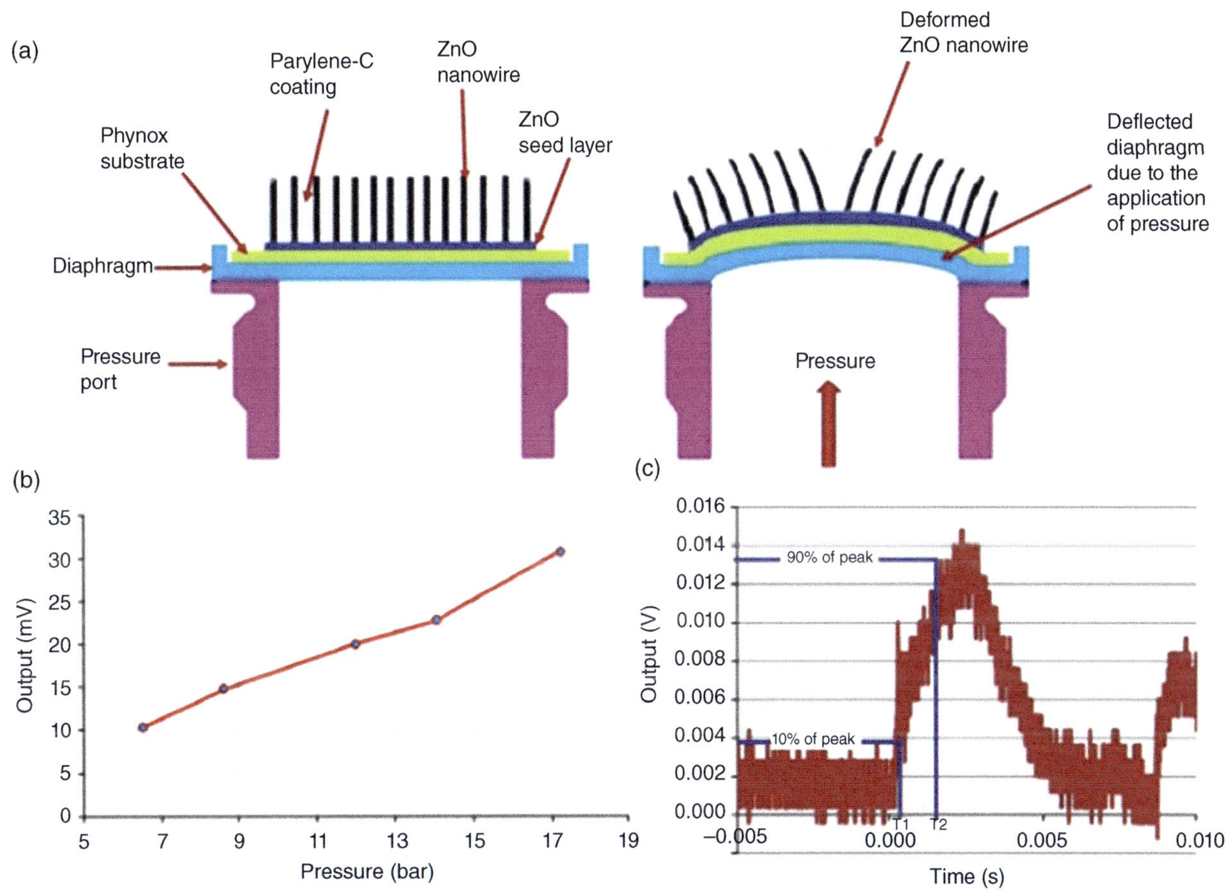

Figure 9.11 ZnO pressure sensors using vertical nanowires formed on membranes. (a) The concept of the sensors; (b) output voltage vs applied pressure; (c) time response.

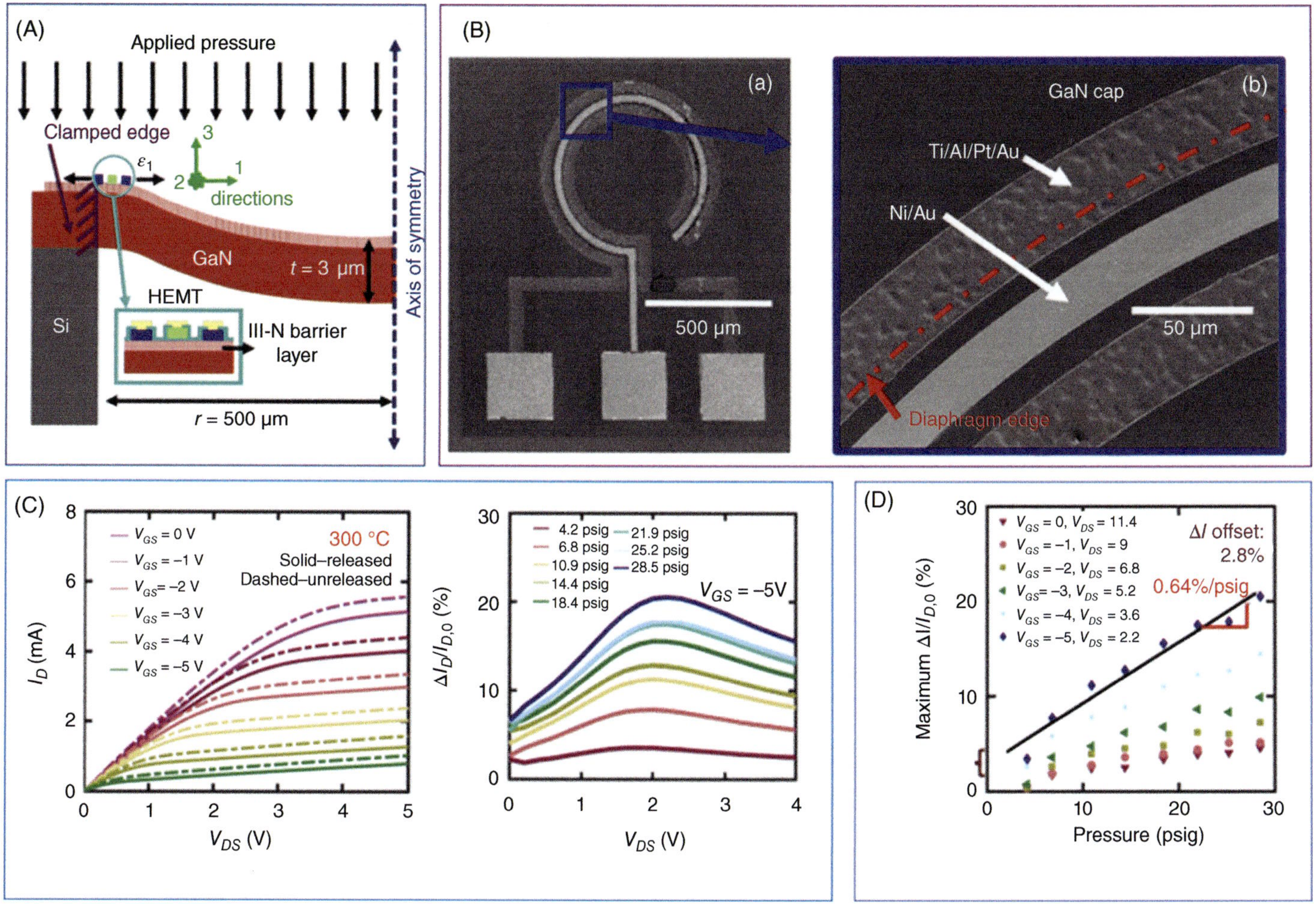

Figure 9.12 GaN pressure sensors using III-nitride materials. (A) Schematic sketch of the InGaN/GaN pressure sensor. (B) SEM image of the top surface of the sensor. (C) $I_{ds}-V_{ds}$ curve in released and non-released devices at 300 °C. The change in current at different drain to source bias for pressures ranging from 0 to 28.5 psig. (D) The relationship between the maximum change in the drain current and applied pressure at different gate biases. *Source:* [33], Chapin, et. al., (2017)/Elsevier.

upon Si removal. The devices exhibit good current modulation function at high temperature of 300 °C. The pressure sensing function was characterized by applying air pressure and monitoring the change in I_{ds}–V_{ds} curves. The authors also developed a theoretical model to predict the behavior of the HEMT sensing element under applied pressure (i.e. strain). The numerical analysis showed that applying a positive pressure to the membrane induces a tensile stress which results in an increase in the 2DEG carrier concentration. The increase in the sheet carrier density is considered to be due to the increase in the total polarization in the piezoelectric materials of the HEMT device. Experimental results agreed well with the theoretical model, where an increase in the I_{ds} current was clearly observed for all applied pressures ranging from 4.2 to 28.5 psig. The current change increased with increasing external pressure. At a constant gate bias of −5 V, the current change initially increased with increasing the drain–source voltage and reached a peak near the saturation voltage. Figure 9.12d shows the maximum current change ($\Delta I/I_{D,0}$) as a function of pressure at different gate bias. A linear relationship between the change in current and the pressure was observed, with a sensitivity of 0.64%/psig. The results demonstrate the potential of III-materials for mechanical sensing at high temperatures.

As well as GaN materials, SiC has been used for pressure sensing applications owing to its significant piezoresistive effect and the capability of working under extreme condition. SiC nanowires have been utilized to enhance the sensitivity of pressure sensors. Specifically, using a stress concentration configuration, Phan et al. developed top-down SiC nanowires pressure sensors on a Si substrate [35]. The sensing elements were formed by etching free standing SiC membranes (5 μm width) into nanowires with a diameter of approximately 300 nm using focused ion beam. The Si substrate was etched from backside using laser machining to form square-shaped membranes with a thickness of 150 μm. A simple numerical analysis model revealed that by locally fabricate SiC nanowires at the center of a free-standing microbridge, the effective strain induced into the nanowires can be significantly amplified with respect to the strain applied to the substrate. Briefly, the relationship between the strain induced into nanowires (ε_{NW}) and the applied strain (ε_{sub}) is given by:

$$\varepsilon_{NW} = \frac{L}{2\frac{K_{NW}}{K_{MF}}L_{MF} + L_{NW}} \times \varepsilon_{sub}$$

where L is the total length of the free-standing bridge; and (K_{MF}, L_{MF}) and (K_{NW}, L_{NW}) are the stiffness and the length of the micro frame and the nanowires, respectively, Figure 9.13a,b. Due to its low dimension, the stiffness of the nanowires is much smaller than that of the micro-frame ($K_{NW} \ll K_{MF}$). Therefore, by choosing a proper ratio between L_{MF} and L_{NW}, it is possible to amplify the effective strain induced into nanowires from the applied surface strain. The numerical model was in a solid agreement with finite element analysis using COMSOL Multiphysics simulation. Figure 9.13c shows the current change of SiC pressure sensor with and without using the strain amplification. Evidently, the strain amplification pressure sensor possesses a much higher response, at least threefold in the current change compared with the conventional design. A pressure ramping-up cycle (10 seconds for each cycle) was applied to characterize the dynamic response of the sensor. The sensor showed excellent time response, which can track the change in pressure in real-time, with consistent repeatability, Figure 9.13d.

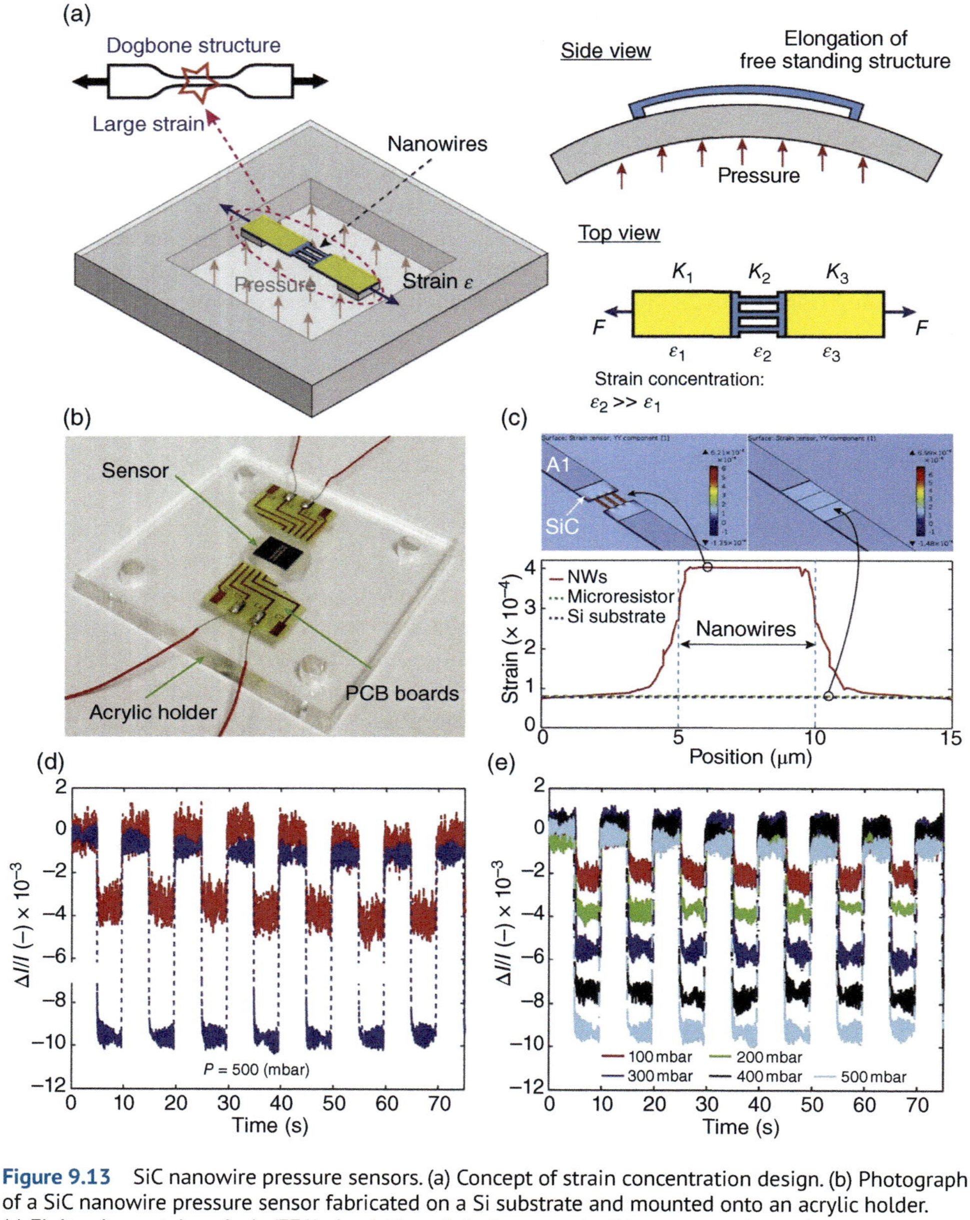

Figure 9.13 SiC nanowire pressure sensors. (a) Concept of strain concentration design. (b) Photograph of a SiC nanowire pressure sensor fabricated on a Si substrate and mounted onto an acrylic holder. (c) Finite elemental analysis (FEA) simulation of strain concentration on nanowires using a smart design. (d) A comparison between the response on the sensors with (blue) and without (red) the stress concentration concept. (e) The repeatability of the pressure sensors over several pressurizing cycles. *Source:* Phan et al. [35]/Elsevier.

9.3.3 Tactile Sensors

Beside air and liquid pressure sensors, the touch or tactile pressure is an important physical parameter, particularly in the field of robotic and haptic control to mimic the tactile sensing capability in human skin [36, 37]. The capability of transferring and integrating ZnO nanowires into soft polymers along with their large piezotronics enabled the development of several tactile sensors [38]. Generally, ZnO nanowires in the form of forest or planar array are embedded inside flexible elastomer such as PDMS or Ecoflex. A direct contact between the sensors and external object causes these elastomer layers to deform, activating the piezotronics effect in each individual nanowire. This generates an output voltage that correlates to the amount of applied touching pressure.

Figure 9.14 illustrates examples of ZnO nanowires tactile sensors. Wu et al. developed a tactile sensor using 3D strain-gated vertical piezotronics transistors (SGVPTs) [39]. The SGVPT array consists of multiple ZnO nanowires. Each pixel of the tactile sensors (i.e. a group of ZnO nanowires) was in contact with line-shaped indium tin oxide (ITO) electrodes at the top and bottom. This configuration allows for reading the output of each individual pixel. The nanowires were uniformly coated with a thin layer of parylene-C to prevent moisture and corrosion, and then the whole device was fully covered in a thick SU8 film. When a pressure is applied to the top surface of the sensor, it induces an axial strain into each nanowire, creating a piezopotential due to the polarization of nonmobile ions distributed at the tip of each wire. By monitor the current response of each SGVPT, the location of the applied pressure and its magnitude can be mapped. The taxel area density of the ZnO sensors was 8464/cm^2 (i.e. 92×92 array), enabling a high spatial resolution of 100 μm. Under a small pressure of 3.5 kPa, the SGVPT experienced a current change of approximately 2 μA, indicating a high sensitivity in the sensor (2.1 μS/kPa) at as low-pressure regime of less than 10 kPa. The high sensitivity in the sensor is attributed to a significant change in the carrier transport in the SGVPT due to the

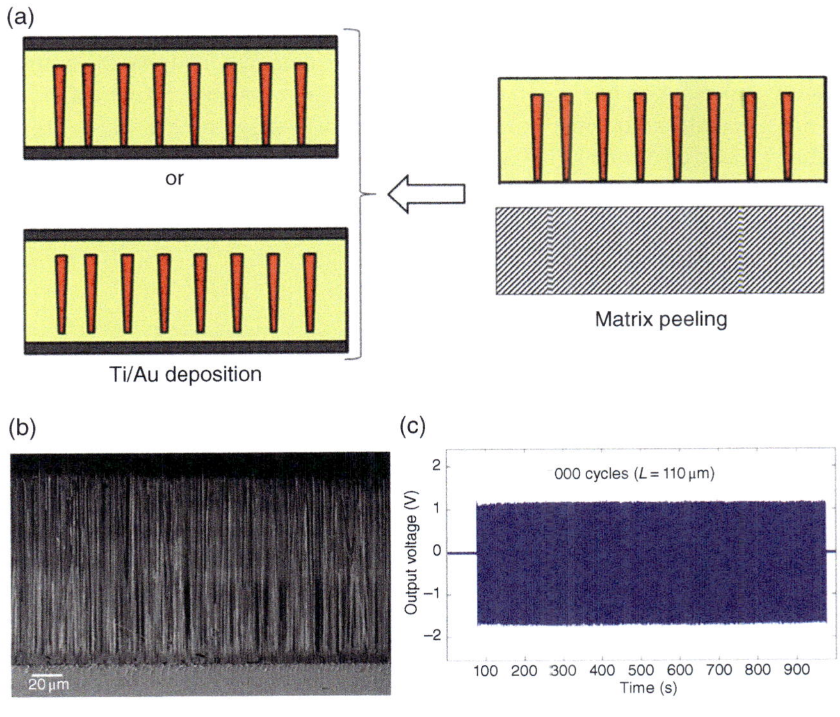

Figure 9.14 Tactile sensors using the piezotronics effect in wide bandgap semiconductor nanowires. (a) Fabrication process for capacitive sensors; (b) SEM image of a GaN nanowire for the capacitive piezotronics tactile sensors. (c) Output voltage of the GaN nanowire tactile sensors under a cyclic pressurizing test at 1.1 Hz. *Source:* [40], El Kacimi, et. al., (2018)/American Chemical Society.

modulation of the Schottky barrier height at the ITO/ZnO junction under pressure. The sensors exhibit a fast response time of 0.15 seconds. The sensors remained its operational functions after 24 hours of immersion into 0.9% saline solution. This indicates the feasibility of the ZnO tactile sensors for biophysiological sensing in complex environments.

Kacimi et al. reported a scalable fabrication technology to form flexible capacitive piezoelectric tactile sensors using vertical GaN nanowires, Figure 9.14d [40]. GaN nanowires with lengths ranging from 10 to 700 μm were synthesized using metal organic vapor phase epitaxy (MOVPE) on a c-sapphire substrate. This process provides a relatively high growth rate of approximately 100 μm/h, enabling the formation of nanowires with lengths approximating 1 mm. PDMS was prepared at a mixture of 10 : 1 (elastomer to curing agent ratio), and spin-coated onto the as-grown vertical GaN nanowires. The PDMS layer was then peeled off from the sapphire substrate and 20 nm Ti/200 nm Al was deposited on top and bottom to form a capacitive configuration sensor. Under an applied pressure, GaN nanowires are deformed in the vertical direction, creating a potential along the principal piezoelectric axis of the wires as a result of the piezotronics effect. This electric potential separates charges to the bottom and top parts of GaN nanowires, resulting in an output voltage between the two metal contacts previously formed on the PDMS film. Here the PDMS layer plays an important role in the device performance, as its elasticity allows for a large deformation under compressive stress, while its thickness and dielectric contact are directly correlated to the output voltage. As such, using finite element analysis (COMSOL Multiphysics), the authors estimated a high potential voltage when the ratio between the PDMS thickness and wire length was 1.5, and it decreased significantly when this ratio reached around 4.0. Experimental data showed a large voltage of 1 V was observed when applying a pressure of 1 N/cm^2 over a 4 cm^2 surface area of a sensor with 110 μm length nanowires. The device also exhibited good repeatability over 1000 cycles at a frequency of 1.1 Hz, Figure 9.14e. Experimental results also demonstrated that the output voltage significantly depended on the length of the nanowires. For instance, under a similar finger touch, 190 μm length nanowires provide a peak output voltage of 2 V, while 70 μm nanowires generate a lower peak voltage of 0.4 V. These results suggest interesting approaches to engineering the sensitivity and performance of piezotronics wide bandgap semiconductors nanowires via their nanoarchitectures and the choice of dielectric materials.

9.3.4 Acceleration and Vibration Sensors

Accelerometer and vibration sensors are used to detect shock and monitor structural conditions of industrial systems [41, 42]. Acceleration sensors using wide bandgap semiconductor nanowires share the same concept design as the conventional devices, in which a proof mass is utilized to convert acceleration into a mechanical force (i.e. $F = ma$) that cause a deformation in the sensing element [43]. Song et al. developed an accelerometer using vertically aligned ZnO nanowires [44]. Figure 9.15a shows the fabrication flow of the acceleration sensor. ZnO nanowires were synthesized using either seedless hydrothermal process or refresh hydrothermal process on a square shaped copper film previously formed on a PDMS handling layer. After the growth process, silver paste was used to form one contact to the tips of ZnO nanowires, while the copper substrate formed another contact pad. Here, the thick copper substrate (1.5 mm × 1.5 mm × 0.5 mm) also functioned as the proof mass of the acceleration sensor. The device was mounted onto a shaker, where the silver electrode side was fixed to the mounting stage, while the copper side can freely move to detect acceleration, Figure 9.15b. A commercial sensor was also attached onto the same setup for calibration.

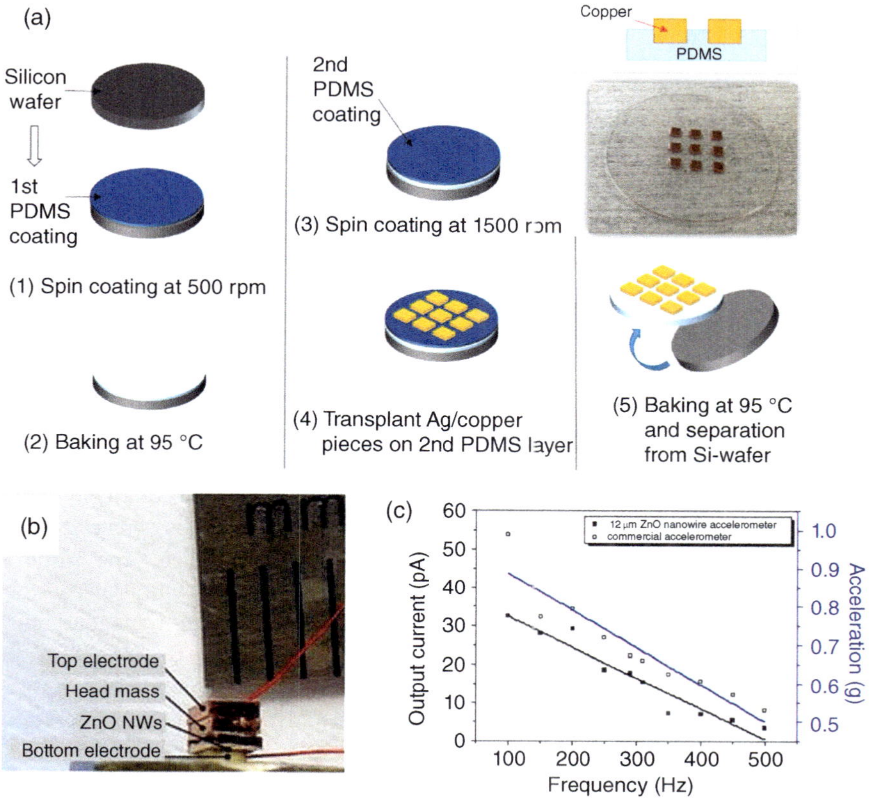

Figure 9.15 Acceleration sensors using wide bandgap semiconductor nanowires. (a) Fabrication process. (b) Photograph of a sensor with a head mass. (c) Output voltage of the sensors at different frequency. *Source:* [44], Song, et. al., (2017)/The Royal Society/CC BY 4.0.

Experimental results showed that the sensor with 6 μm length nanowires generated an output current of 1.5–14.9 pA under input acceleration of 0.1–1.0g, Figure 9.15c. The sensor with longer wires (12 μm) provided a higher output, which was approximately double of that of the short wires. This result is considerable as the generated potential is proportional to the length of nanowires. These sensors demonstrated their capability of detecting acceleration at low frequencies ranging from 100 to 500 Hz with a similar trend observed in the commercial device.

9.3.5 Energy Harvesting Devices

Mechanical energy is an abundant resource, which is available almost everywhere in our daily life, from mechanical impacts, vibrations, sonic waves, air/liquid flows, ocean waves, and even from human motions. Harvesting these mechanical energy resources can be an effective approach to power several electronics and sensing devices [45–47]. In this regard, ZnO nanowires are attractive for nano-energy harvesters due to their scavenging mechanical energy and versatile fabrication processes. Wang et al. developed a direct-current ZnO nanogenerator driven by ultrasonic wave [48]. The nanogenerator was fabricated using a

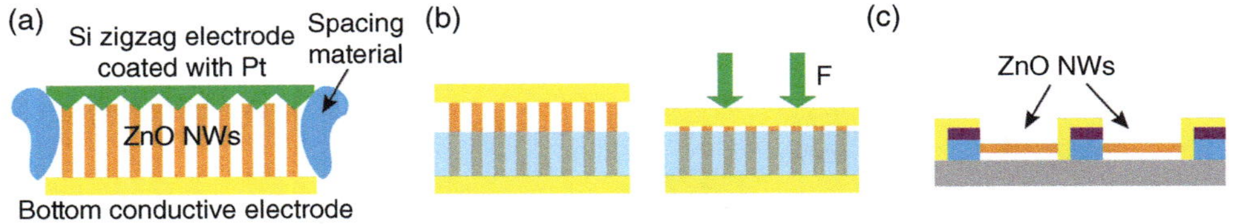

Figure 9.16 Energy harvesting using piezoelectric effect in ZnO nanowires. (a) DC energy harvester from ultrasonic wave using zig-zag Pt electrode. *Source:* Wang et al. [48]/American Association for the Advancement of Science – AAAS. (b) Nanogenerator with VING configuration using vertical nanowire. Right: The output voltage of a VING device at different applied loads. (c) Schematic sketch of a LING device with lateral nanowires. Right: Output voltage of a LING nanogenerator at cyclic loads. *Source:* Xu et al. [49]/Springer Nature.

vertical ZnO nanowire forest grown on a sapphire or GaN substrate that was coated by a thin layer of conductive ZnO film. This bottom ZnO layer serves as an electrode to connect the nanowire to external circuit. The top surface of the vertical nanowire array was connected to a zig-zag shaped Si substrate coated by Pt, Figure 9.16a. This configuration is relatively similar to the AFM tip used in the first discovery by the Wang group, where the zig-zag electrode forced the nanowires to bend in response to external mechanical impacts. The authors demonstrated that ultrasonic waves induced elastic deformation and vibration of the ZnO nanowires and could generate an output voltage of ~1 mV from a 2 mm^2 nanogenerator. A tradeoff in this concept was the relative scrubbing and sliding between the top electrode (Pt) and ZnO nanowires that may lead to wearing issues and make the contact resistance unstable. To overcome this problem, Xu et al. developed alternative-current nanogenerators using a solid contact between the electrode and the top end of a ZnO nanowire array. Two types of configurations were proposed, vertical nanowire arrange integrated nanogenerators (VINGs) and lateral-nanowire-array integrated nanogenerator (LING), both of which provide a high-power density. Figure 9.16b illustrates the concept of VING with a solid top and bottom electrodes. Specifically, ZnO nanowires with diameters of approximately 300 nm, and lengths of 4 μm were grown on an Au coated surface at 100 °C. A layer of PMMA was spin coated onto the ZnO nanowire array to fully cover them and to prevent short-circuiting between the top electrode to the bottom Au surface. Slightly etching back the PMMA layer left the tip of ZnO nanowires exposed, enabling the deposition of a 300 nm Pt film to form a top contact with the nanowires. Figure 9.16b shows the output voltage of a VING nanogenerator at compressive stress applied at 2 Hz. The results demonstrate a significant output voltage of over 50 mV at 6.25 MPa stress in a 4 mm^2 VING device. Integrating multiple VING devices in a series can further enhancing the output power.

Figure 9.16c shows the concept of the LING configuration in which Au and Cr are employed as the Schottky and Ohmic contact at the two ends of lateral ZnO nanowires. The ZnO nanowires were grown parallelly to the substrate using a chemical approach below 100 °C on a flexible substrate. All ZnO nanowires were crystallographically aligned to make sure that the generated piezoelectric potentials were in the same direction to maximize the output. A periodic force was applied to induce cyclic stretching and deformation process in ZnO nanowires. Enhancement in the interconnection of the electrodes and nanowires, and integration a large number of nanowires into a single LING device play an important role in boosting the output voltage and current of a LING nanogenerator. The authors demonstrated that a LING device with 700 rows, and 20 000 nanowires in each row can generate a significant voltage of ~1.2 V and current of ~26 nA at a strain of 0.19% and a straining rate of

2.13%/s. Furthermore, integration of layer-by-layer ZnO nanowire arrays could enable the development of three-dimensional LING nanogenerator that could provide sufficient output voltage and current to power miniaturized electronic and sensing components.

9.4 Applications of Mechanical Sensors Using Wide Bandgap Materials

9.4.1 Structural Health Monitoring

Being recognized as one of the top five high technologies, aerospace still faces several technical challenges to maintain a sustainable growth, reduce transportation cost, and generate cleaner solution [50, 51]. The push for new missions such as space exploration requires new technologies with enhanced energy efficiency, light weight, and improved safety. These needs have promoted research into the development of distributed, miniaturized sensor networks that can provide real-time structural health monitoring of airspace systems. Wide bandgap semiconductors play an important role due to their robustness that can with stand extreme conditions presented in aerospace systems [52].

In recent years, several types of composites have been utilized to develop structural components in aircraft such as fuselage, wing skins, and tail skins. Composite materials offer strong mechanical properties and reduced total weight. However, as these materials are typically subjected to long-term thermal expansion, and corrosion due to humidity conditions, damage assessment to identify potential of crack, delamination, and degradation is imperative. Strain sensors embedded onto the surface of structural components can provide real time monitoring of mechanical deformation.

Another dominant cause in aerospace systems is the failure of landing gear. The main failure mechanism is the fracture of the outer cylinder attachment lug of the landing gear resulted from the rupture of the connections caused by fatigue and/or crack [53, 54]. Furthermore, the deviations in the set point of tire inflation pressure can cause explosion, leading to catastrophic accidents. Real-time and accurate assessment of the forces applied on gear systems as well as tire pressure could help preventing these problematic issues. In this regards, wide bandgap material nanowires with their excellent tolerance to high strain and stress along with excellent sensitivity to mechanical impacts (e.g. via the piezoresistive, piezoelectric, and piezotronics effect) can be utilized to construct force and pressure sensors for gear system monitoring. These sensors can be distributed at different locations of a landing gear to improve the safety of the system.

Enhancing the performance of jet engines of aerospace systems is critically important to reduce the hazardous emission to environments as well as lower transportation costs. Particularly, monitoring of the air to fuel ratio and real-time measurement of the temperature inside combustion chambers can enhance engine efficiency. SiC and GaN nanowires are good candidates for these applications for monitoring the pressure inside engine chambers owing to their stable electrical properties and extremely low oxidation rate. SiC pressure sensor has been demonstrated to function at high temperature up to 800 °C with an output voltage of 40 mV under a high pressure of 200 psi [8].

A similar concept using wide bandgap sensors for structural health monitoring of automobile systems has also been proposed owing to the sustainable growth of wide bandgap semiconductor devices (e.g. high-power converters) [55]. For instance, Kroetz et al. demonstrated 3C-SiC pressure sensors embedded inside a combustion engine. The sensors were developed from a 3C-SiC thin film deposited on a silicon on insulator (SOI) wafer. Experimental studies showed that the sensor can continuously monitor the variation of the chamber pressure in a full engine cycle at high temperatures. The data also demonstrated the relationship between chamber pressure and the amount of the loaded fuel [56]. The data obtained from the pressure sensors can be used as a feedback signal to improve the design of engines.

9.4.2 Advanced Health Care

Wide bandgap nanowires transferred onto a soft substrate such as PDMS show promising opportunities for wearable device for personal health care applications, Figure 9.17.

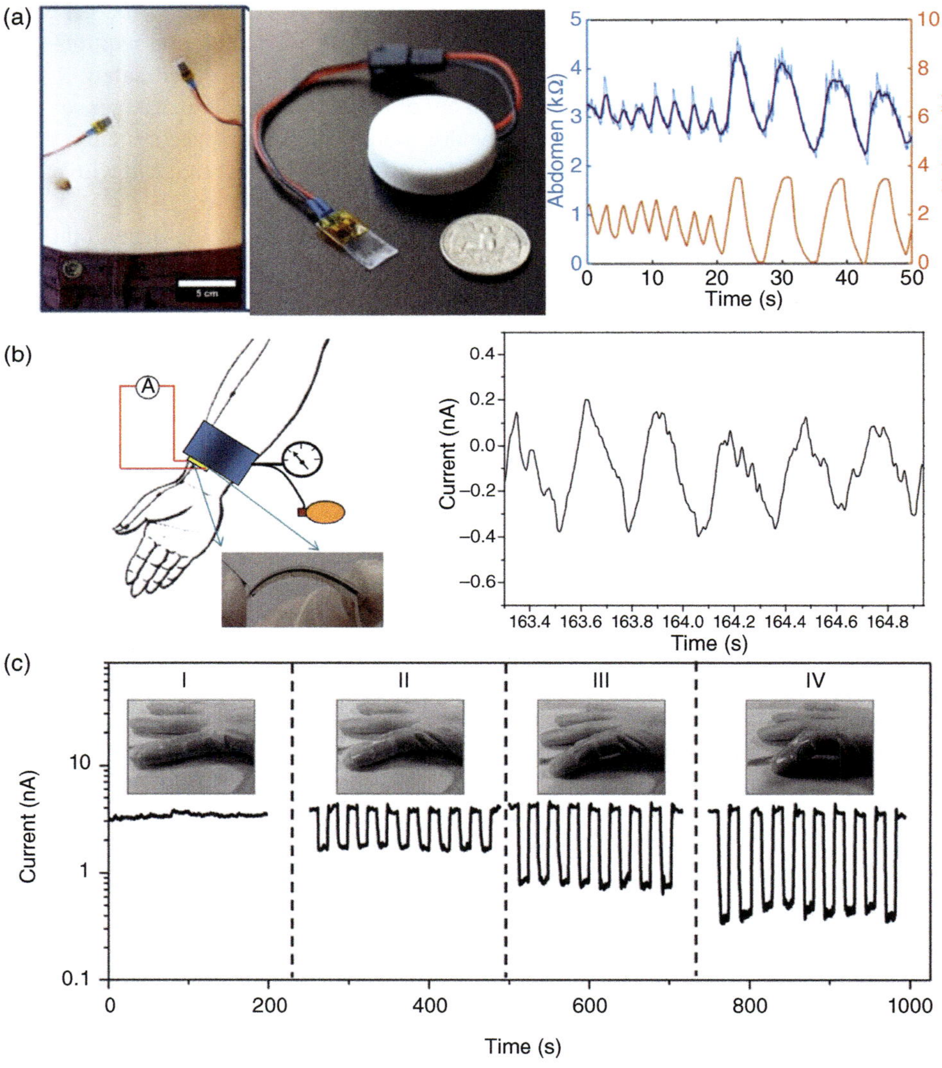

Figure 9.17 Mechanical sensors for advanced health care applications. (a) Strain sensors for respiration measurement. *Source:* Chu et al. [57]/Nature Publishing Group. (b) Piezotronics nanowires for heart rate monitoring. *Source:* Li et al. [58]/John Wiley & Sons, Inc. (c) Tactile sensors for joint angle measurement and gesture monitoring. *Source:* Xiao et al. [29]/John Wiley & Sons, Inc.

The piezoresistive, piezoelectric, and piezotronics effect in wide bandgap nanowires SiC, GaN, and ZnO can be utilized to monitor different biophysiological signals from users [29, 57, 58]. Furthermore, materials with capability of converting mechanical energy into electrical power (e.g. piezoelectric and piezotronics effects) can eliminate the requirement for external battery, which enhances the flexibility, lightweight, and compactness of wearable systems.

Examples of mechanical sensors used for advanced health care include respiration sensors, lung expansion sensors, heartbeat sensors, and skin strain sensors [57]. For respiration measurement, the sensors are required to detect small-scale motions, such as strain and pressure induced from air flow and the movement of ribcage and abdomen. Several wide bandgap nanowire devices have been employed in respiration measurement owing to their excellent sensitivity and stretchability. These devices share relatively similar concept where a network of nanowires is embedded into an elastomer and then mounted onto nostril of a wear mask. During each inhalation and exhalation cycle, the nostril or mask will deflect, causing a mechanical deformation in the sensing element. Measurement of the respiration rate (i.e. the number of respiration pulses in a minute, typically 14–18 ppm for a healthy adult) and the respiration depth (i.e. the peak-to-peak amplitude) can determine abnormality in breathing patterns, providing valuable information for subsequent intervention or diagnosis. Another method to monitor the respiration rate is the measurement of lung expansion using a strain sensor attached on chest and abdomen. The change in the lung volume during inhalation and exhalation cause a significance strain on the skin surface around chest and abdomen. Therefore, highly sensitive nanowire strain sensors attached on these parts of human body can be used to diagnose sleep apnea, lung functions, and possible injuries in lung.

Continuous monitoring of heart rate is of particular interest in diagnosis of cardiac diseases as well as in performance enhancement in sport science [58]. Recent studies suggest a significant research interest from research community in developing wrist-worn smart pads capable of detecting the heart rate from user. Nanowire strain sensors can be integrated into wrist-worn smart pads and convert the small mechanical impact resulted from heartbeat into an output voltage spike. The magnitudes and frequency of the output voltage signal of nanowire sensors can help assessing the behavior of heart functions, offering useful information for medical diagnostics.

Nanowire strain sensors with good flexibility and stretchability can form conformal contact with human tissues to accurately detect skin strain [29]. Specifically, skin close to human joints experiences large strains due to large angle movement. This large strain also generates a significant electrical potential in piezotronics nanowire networks, offering sufficient power for wearable devices to fully function without the need of external power sources (e.g. battery). Information obtained from skin strain and joint angles provide valuable information for designing and assessing prothesis as well as improving physical therapy. For example, the degree of joint angle with joint flexion and extension can be used in the diagnosing progression of muscular diseases.

Other types of wearable mechanical sensors used in health monitoring include blood flow sensors and vocal fold vibration sensors. Blood flow pressure measurement is useful for the diagnosis and treatment for patients with vascular conditions such as diabetes and high blood pressure [59]. Vocal vibration sensors operate by converting acoustic pressure from throat to an electrical output (e.g. resistance change). These sensors are of significant interest to evaluate the performance of vocal fold.

9.4.3 Robotics

Power assist devices (e.g. exoskeleton robots) and humanoid robots have been active research topics over the last two decades. As these types of robots are required to interact with human, they could pose potential safety risk if they lack knowledge of surrounding environments. Tactile sensors, which mimic the sense of human touch, have been proven as a valuable sensing platform that can help robot to understand physical properties of surrounding objects and environments [29, 60]. The use of tactile sensors attached on the fingertip of robot hands can significantly enhance the control of robotic manipulation, Figure 9.18. The two main functions of tactile sensors include (i) the measurement of contact forces in 3D directions (i.e. shears forces and normal pressure) for grasping tasks and (ii) characterization of different surface textures. To enable the integration of tactile sensors into robot finger and to sense gentle touching forces, small thickness, high stretchability, and high sensitivity are imperative properties in tactile sensors. In this regard, wide bandgap semiconductors

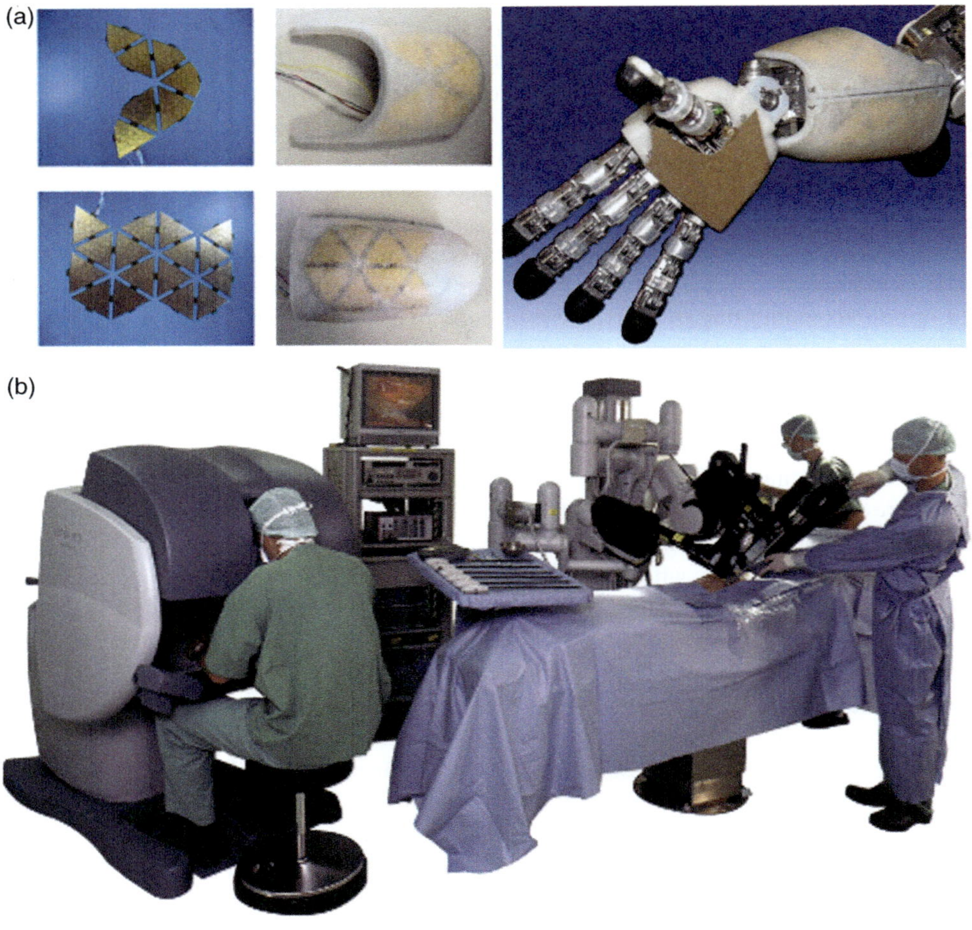

Figure 9.18 Application of mechanical sensors for robotics and haptic control. (a) Sensors are embedded onto the fingertips of robot hand for holding and manipulation. *Source:* [61], Schmitz, et. al., (2011)/IEEE. (b) The da Vinci surgical system where operators can control robot tool by translating their hand, wrist, and finger. *Source:* [62], Tiwana, et. al., (2012)/Elsevier.

nanowires have risen as a promising candidate to construct low cost and high-performance tactile sensors due to their significant strain effect, and capability to be integrated into a complex 3D nanoarchitecture for multi-axis and multi-variable measurement. The use of wide bandgap nanowires has demonstrated to offer robotics the capability to perform multiple tasks such as grasping and holding without dropping or damaging objects.

Tactile sensors can also be applied in medical robots, particularly to enhance the haptic control interface of surgery robotics [61, 63]. Several surgery using robot-assisted systems have been successfully commercialized and employed in clinical applications. Medical robots offer higher precision compared with conventional manual operation and minimize invasive procedures. A well-known example is the da Vinci System that has been Food and Drug Administration (FDA) approved and implemented for several cardiac surgeries and have save many lives [62]. A challenge in medical robots is to work in narrow and complex environments, where the robots need to understand the geometry, and physical properties of tissues as well as the force applied to sutures to precisely perform surgery. In this regard, flexible mechanical sensors can be used to provide haptic feedback to operators to enhance surgeon's performance. Haptic feedbacks, in a broad term definition, are the tactile and force information, which are imperative to form the typical sensations felt with human hand. These haptic feedback signals can enhance the robotic control and visualization of targeted objects or organs, leading to significant improvement in patient outcomes. As the piezoelectric and piezotronics effects in wide bandgap nanowires are a two-way phenomenon, where they can sense mechanical input, and can also generate mechanical motions (e.g. vibrations). These features make piezoelectric and piezotronics materials an attractive choice for the development of haptic devices.

References

1 Phan, H.P., Dao, D.V., Nakamura, K. et al. (2015). *Journal of Microelectromechanical Systems* 24 (6): 1663–1677.

2 Barlian, A.A., Park, W.T., Mallon, J.R. et al. (2009). *Proceedings of the IEEE* 97 (3): 513–552.

3 Fathalilou, M., Soltani, K., Rezazadeh, G., and Cigeroglu, E. (2020). *Measurement* 153: 107428.

4 Giurgiutiu, V., Redmond, J.M., Roach, D.P., and Rackow, K. (2000). *Proceedings of the Society for Photo-Instrumentation Engineers* 3985: 294–305.

5 Jeong, H., Kim, H., Lee, S., and Dornfeld, D. (2006). *CIRP Annals* 55 (1): 325–328.

6 Fraga, M.A., Furlan, H., Pessoa, R.S., and Massi, M. (2014). *Microsystem Technologies* 20 (1): 9–21.

7 Phan, H.-P., Dao, D.V., Tanner, P. et al. (2014). *Applied Physics Letters* 104 (11): 111905.

8 Okojie, R.S., Lukco, D., Nguyen, V., and Savrun, E (2014). *IEEE Electron Device Letters* 36 (2): 174–176.

9 Phan, H.-P., Dinh, T., Kozeki, T. et al. (2016). *IEEE Electron Device Letters* 37 (8): 1029–1302.

10 Phan, H.-P., Dinh, T., Kozeki, T. et al. (2016). *Applied Physics Letters* 109: 123502.

11 Shao, R., Zheng, K., Zhang, Y. et al. (2012). *Applied Physics Letters* 101 (23): 233109.

12 Bi, J., Bi, J., Wei, G., and Wang, L. (2013). *Journal of Materials Chemistry C* 1 (30): 4514–4517.

13 Gao, F., Zheng, J., Wang, M. et al. (2011). *Chemical Communications* 47 (43): 11993–11995.

14 Nakamura, K., Toriyama, T., and Sugiyama, S. (2011). *Japanese Journal of Applied Physics* 50 (6S): 06GE05.

15 Rowe, A.C. (2008). *Nature Nanotechnology* 3 (6): 311–312.

16 Wang, X. (2012). Synthesis fabrication, characterization and application of ultrananocrystalline diamond micro- and nanostructures. PhD thesis, University of Puerto Rico, San Juan, PR, USA.

17 Zhou, A.F., Wang, X., Pacheco, E., and Feng, P.X. (2021). *Materials* 14 (3): 661.

18 Sappati, K.K. and Bhadra, S. (2018). *Sensors* 18 (11): 3605.

19 Ballas, R.G. (2007). *Piezoelectric Multilayer Beam Bending Actuators: Static and Dynamic Behavior and Aspects of Sensor Integration*. Springer Science & Business Media.

20 Jenkins, K., Nguyen, V., Zhu, R., and Yang, R. (2015). *Sensors* 15 (9): 22914–22940.

21 Wang, Z.L. and Song, J. (2006). *Science* 312 (5771): 242–246.

22 Zhao, M.H., Wang, Z.L., and Mao, S.X. (2004). *Nano Letters* 4: 587.

23 Wang, Z.L. (2010). *Journal of Physical Chemistry Letters* 1 (9): 1388–1393.

24 Minary-Jolandan, M., Bernal, R.A., Kuljanishvili, I. et al. (2012). *Nano Letters* 12 (2): 970–976.

25 Gogneau, N., Chrétien, P., Galopin, E. et al. (2014). *Physica Status Solidi Rapid Research Letters* 8 (5): 414–419.

26 Chen, Y.S., Hsieh, G.W., Chen, S.P. et al. (2015). *ACS Applied Materials & Interfaces* 7 (1): 45–50.

27 Zhou, J., Gu, Y., Fei, P. et al. (2008). *Nano Letters* 8: 3035–3040.

28 Jenkins, K. and Yang, R. (2017). *Semiconductor Science and Technology* 32 (7): 074004.

29 Xiao, X., Yuan, L., Zhong, J. et al. (2011). *Advanced Materials* 23 (45): 5440–5444.

30 Lin, L. and Yun, W. (1998). MEMS pressure sensors for aerospace applications. In: *1998 IEEE Aerospace Conference Proceedings (Cat. No. 98TH8339), 28 March 1998, Snowmass at Aspen, CO, USA*, vol. 1, 429–436. IEEE.

31 Song, P., Ma, Z., Ma, J. et al. (2020). *Micromachines* 11 (1): 56.

32 Nallathambi, A., Shanmuganantham, T., and Sindhanaiselvi, D. (2018). *Materials Today: Proceedings* 5 (1): 1897–1903.

33 Chapin, C.A., Miller, R.A., Dowling, K.M. et al. (2017). *Sensors and Actuators A: Physical* 263: 216–223.

34 Suma, M.N., Gaddam, V., Prasad, M.V.N. et al. (2020). *SN Applied Sciences* 2 (10): 1–8.

35 Phan, H.-P., Dowling, K.M., Nguyen, T.-K. et al. (2018). *Materials & Design* 15: 16–21.

36 Wan, Y., Wang, Y., and Guo, C.F. (2017). *Materials Today Physics* 1: 61–73.

37 Yang, T., Xie, D., Li, Z., and Zhu, H. (2017). *Materials Science & Engineering R: Reports* 115: 1–37.

38 Ha, M., Lim, S., Park, J. et al. (2015). *Advanced Functional Materials* 25 (19): 2841–2849.

39 Wu, W., Wen, X., and Wang, Z.L. (2013). *Science* 340 (6135): 952–957.

40 El Kacimi, A., Pauliac-Vaujour, E., and Eymery, J. (2018). *ACS Applied Materials & Interfaces* 10 (5): 4794–4800.

41 Eddy, D.S. and Sparks, D.R. (1998). *Proceedings of the IEEE* 86 (8): 1747–1755.

42 Currano, L.J., Bauman, S., Churaman, W. et al. (2008). *Sensors and Actuators A: Physical* 147 (2): 490–497.

43 Kim, H.C., Song, S., and Kim, J. (2016). *Sensors* 16 (9): 1499.

44 Song, S., Kim, H.C., Kim, J.W. et al. (2017). *Royal Society Open Science* 4 (9): 170557.

45 Wang, Z., Pan, X., He, Y. et al. (2015). *Advances in Materials Science and Engineering* 2015: 165631.

46 Zhou, Z., Tang, H., and Sodano, H.A. (2014). *Advanced Materials* 26 (45): 7547–7554.

47 Alexe, M., Senz, S., Schubert, M.A. et al. (2008). *Advanced Materials* 20 (21): 4021–4026.

48 Wang, X., Song, J., Liu, J., and Wang, Z.L. (2007). *Science* 316 (5821): 102–105.

49 Xu, S., Qin, Y., Xu, C. et al. (2010). *Nature Nanotechnology* 5 (5): 366–373.

50 Park, G., Farrar, C.R., di Scalea, F.L., and Coccia, S. (2006). *Smart Materials and Structures* 15 (6): 1673.

51 Giurgiutiu, V., Zagrai, A., Bao, J. et al. (2003). *International Journal of COMADEM* 6 (1): 3.

52 Senesky, D.G., Jamshidi, B., Cheng, K.B., and Pisano, A.P. (2009). *IEEE Sensors Journal* 9 (11): 1472–1478.

53 Azevedo, C.R.F., Hippert, E. Jr., Spera, G., and Geradi, P. (2002). *Engineering Failure Analysis* 9: 1–15.

54 Boller, C. (2001). *Smart Materials and Structures* 10: 432–440.

55 Eftekhari Shahroudi, K. (2008). Robust design evolution and impact of in-cylinder pressure sensors to combustion control and optimization: a systems and strategy perspective. Doctoral dissertation, Massachusetts Institute of Technology.

56 Kroetz, G., Wondrak, W., Obermeier, E., and Cavalloni, C. (1998). Silicon carbide on silicon-an ideal material combination for harsh environment sensor applications. In: *IEEE International Symposium on Industrial Electronics. Proceedings. ISIE'98 (Cat. No. 98TH8357), 7–10 July 1998, Pretoria, South Africa*, vol. 2, 732–736. IEEE.

57 Chu, M., Nguyen, T., Pandey, V. et al. (2019). *npj Digital Medicine* 2 (1): 1–9.

58 Li, Z. and Wang, Z.L. (2011). *Advanced Materials* 23 (1): 84–89.

59 Li, L., Zheng, J., Chen, J. et al. (2020). *Advanced Materials Interfaces* 7 (17): 2000743.

60 Dahiya, R. (2019). *Proceedings of the IEEE* 107 (2): 247–252.

61 Schmitz, A., Maiolino, P., Maggiali, M. et al. (2011). *IEEE Transactions on Robotics* 27 (3): 389–400.

62 Tiwana, M.I., Redmond, S.J., and Lovell, N.H. (2012). *Sensors and Actuators A: Physical* 179: 17–31.

63 Dargahi, J. and Najarian, S. (2005). *Industrial Robot: An International Journal*.

10

Gas Sensors

10.1 Introduction

Detection of toxic gases is of interest for several fields, from mining, energy, to security [1]. The emission of hazardous gas from industrial systems over a long period of time can lead to negative environmental impacts that could subsequently affect the quality of life [2, 3]. The implementation of a distributed gas sensing devices can assess air quality and help predicting the possibility of environmental pollution and catastrophic failures.

The capability to adsorb and react to different types of gas in semiconductors provides an engineering route to detect and quantify the amount of gas emitted from industrial processes to environments [4, 5]. This property is further enhanced using nanowire architectures with large surface areas to boost the adsorption capacity as well as sensitivity of gas sensors [6–8]. Wide bandgap semiconductor nanowires such as metal oxide, SiC, group-III nitrides, and diamond have been employed for the development of numerous gas sensors owing to their significant surface to volume ratio, excellent electrical conductance properties, and long-term stability [9]. Remarkable progress in the synthesis and micro-/nanofabrication technologies in these semiconductor nanowires shows promising pathway toward commercialization and practical use of low-cost, large-network density gas sensors.

This chapter provides an overview on gas sensors using wide bandgap nanowires, covering functioning principles, important parameters such as selectivity and time response, and practical applications.

10.2 Principle of Gas Sensing

The fundamental working principle of gas sensors using semiconductor nanowires is based on the change in the electrical conductance of nanowires upon exposure to various organic and inorganic gases. Compared with bulk materials, the electron and hole transportation within nanowires is significantly influenced by their surface state due to the low dimension.

10.2.1 Transconductance Sensing Mechanism

Adsorption of gas from either natural physical diffusion or chemical bonding (e.g. with the help of surface functionalization) leads to a dramatic change in surface states that alters

Wide Bandgap Nanowires: Synthesis, Properties, and Applications, First Edition. Tuan Anh Pham, Toan Dinh, Nam-Trung Nguyen, and Hoang-Phuong Phan.

the electrical conductance of nanowires [10, 11]. The conductance of a nanowire is given by:

$$G = \frac{ne\mu\pi D^2}{4L} \tag{10.1}$$

where n is the carrier concentration, q is the unit charge, μ is the electron mobility, and D and L are the diameter and the length of the nanowire, respectively. The adsorption of gas into nanowires due to either reduction or oxidation reaction leads to the formation of trapped charges on the surface, generating a small electric field. This electric field alters the concentration of free charges in nanowires (Δn), leading to a change in their conductance:

$$\Delta G = \frac{\Delta ne\mu\pi D^2}{4L} \tag{10.2}$$

Assume that the mobility of charge carriers is not affected by gas adsorption; the change in electrical conductance of nanowires is proportional to the change in the charge carrier concentration:

$$\frac{\Delta G}{G} = \frac{\Delta n}{n_0} \tag{10.3}$$

Figure 10.1a illustrates the sensing mechanism of a CO gas sensor using SnO_2 nanowires. Initially, due to the presence of O_2 in ambient condition, the vacancies on the surface of SnO_2 nanowires are repopulated, creating an ionized surface oxygen in the form of O^-. When the SnO_2 nanowires are subjected to CO gas, a reduction reaction takes place, releasing CO_2 to ambient and generating free electrons in the nanowires: $CO + O^- = CO_2 + e^-$. The increase in free electrons in SnO_2 nanowires results in an enhancement in their electrical conductivity, Figure 10.1b. As a consequence, measurement of SnO_2 electrical conductivity can quantify the concentration of CO gas.

10.2.2 Field Effect Transistor-Based Gas Sensors

Due to its low dimension, nanowires provide an ideal platform for field effect transistors (FETs), with enhanced rectification ratio and reduced electric field. The FET configuration has also been deployed in gas sensors where a back gate is applied to amplify sensitivity [12, 13]. Figure 10.2a,b shows the concept of a SnO_2 FET sensors to detect CO gas [14]. SnO_2 nanowires were synthesized by fabricating β-Sn nanowire using an electrochemical method, followed by thermal annealing in air to turn Sn into an oxide form (i.e. SnO_2). The nanowires were then placed onto a SiO_2/Si substrate, where a 300 nm-thick SiO_2 layer functions as the dielectric layer, and p-doped Si functions as the gate electrode. The oxidation and reduction reactions between the surface of SnO_2 and CO gas species can be tuned by modulating electron concentration in the nanowires using a bias voltage at the Si gate. Figure 10.2c plots the output current of SnO_2 nanowire FETs under a gas flow mixture of 100 ml/min N_2, 10 ml/min O_2, and 5 ml/min CO. The data suggested that the current change in the FET depends on the gate bias, where a positive bias enhances the sensitivity while a negative bias reduces the response of the gas sensors. Briefly, electrons are pushed out of the nanowire into the source and drain under a negative gate voltage. This reduces the number of electrons available for oxygen surface chemistry, thus lowering the oxidation rate and diminishing the

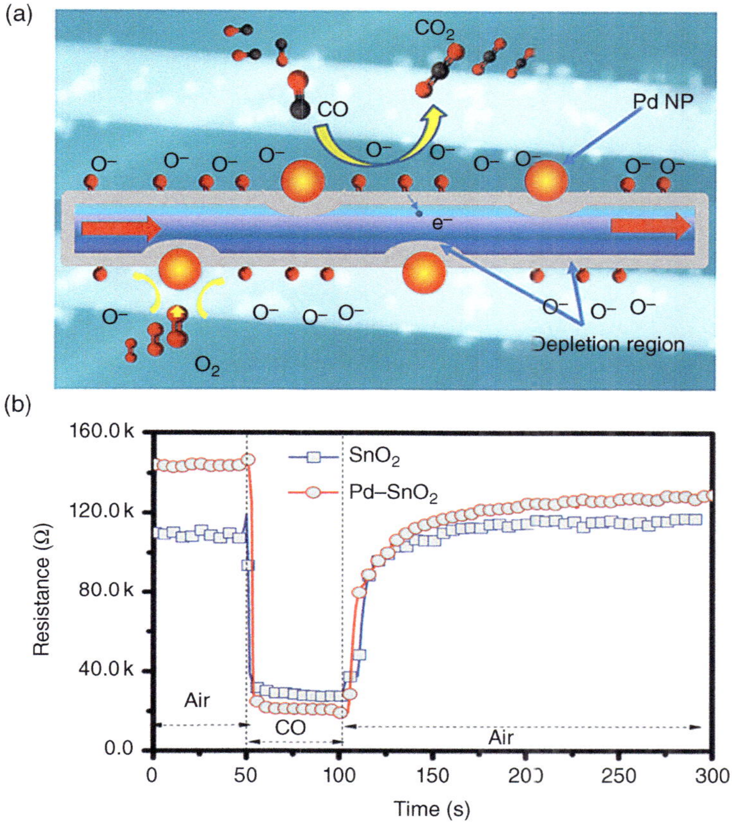

Figure 10.1 CO sensor using SnO_2 nanowire. (a) A schematic sketch of CO adsorption mechanism; (b) an increase in electrical conductance upon CO adsorption.

sensitivity. On the other hand, a positive bias attracts more electrons toward the channel that directly increases the rate of oxidation and resulting in a higher sensitivity. The enhancement in the sensitivity saturated around 4 V, which is attributed to the saturation of electron density and the full adsorption of vacancies.

10.2.3 Metal–Semiconductor Schottky Contact-Based Gas Sensors

Due to a small OFF-state current, the Schottky diode formed at the interface of metals and semiconductors provides a highly sensitive platform for gas sensing [15, 16]. The metal layer in a Schottky diode functions as the electrical contact and at the same time as a catalyst to modulate the conductance characteristic under gas adsorption. Figure 10.3a,b illustrates a typical concept of Schottky diode gas sensors, developed from MoO_3 nanoplatelets (with a bandgap of 2.75–2.95 eV) thermally deposited on a 6H-SiC substrate [17]. A bi-layer of metals (100 nm Pt/400 nm Ti) was deposited on the backside of 6H-SiC and annealed at 500 °C in Ar to form the Ohmic contact of the sensors. The Schottky junction was formed by sputtering a thin film of 20 nm Pt on MoO_3. Experimental results showed a significant increase in the reverse current of the Pt/MoO_3 Schottky junction when exposed to 1% H_2 gas.

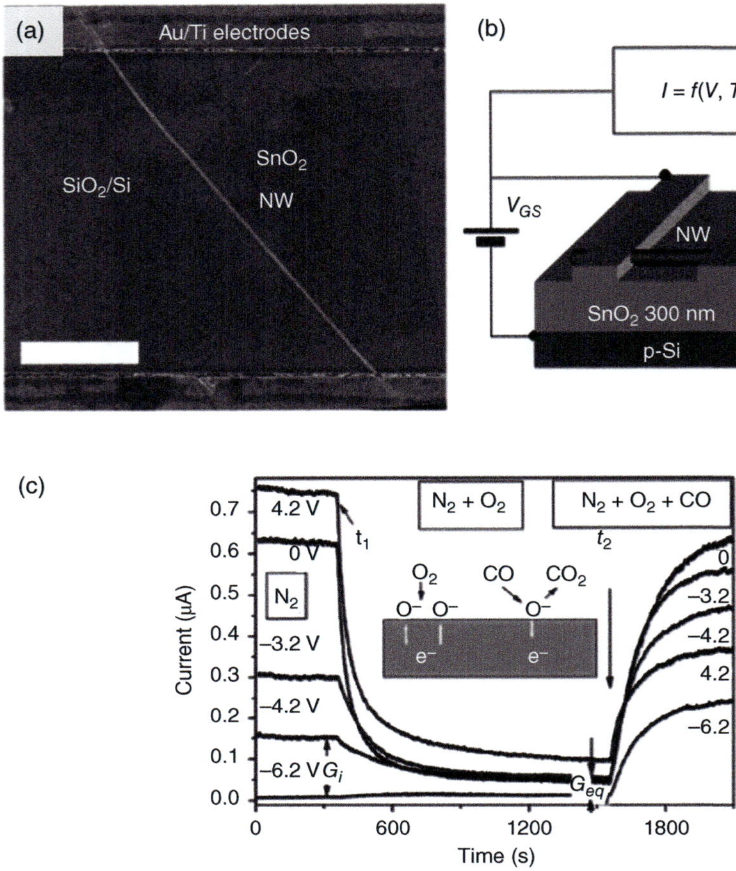

Figure 10.2 Gas sensor using field effect transistor nanowires. (a) SEM images of an SnO_2 nanowire-based FET for CO sensing; (b) sketch of experimental setup; (c) sensor response to CO gas. *Source:* [14], Zhang, et. al., (2004)/American Chemical Society.

The device operated in a wide temperature range from 22 to 250 °C, with the most stable performance at 180 °C. Figure 10.3c illustrates the principle of the Schottky gas sensors, in which the diffusion of the H_2 gas induces dipole-like charges at the Pt/MoO_3 interface. These charges reduce the barrier height, resulting in an increase in the current and a shift in the threshold voltage in the reverse regime.

10.2.4 Integration of Nanowires with Microheaters

Typically, adsorption of gases on nanowire surfaces requires a sufficient activation energy to trigger the chemical reaction (e.g. reduction or oxidation). Thermal energy is widely employed to facilitate surface reactions that lead to a higher oxidation/reduction rate as well as lower the decay time in gas sensors. To provide a sufficient thermal energy, nanowire gas sensors are typically integrated with a built-in microheater that can increase the surface temperature [18, 19]. Due to its stability and good electrical conductivity, Pt thin films are widely used to develop these microheater. Applying a voltage to the heater (i.e. with a two-terminal resistance configuration) increases device temperature through the Joule heating

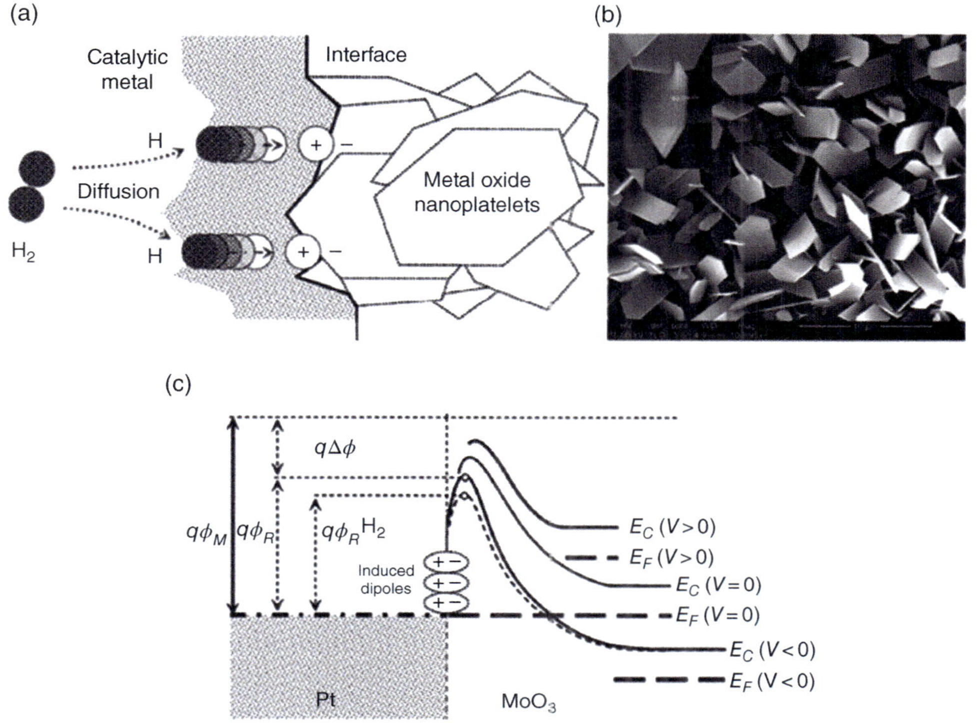

Figure 10.3 Schottky diode gas sensor. (a) Schematic sketch of a H_2 sensor functioning based on the Schottky junction between Pt and MoO_3 nanoplatelets; (b) an SEM image of MoO_3 nanoplatelets; (c) schematic sketch of Schottky barrier change upon H_2 gas adsorption. *Source:* (a, c) Yu et al. [17]/ American Institute of Physics, (b) [17], Yu, et. al., (2009)/A P Publishing LLC.

effect. The temperature of the heater can be monitored using an integrated thermocouple or based on the calibration data between applied powers and temperatures. In several cases, to reduce the power consumption, microheaters are released from substrates (i.e. suspended configurations) to minimize the heat loss due to thermal conductance.

Figure 10.4a shows a conductive type ZnO nanowire sensor developed on a free-standing ITO heater. A 1 μm-thick indium tin oxide (ITO) film was sputtered onto a SiO_2/Si substrate and then patterned into a microheater configuration [20]. ZnO nanowires were then grown on the surface of ITO heater using a hydrothermal method. Figure 10.4b shows a high-resolution scanning electron microscopy (SEM) image of the ZnO nanowires on ITO sensors, in which the ZnO nanowires were randomly orientated and distributed on both the conducting finger regions and the spacing regions (i.e. SiO_2). The temperature of the heater was monitored using an infrared (IR) camera, showing an increase in the temperature from 150 to 350 °C when raising the applied voltage from 6.73 to 11.55 V, Figure 10.4c. The ZnO nanowire sensors were then subjected to several types of gases such as C_2H_5OH, HCHO, H_2S, and NO. Experimental results demonstrated that the sensors obtained higher sensitivity at 300 °C for C_2H_5OH and HCHO, while its response to H_2S and NO gas showed a peak at 250 and 200 °C, respectively, Figure 10.4d. The variation in response peaks between different gases suggests an interesting approach to enhance sensing selectivity by adjusting the operating temperature of devices.

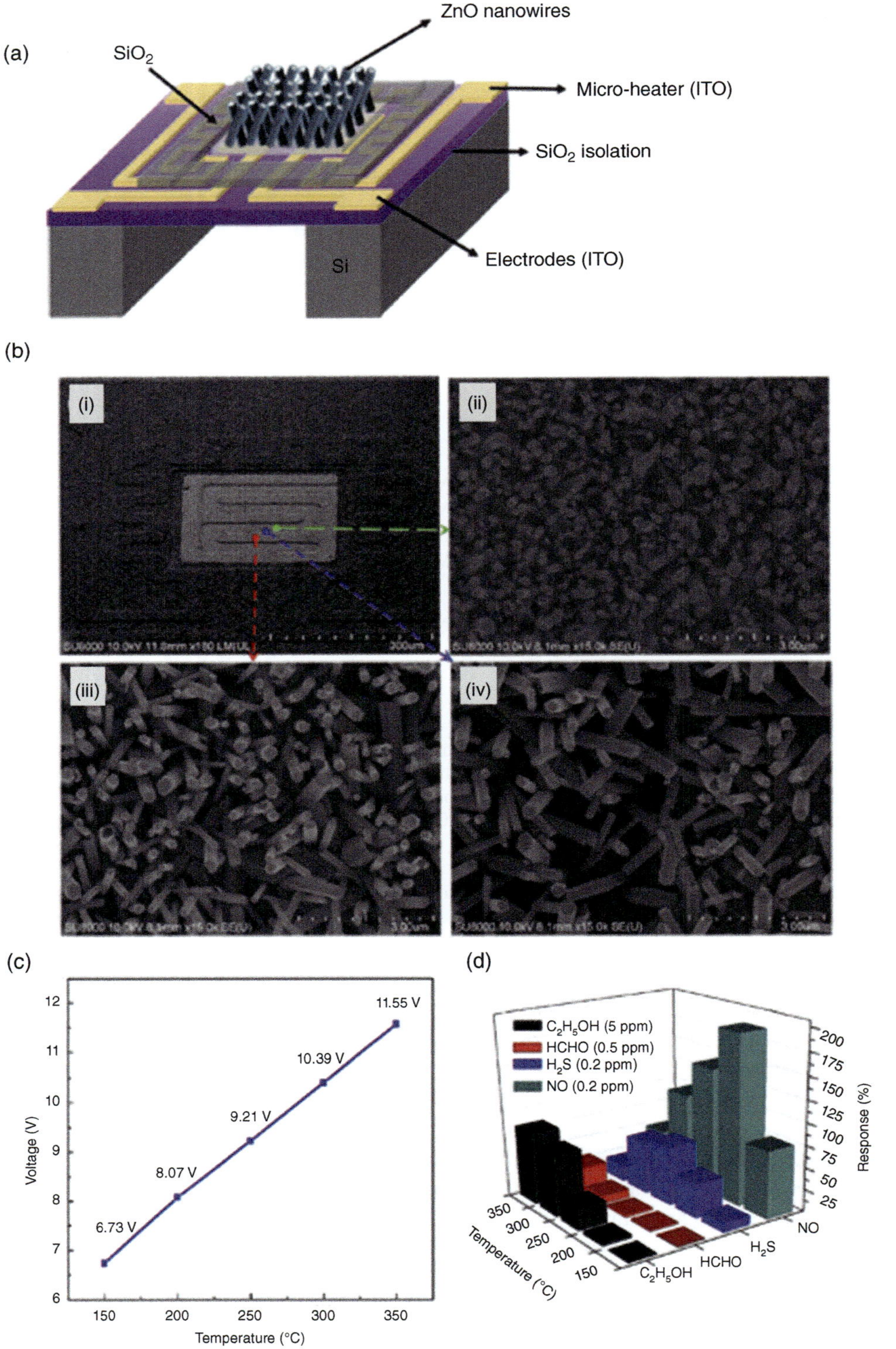

Figure 10.4 Implementation of integrated heater to enhance the performance of gas sensors. (a) The configuration of a gas sensors using ITO as heater and ZnO nanowires as the sensing element; (b) SEM images showing ZnO nanowires were grown at different location of the interdigitated electrodes; (c) the temperature of the ITO heater monitored using an IR thermal camera; (d) response of the sensor to different gases and at different temperatures. Source (a, c, d) Hsueh et al. [20]/with permission of Elsevier, (b) [20]. Hsueh, et. al., (2020)/Elsevier.

10.3 Standard Physical Parameters for Gas Sensors

10.3.1 Sensitivity

Sensitivity is defined by the ratio between resistance change (i.e. conductance change) and the gas concentration [6]. This parameter is critically important for applications where a tiny variation in gas concentration could bring significant impacts to environments, safety, and/or health issue.

An effective approach to enhance sensitivity is the use of high density of nanowire networks distributed on a substrate either planarly or in a forest-shaped architecture. A large number of nanowires enable large surface area that can simultaneously adsorb the targeted gas. Another method is engineering of the nanowire morphology to create more reactive sites. Specifically, the grain size in nanowires can be reduced to allow gas adsorption at each grain boundary, thereby markedly modulating the nanowire conductance. Note that, for poly- or nanocrystalline nanowires, the electrical conductance is mainly dominated by the physical parameters at grain boundaries. For instance, a thicker boundary typically leads to a higher potential barrier between two adjacent grains, impeding the current flow within the nanowires. Gas adsorption at the boundary leads to an energy deformation, in some case lowering the potential barrier that results in an increase in conductivity.

Inducing impurities such as noble metals into nanowires has also been experimentally demonstrated to boost the gas sensitivity [21, 22]. However, the physical chemistry process that leads to enhancement in sensitivity due to dopants is not fully understood. Some hypotheses suggested that spill-over process and electronic mechanism could be the main reasons behind the enhancement of gas sensors with the introduction of impurities.

10.3.2 Selectivity

Perhaps, selectivity is one of the most challenging research questions in the development of gas sensors [23] that is due to that fact that nanowires can actively react to a large number of different gases, which could cause errors in reading the target gas species. Different techniques have been proposed to improve the selectivity in gas sensors. One of the bottom-up approaches is to engineer the chemical properties in materials during the synthesis process to enable selective chemical reaction to targeted gases. These materials are required to exhibit negligible response to other type of gases, to avoid cross sensitivity. Another common method employs top-down methods, where an integrated heater is employed to raise the temperature of the sensing element so that it only adsorbs the targeted gas at a certain temperature.

Implementation of an array of different types of nanowires can further enhance selectivity. Briefly, each nanowire array reacts with a certain gas at a different rate. Based on the responsive matrix from the sensors array, it is possible to quantify the concentration of the targeted gas from non-specific gas species. For instance, Ng. et al. developed a SnO_2 gas sensor integrated into a single complementary metal-oxide-semiconductor (CMOS) chip, which can selectively detect different types of gases [24]. Each unit of the sensor contains a 4×4 array of SnO_2 sensing elements which were decorated with different metals (e.g. Au, Ag, Pt) and various dopants (e.g. P, B, H). A serpentine-shaped Pt heater was formed on each sensing element to allow the device to operate at a high temperature of 300 °C.

The sensor array was then subjected to several gases such as CO, C_2H_5OH, and C_3H_8 with concentrations varying from 20 to 200 ppm. Over 800 data samples were collected, among which 10% of data was used for computational training, and the remaining 90% was employed to validate the accuracy of the sensor. A two-dimensional rank order gas recognition scheme was developed which exhibits an excellent detection rate of approximately 95%.

10.3.3 Response Time

Response time represents the reaction rate of a sensor to a variation in gas concentration. Response time includes the raising time (when the gas concentration starts to change) and the recovery time (when the targeted gas is removed) [25, 26]. Briefly, response time is the time needed for a sensor to reach 90% of the output signal upon gas exposure. On the other hand, the recovery time indicates the time required for the sensor to return to 90% of its baseline signal when the concentration of gas returns to nearly zero. In several cases, these parameters are critically important to timely detect the gases of interest (e.g. hazardous, toxic, or explosive gases).

The time response of a sensor depends on the diffusivity of gas into the material and the reaction rate between the constructing compound and the targeted gas. Typically, the response time of thin films or bulk materials is in a range of several minutes. This parameter can be significantly reduced using nanoarchitectures such as nanowires, ribbons, or mesoporous structures, due to their large surface area that allows simultaneous reaction at different sites of the sensors that rapidly reflects in the sensor output (i.e. electrical resistance). The use of integrated heaters to raise the reaction rate also plays an important role in reducing the raising time of gas sensors. Several studies reported relatively fast response time in wide bandgap semiconductors, in an order of several milliseconds.

Figure 10.5 presents the time response of ZnO gas sensors to H_2 at 300 °C [27]. Two types of ZnO nanowires were developed, one without intentional dopant, and the other one co-doped with 8 mol% Co content (i.e. Co:ZnO nanowires). Both sensors showed an increase in the current subjected to the change in H_2 concentration from 1000 to 3000 ppm. The pristine ZnO nanowires exhibited a rising time of over 240 seconds, while that of the Co:ZnO nanowires had a faster raising time of less than 80 seconds. A similar trend was also observed in the recovery time, where the Co:ZnO nanowire sensors can return to 90% of its baseline within 40 seconds compared with 120 seconds in ZnO nanowires. This observation was attributed to a larger adsorption sites due to oxygen vacancy defects resulted from Co dopants.

10.4 Materials for Different Types of Gases

Common gas sensors are used for detecting oxygen, carbon dioxide, organic gases (methane, ethylene, toluene), and hydrogen. Detection of these gases is critically important to ensure system operation, avoid catastrophic failures, and minimize environmental impacts. For instance, organic substances such as methane and ethylene are highly flammable materials; therefore, timely detection of these gases can avoid potential risk such as fire and asphyxiation hazard. Each type of gas requires a different semiconductor as the sensing element for optimal selectivity, sensitivity, and timely detection.

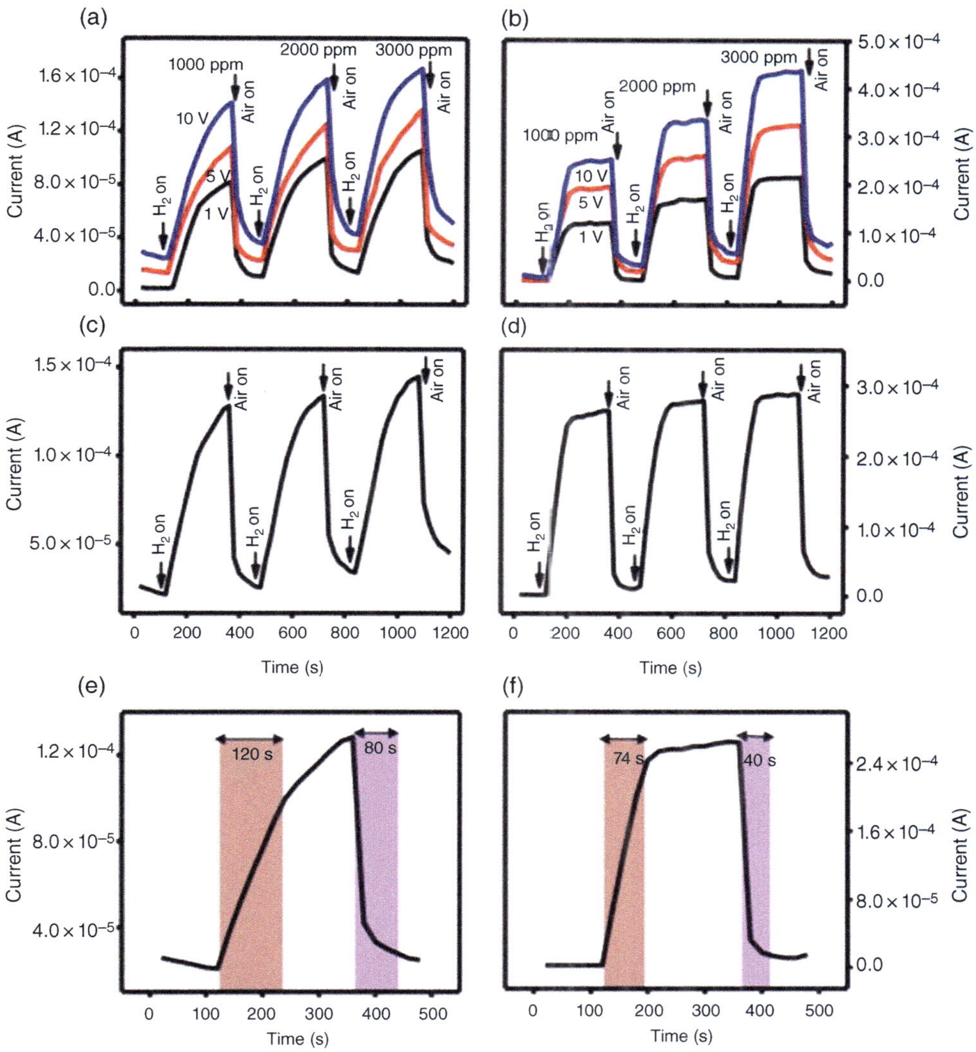

Figure 10.5 Transient gas response of (a, c, e) ZnO nanowire sensors; and (b, d, f) ZnO nanowires co-doped with 8 mol% CO content. *Source:* (a, b) Sett and Basak [27]/with permission of Elsevier.

10.4.1 Oxygen Sensors

An oxygen sensor is commonly referred to as oximetry, a device that can measure the concentration of oxygen in ambient to ensure respiration sufficiency. In this regard, wide bandgap materials such as TiO_2, Ga_2O_3, and $SiTiO_3$ have been proven to be sensitive to O_2 adsorption [28, 29]. The electrical conductivity in oxide semiconductors is dominated by the electron hoping between crystal defects formed by the oxygen vacancies during the growth process. This means a higher density of oxygen vacancies typically leads to a better electrical conductivity in oxide semiconductors. Upon oxygen adsorption, the vacancies are gradually filled up, which impedes electron hoping paths, thereby reducing the electrical conductivity of the semiconductor. Oxide nanowires and nanorods with much larger surface areas (corresponding to higher surface vacancies) generally offers higher sensitivity compared with thin film-based sensors.

Minaee et al. compared the oxygen sensing performance of ZnO with different nanoarchitectures of nanowires, nanorods, and nanoflowers, Figure 10.6a [30]. The authors employed a chemical vapor deposition (CVD) method with Zn shots as source materials, Ar and oxygen as the carrier and active gases, and alumina and silicon wafers as the substrates. Oxygen sensing was performed at a constant flow rate with varied temperature ranging from 25 °C up to 150 °C, Figure 10.6b,c. The ZnO sensors show the highest sensitivity at 50 °C. When varying the flow rate (from 300 to 900 cc/min) and keeping the operating

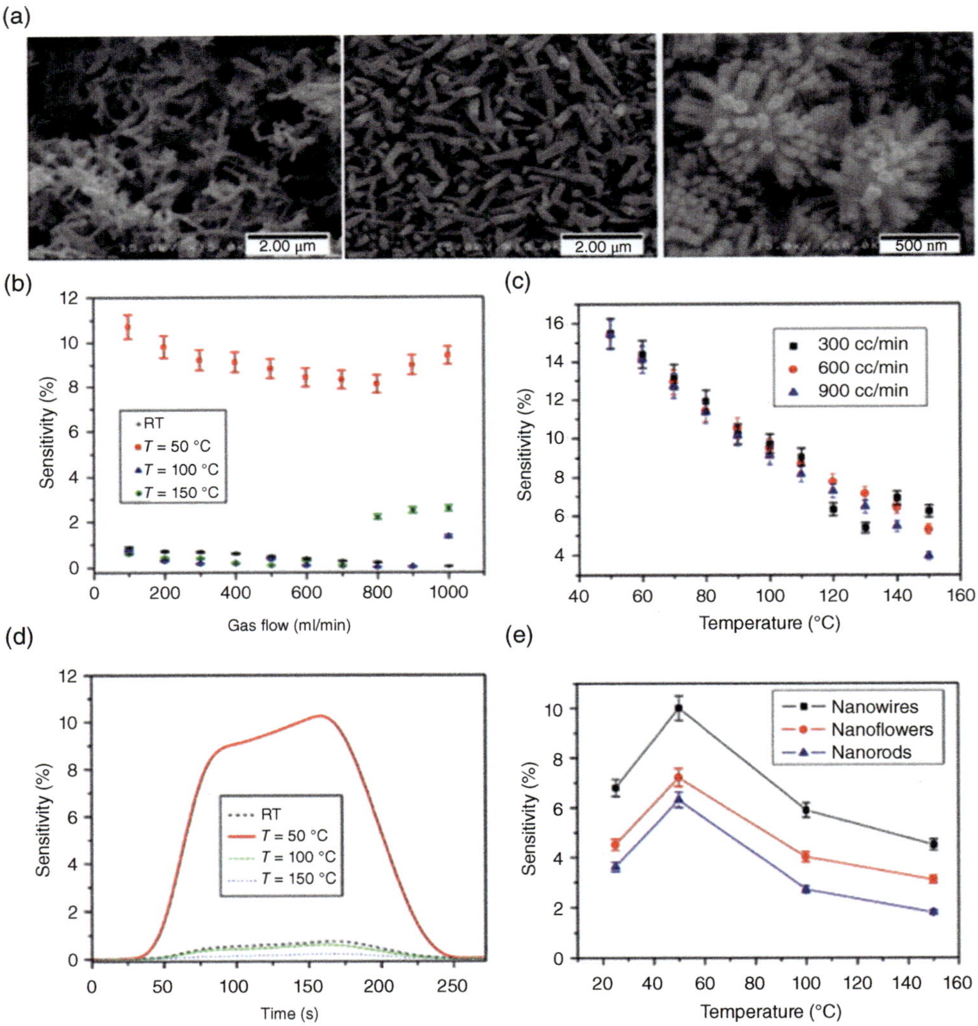

Figure 10.6 An example of oxygen sensors using ZnO nanoarchitectures. (a) From left to right: SEM images of ZnO nanowires, nanorods, and nanoflowers; (b) sensitivity at different temperatures. (c) Sensitivity at different gas flow rate. (d) Time response. (e) A comparison between different morphologies. *Source:* (a) [30], Minaee, et. al., (2013)/Elsevier, (b–e) Minaee et al. [30]/with permission of Elsevier.

temperature constant, the sensitivity exhibits a flow rate independent property, Figure 10.6c. The sensors have a relatively fast raising time and recovery time of 35 and 50 seconds at 50 °C, respectively, Figure 10.6d. A comparison of the sensing response between different morphologies shows that ZnO nanowires grown on alumina possess a higher sensitivity than other types of architectures (nanorods and nanoflowers), Figure 10.6e. This indicates that the performance of oxide nanowire-based oximetry devices can be engineered by changing morphologies and operating temperatures.

10.4.2 Carbon Dioxide

Measurement of carbon dioxide (CO_2) is imperative in several industrial process controls, environmental monitoring systems (e.g. with the concern of the global warming issue), and health care applications [31, 32]. For instance, real-time and continuous monitoring of carbon dioxide from exhaled breadth provides informative indicators for lung and stomach related diseases [33].

Figure 10.7a presents a CO_2 gas sensor using ZnO nanorods deposited on an AlGaN/GaN high electron mobility transistor (HEMT). The AlGaN/GaN (30 nm $Al_{0.3}GaN$ and 2.5 μm GaN) heterostructure was grown using a metal organic chemical vapor deposition method (MOCVD) on a sapphire substrate. Metal deposition and etching processes were applied to form AlGaN/GaN HEMT devices. An array of ZnO nanorods was selectively grown onto the gate area using a combination of (i) photolithography, (ii) hydrothermal ZnO nanorod growth, and (iii) a lift-off step to remove unnecessary nanorods. The diameter and length of

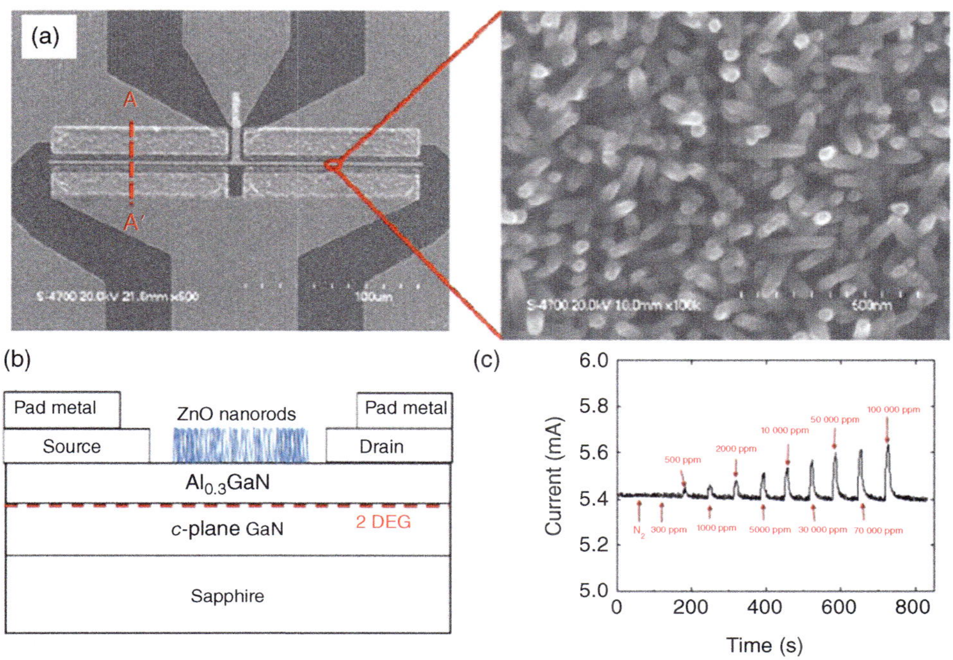

Figure 10.7 CO_2 gas sensors using ZnO nanorods selectively grown on the gate are of an AlGaN/GaN HEMT device. (a) SEM image of devices. (b) Schematic sketch of the gas sensors. (c) Current change under applying CO_2 gas. *Source:* [34], Baik, et. al., (2020)/MYU K.K./CC BY 4.0.

ZnO nanowires are approximately 50 nm and 1 µm, respectively. The ZnO nanowire array functions as a gas-sensitive gate layer, which modulates the electrical properties of the two dimensional electron gas (2DEG) layer at the AlGaN/GaN interface, Figure 10.7b.

To characterize the performance of the sensors, the drain–source bias was fixed at 1 V, and the operating temperature was varied from room temperature to 300 °C. Experimental data shows that the sensors exhibit an insignificant change in the current when the operating temperature was set below 150 °C. However, at 300 °C, a marked drain–source current was observed with a flow rate of 500 ppm, Figure 10.7c. This indicates that to effectively modulate the current in the AlGaN/GaN device, high temperature is required to sufficiently adsorb CO_2 and thus creating an electric field that can influence the conducting channel. The current returned to its original value when the gas flow was stopped. Further increasing gas flow rate shows an increase in the response current. These results demonstrate the interesting possibility in integrating multiple layers of wide bandgap materials for highly sensitive CO_2 sensing applications.

10.4.3 Organic Gases

Detection of organic gases such as methane, ethylene, and toluene is of particular interest to ensure the safety (e.g. in mining industry), prevent potential of fire (e.g. due to their highly flammable property), and minimize health issues (e.g. skin contact and exhale of toluene) [35–37]. In mining industry, detection of CH_4 is important as this colorless and odorless gas can cause explosion at a concentration ranging from 5% to 15%. Exposure to high concentration CH_4 can lead to health issues to underground miners. Tin oxide (SnO_2) with an energy bandgap of 3.6 eV is one of the most common choice for methane sensing due to its high specific surface area, nontoxic, and low cost [38]. The electron transfer in SnO_2 can be markedly improved through the introduction of nobble metals such as Pt, Pd, Co, Ni, and Cr. For instance, Das et al. developed undoped SnO_2 nanoparticles using sonication-assisted simultaneous precipitation [39]. The undoped SnO_2 sensors exhibited good response at an operating temperature of 350 °C to CH_4 gas with concentration ranging from 10 to 1000 ppm. Vuong et al. used Ni_2O_3 as the catalyst for a SnO_2 film, which shows high response to CH_4 with a low concentration of 2.2 ppm at 400 °C [40]. The use of 2 wt% Pd loaded SnO_2 synthesized by simultaneous precipitation can detect low concentrations of CH_4 down to 6.7 ppm at 350 °C.

Figure 10.8 presents an example of CH_4 gas sensors using Cr doped SnO_2 nanoparticles [38]. The SnO_2 nanoparticles were synthesized by evaporating and combusting precursors to form nucleation of metal oxide species that subsequently grew into clusters and coagulated into the form of SnO_2 nanoparticles. Various Cr contents ranging from 0.1 to 0.2 wt% were *in situ* doped by introducing chromium(III) 3-acetylacetonate hydrate during the synthesis process. The SnO_2 particle sensors were then fabricated using powder pasting and spin coating. High resolution transmittance electron microscopy images of Cr-doped SnO_2 nanoparticles show high crystallinity, Figure 10.8a. The SnO_2 sensors with different Cr concentration were then tested with various gases such as CH_4, C_2H_2, NO_2, NO, N_2O, NH_3 at the same operating temperature of 350 °C. In the experiment, NO_2, NO, and N_2O were set at a concentration of 5 ppm to represent typical environmental condition (where the concentration of these gases is below 10 ppm). The other gases were set at a concentration in a range of 0.01–1 vol%. As shown in Figure 10.8b, undoped SnO_2 nanoparticles show high response to C_2H_5OH, but lower outputs with CH_4 and other gases. On

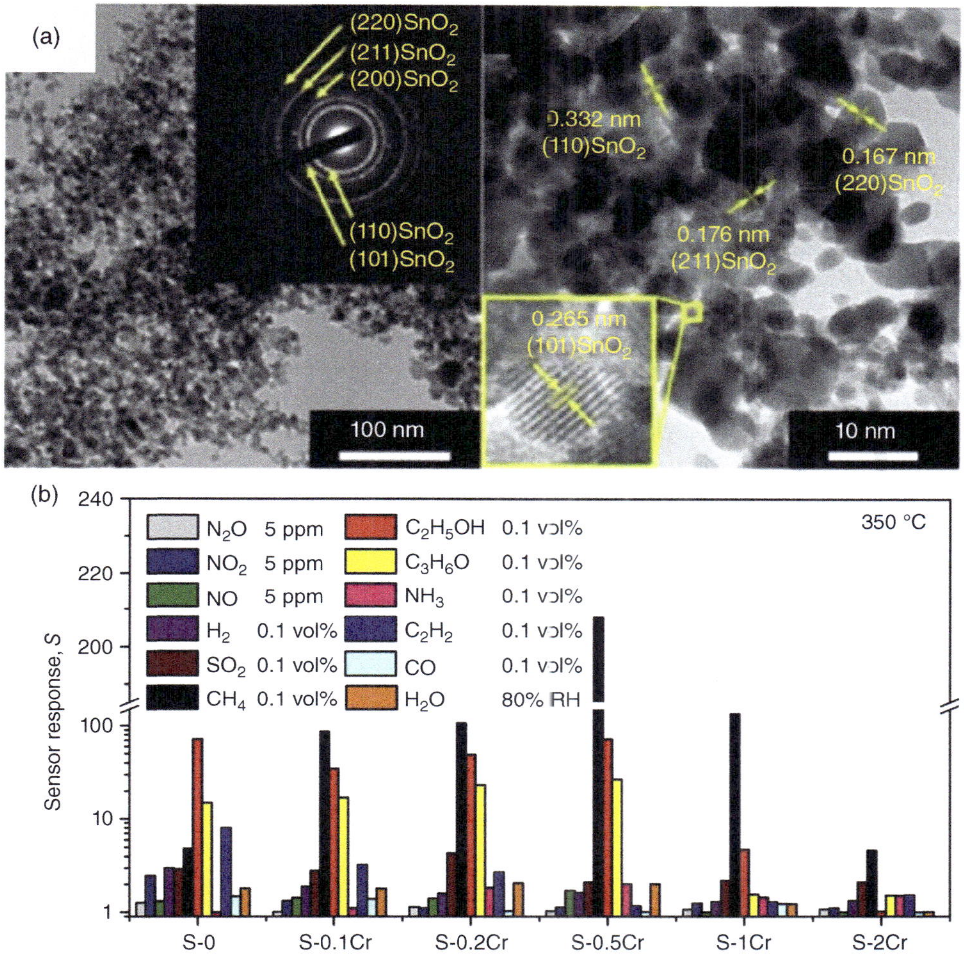

Figure 10.8 An example of CH_4 sensors using SnO_2 nanoparticles. (a) Bright field transmission electron microscopy (BF-TEM) with selected area electron diffraction (SAED) patterns (insets) and high resolution transmission electron microscopy (HR-TEM) images of 2 wt% Cr; (b) The selectivity histogram of the Cr doped SnO_2 sensor for different types of gases such as CH_4, H_2, C_2H_2, NO_2, NO, N_2O, CO, NH_3, SO_2, C_2H_5OH, C_3H_6O, and H_2O at 350 °C in dry air. *Source:* [38], Bunpang, et. al., (2019)/Elsevier.

the other hand, introducing Cr dopants into the particles significantly enhanced CH_4 sensitivity, and markedly reduce the response to other gases including C_2H_5OH. In particular, SnO_2 sensors with 0.5 wt% Cr content exhibits the highest sensitivity to CH_4 at 350 °C and a good selectivity of over 2.98 against a wide range of gases. These experimental results demonstrated the effectiveness of metallic dopants in enhancing the selectivity and performance of gas sensors.

A similar concept using metallic dopants was applied to enhance the sensitivity to other organic gases such as trimethylamine and butane. For instance, by doping Cr_2O_3 into SnO_2 nanowires grown by thermal evaporation, Kwak et al. demonstrated highly sensitive trimethylamine sensors that can detect low concentration down to 5 ppm at 450 °C. The sensors exhibited good selectivity against C_2H_5OH, HCHO, CO, H_2, and benzene [41].

Besides metal oxide materials, other wide bandgap semiconductors such as GaN and SiC have also been employed to construct organic gas sensors. A common configuration in these devices is the use of metals such as Pd or Pt as the catalyst to modulate the channel current in nanowires upon gas adsorption. For instance, Li et al. developed a Pd decorated GaN nanowire-based sensors using a MOCVD process [42]. The authors synthesized the GaN nanowires that directly bridged a microtrench of sapphire coated by a thin layer of GaN, Figure 10.9a. Pd was then deposited on the as-grown nanowire using sputtering with different thicknesses ranging from 2 to 10 nm, Figure 10.9b. The sensors were then subjected to CH_4 gas with concentration ranging from 0 to 500 ppm. The photoelectric response of the sensors was measured under ultraviolet (UV) light illumination with an optical power of 1400 μW/cm^2. Experimental result showed that compared with pure GaN nanowires, the introduction of Pd particles significantly enhanced the change in the photocurrent, Figure 10.9c,d. The sensor with a Pd thickness of 10 nm can detect low concentration of CH_4 down to 20 ppm. The data also revealed that, compared with dark condition, UV illumination markedly improved the response of Pd decorated GaN nanowires, indicating the potential of coupling photoelectronic with chemical adsorption for gas sensing. The time response measurement shows that increasing the thickness of Pd leads to a decrease in the raising time but causes an increase in the recovery time of the sensors.

10.4.4 Hydrogen Gas

Hydrogen has been used as an energy source for combustion engines for several decades. Recent years have overseen the raise of hydrogen for clean energy production through the water splitting process [43, 44]. As it is an odorless, colorless, and highly combustible, hydrogen poses several safety issues such as flame and explosion. In addition, in several cases, hydrogen is stored in a liquid form (cryogenic); therefore, sensing devices require extra robustness to withstand extreme industrial processing conditions [45]. In this regard, wide bandgap semiconductors raise as an excellent choice that offer excellent sensitivity and at the same time provide long-term reliable performance. There are several types of hydrogen sensors employing wide bandgap materials. These include metal oxide (MO_x) nanowires, III-nitride based HEMT devices, GaN, and SiC Schottky diodes, GaN nanowires, and InN nanobelts.

Metal oxide sensors operate based on the adsorption of oxygen followed by the desorption of ionic species (e.g. O^-, O^{2-}) subjected to H_2 gas [46, 47]. The oxygen adsorption in metal oxide acquires electrons from the conduction band ($O_2 + e^- \rightarrow O_2^-$; $O_2^- + e^- \rightarrow 2O^-$), thereby reducing the electrical conductivity of the sensors. When the metal oxide are subjected to hydrogen gas, the adsorption of H_2 causes the desorption of O^- and O^{2-}, releasing electrons back to the conduction band ($H_2 + O^- = H_2O + e^-$). This causes an increase in the electrical conductivity of the sensors. The hydrogen adsorption reaction typically occurs at high temperatures of above 250 °C and the large surface area of the nanowire enables a fast reaction rate and a quick release of H_2O from the surface of nanowires, offering excellent time response. Table 10.1 lists some examples of metal oxide-based hydrogen sensors include ZnO, WO_3, and SnO_2 nanowires with excellent detection range down to 10 ppm and time response in a range of a few seconds [46]. Metal loading effect (e.g. Pt and Pd) has also been proven to enhance the sensitivity of sensors due to the high hydrogen uptake rate in Pt and Pd. For instance, Zhu et al. reported H_2 sensing using a Pt coated WO_3 nanowire film without the requirement for integrated heater [56]. The WO_3 nanowires were synthesized using a two-step process of (i) thermal evaporation and (ii) oxidation. A thin Pt layer that functioned as the catalyst for H_2 gas sensing was sputtered on to the surface of WO_3 nanowires. Gas sensors were constructed from these Pt coated nanowires using an interdigitated configuration where the distance between two adjacent electrodes was 50 μm, and the sensing area was

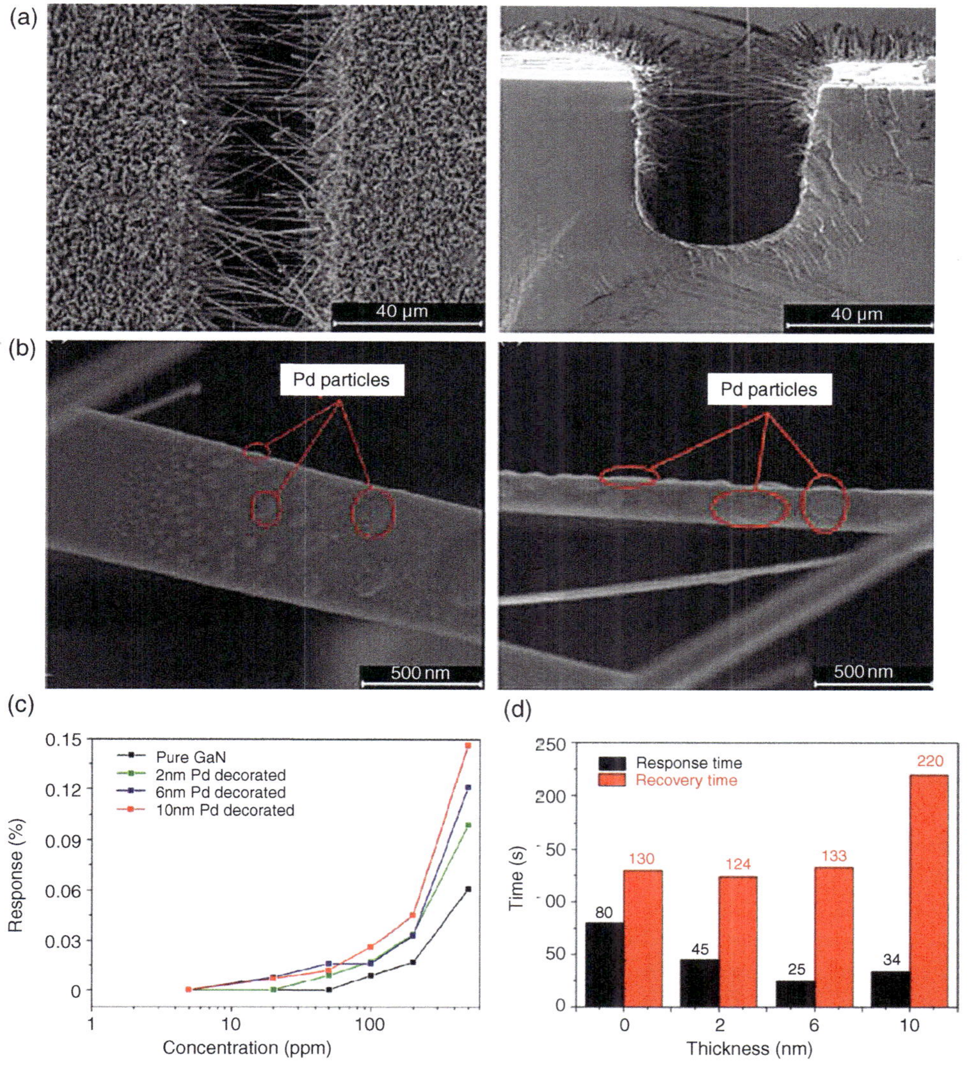

Figure 10.9 CH_4 gas sensors using Pd doped GaN nanowires. (a) SEM images of a GaN/sapphire micro trench bridged by a high density GaN nanowire array; (b) SEM images of GaN nanowire decorated by Pd particle with thicknesses of 6 and 10 nm; (c, d) sensor response at different gas concentrations and different thicknesses of Pd nanoparticles. *Source:* [42], Li, et. al., (2020)/IOP Publishing/CC BY 3.0.

Table 10.1 Examples of H_2 gas sensors using metal oxide semiconductors.

H_2 concentration	Loaded catalytic metal concentration	Semiconductor	Sensor response	References
0.01–1.0 vol%	1.0 wt% Pt	WO_3	1.34×10^5–1.0 vol% at 150 °C	[48]
0.02–1.0 vol%	0.2 wt% Pd	SnO_2	~10^4–1.0 vol% at 200 °C	[49]
0.02–1.0 vol%	0.2–2.0 wt% Ru	SnO_2	27–1.0 vol% at 350 °C	[50]
0.02–1.0 vol%	0.2–2.0 at% Pt	ZnO	164–1.0 vol% at 300 °C	[51]
0.125–1.0 vol%	Pt-loaded	WO_3	1600–1 vol% at 70 °C	[52]
30–200 ppm	0.5 at% Pt	WO_3	26–200 ppm at 95 °C	[53]
0.01–1 vol%	1.0 wt% Pt	WO_3	28–0.01 vol% at 250 °C	[54]
20 ppm	—	WO_3	3.1 at 200 °C	[55]

Source: Adapted from Phanichphant [47].

1.5 mm × 1.5 mm, Figure 10.10a. The hydrogen sensing performance of the Pt–WO_3 nanowire sensors was first characterized in atmosphere and at room temperature. Figure 10.10b shows that the current of the Pt coated WO_3 nanowires increased with increasing the H_2 concentrations. The device can detect H_2 with a diluted ratio of 0.1%. The sensors exhibit a relatively slow recovering time of 15 minutes, which is considerable for devices operating at room temperature. The authors also investigated the influence of co-adsorption of O_2 during hydrogen sensing process, by observing device response under pure H_2 and under a mix of H_2 and O_2 gas. Figure 10.10c plots the resistance change of the sensors under varying pure H_2 concentration, showing a linear increase with a slope of 2.7×10^{-1}. The response of the sensors decreased by 3 orders of magnitude when O_2 gas was applied, Figure 10.10d. The results indicate that co-adsorption of O_2 weakens the electrical response of the H_2 sensors. Modelling of co-adsorption of unspecific gas besides the targeted gas can improve the overall design and performance of H_2 gas sensors.

Metal oxide combined with III-nitride based HEMT devices have also been demonstrated for hydrogen gas sensing. Figure 10.10 illustrates the concept of a AlGaN/GaN HEMT hydrogen gas sensor [57]. The HEMT device includes a 2 μm-thick undoped GaN buffer layer, and a 35 nm-thick unintentionally doped $Al_{0.28}Ga_{0.72}N$ layer, which were grown on a C-plane Al_2O_3 substrate using a MOCVD process. Ohmic contacts were formed at the source and drain using a metal stack of Ti/Al/Pt/Au annealed at 850 °C. A Sc_2O_3 film with a thickness of 40 nm was deposited as a gate dielectric through a silicon nitride masking layer. Pt Schottky contact was then deposited on the Sc_2O_3 dielectric layer followed by Ti/Au deposition for subsequent wire bonding, Figure 10.11a. Figure 10.11b shows the *I–V* characteristic of the AlGaN/GaN HEMT device at 25 °C under pure N_2 and 10% H_2/90% N_2 atmosphere. Clearly, the introduction of H_2 causes an increase in the current. This is attributed to the reduction in the effective barrier height when H_2 decomposes on the Pt electrode and rapidly diffuses through the underlying Sc_2O_3 dielectric, forming a dipole layer. The formation of this dipole layer at the oxide/semiconductor interface (i.e. Sc_2O_3/AlGaN) induces more charge carriers in the HEMT channel, resulting in an increase in the channel current. The dynamic response of the sensors was characterized by applying 10% H_2/90% N_2

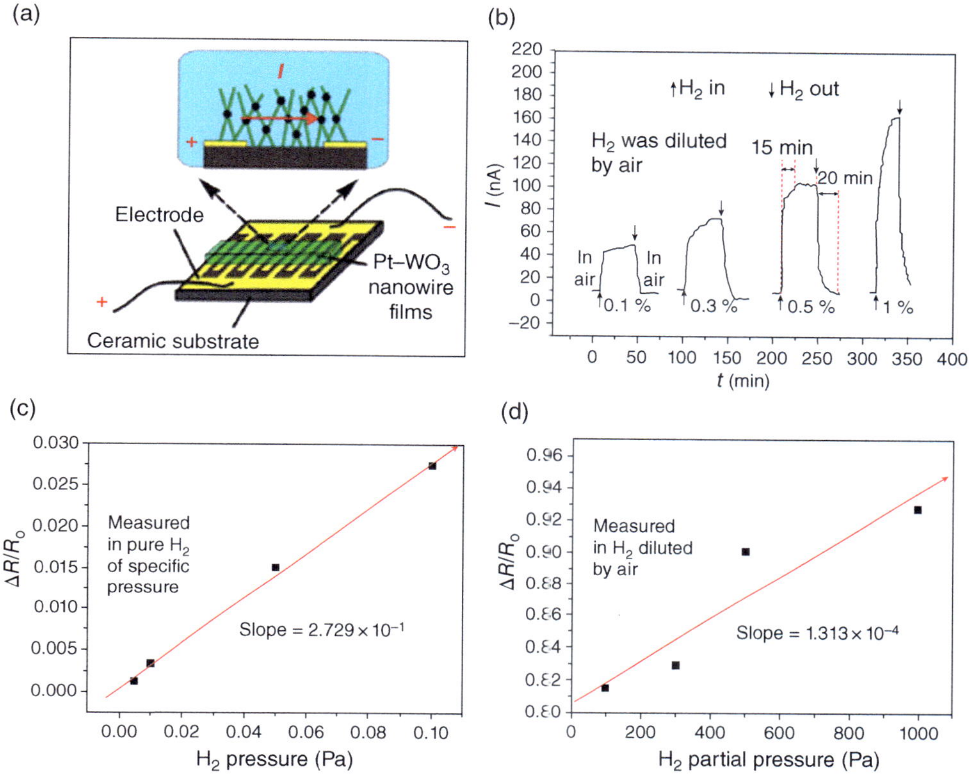

Figure 10.10 (a) Gas sensor using Pt coated WO_3 nanowires; (b) current change of the sensors at different H_2 concentrations; (c, d) the current change of the sensing measured in pure H_2 and measured in H_2 diluted by air. *Source:* Zhu et al. [56]/American Chemical Society.

flow for 10, 20, and 30 seconds, respectively, Figure 10.11c. While the sensors showed immediate response upon applying H_2 gas, it took several hundred seconds for the sensors to return to the original current level when the gas was removed. As the experiment was performed at room temperature (25 °C), increasing device temperature could further enhance the time response of the HEMT sensors.

Along with metal oxide materials and III-nitride, silicon carbide nanostructures have also attracted great deal of interest for gas sensing. Sanger et al. demonstrated a hydrogen gas sensing application employing Pd coated SiC nanocauliflowers [58]. In this configuration, porous anodic aluminium oxide (AAO) was employed as the substrate of the sensors. A thin layer of Ag was sputtered on the porous AAO to define the bottom electrode of the sensor. Silicon carbide were then sputtered on to Ag coated AAO, forming a cauliflower nanoarchitecture, Figure 10.12a,b. The sensing element (SiC) was in an amorphous form characterized by using X-ray diffraction (XRD) measurement. The SiC nanocauliflowers were decorated by a thin layer of Pd with an atomic concentration of 0.88% measured using energy dispersive X-ray (EDX).

The sensing performance was characterized under 100 ppm H_2 in dry air at an operating temperature of 300 °C. Figure 10.12c shows the response of the sensors (i.e. $\Delta R/R$),

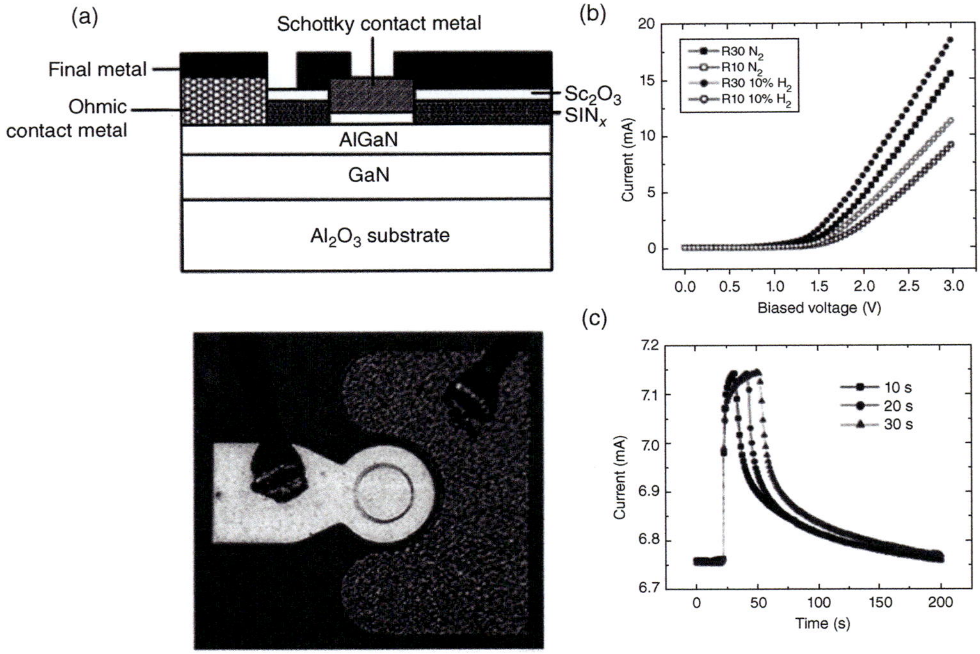

Figure 10.11 (a) Schematic sketch of the AlGaN/GaN hydrogen gas sensor and an optical microscope image of the fabricated device; (b) forward characteristic of HEMT device of two different dimensions under pure N_2 or 10% H_2/90% N_2 ambient; (c) time response at 25 °C to 10% H_2/90% N_2 at a fixed bias of 2 V. *Source:* [57], Kang, et. al., (2004)/AIP Publishing LLC.

indicating a significant change of approximately 41% in the resistance. The experimental data also revealed fast raising time and recovery time of 7 and 13 seconds, respectively, Figure 10.12c, inset. Figure 10.12d plots the output of the sensor under varying concentration of H_2 at the same operating temperature (i.e. 300 °C), which clearly shows a monotonic relationship between resistance change and the H_2 content. The sensors can detect low concentration of 2 ppm. Interestingly, the measurement of the response time shows that higher concentrations led to a faster response time but also a longer recovery time, Figure 10.12e. The high surface area in SiC nanocauliflowers allows for the large number of catalytic sites for H_2, which was accounted for the rapid response in the sensors. Figure 10.12f plots the resistance change against the operating temperature ranging from 25 to 500 °C. Evidently, the H_2 gas sensor exhibited a higher sensitivity at a higher temperature. High temperatures offer larger thermal energy, which promotes the interaction between H_2 and the sensing element, thereby significantly enhancing the sensitivity. The utilization of Pd as a catalyst promotes the reaction between SiC nanocauliflowers and dissociated H atoms, leading to a faster response along with a higher resistance change. Briefly, with the help of Pd, H_2 gas was rapidly dissociated into H atoms (i.e. the spill over mechanism). These H atoms react with the adsorbed oxygen on the surface of SiC nanocauliflowers that create water molecules. This chemical reaction also leaves charge carriers and oxygen vacancies that attribute to an increase in electrical conductivity of SiC devices. These results demonstrated a combination of wide bandgap nanoarchitectures and metallic catalysts can enhance gas sensing performance.

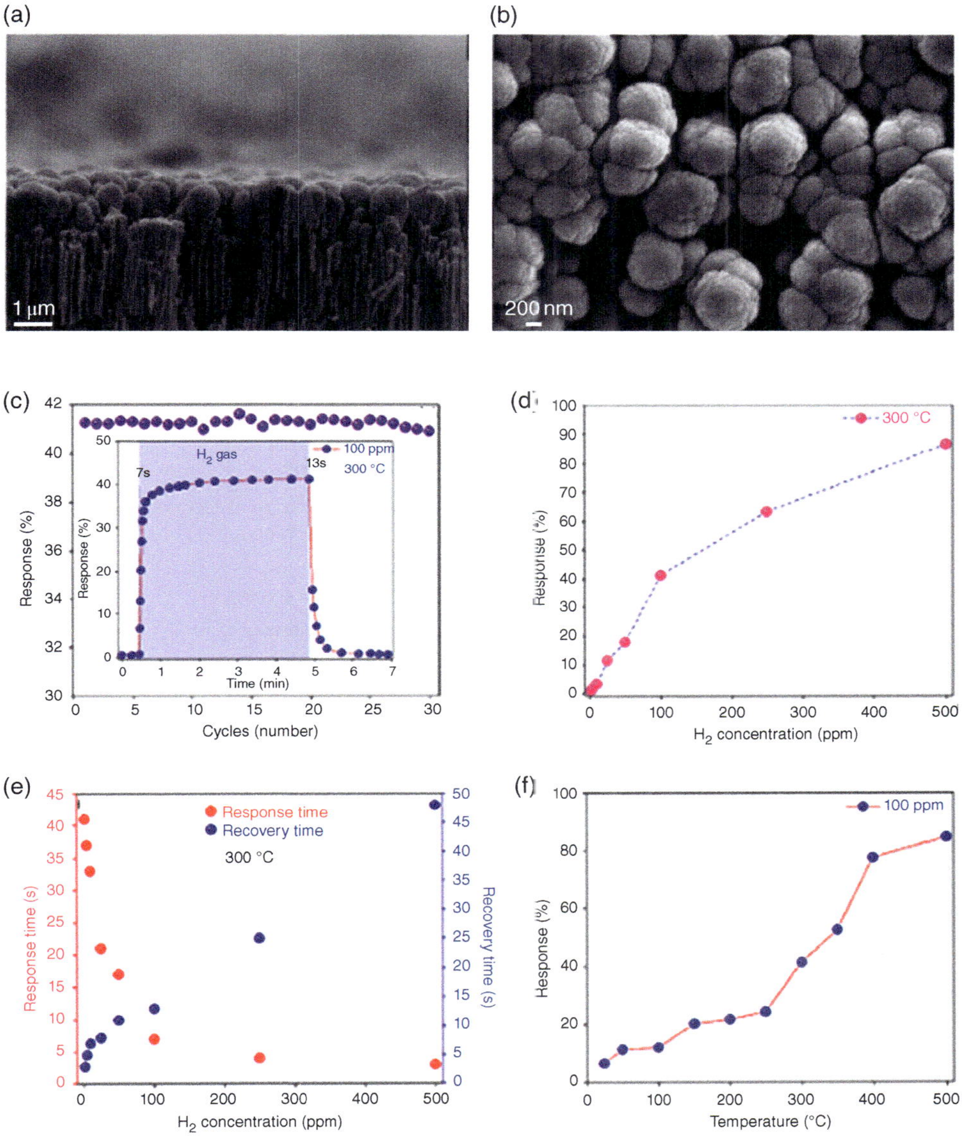

Figure 10.12 (a, b) SEM images of the SiC nanocauliflower architecture; (c) resistance change at an H_2 concentration of 100 ppm, and at 300 °C; (d) resistance change at different H_2 concentration and at 300 °C; (e) time response vs H_2 concentration; (f) resistance change vs temperatures. *Source:* [58], Sanger, et. al., (2016)/American Chemical Society.

10.5 Integration of Nanowires into Flexible Platform for Advanced Health Care Applications

Real-time and continuous measurement of human health using wearable and implantable electronics holds promise for advanced diagnosis and treatment. In this regard, gas sensors integrated into a wearable platform is the key toward timely detection of hazardous and toxic

Table 10.2 Common gas sensors for health care applications.

Sensing gases	Gas properties	Threshold concentration	Applications in healthcare	References
H_2	Colorless, odorless, highly flammable, explosive	20 ppm	Diagnosis of metabolic disease	[59]
NO_2/NO	Brown/colorless, toxic, oxidizer	206 ppt	Diagnosis of lung injury	[60]
CO	Colorless, odorless, toxic	0.1 ppm	Hemolytic disease	[61]
H_2S	Colorless, smell like rotten eggs, flammable, toxic	534 ppt	Asthma, lung cancer	[60]
C_2H_5OH	Colorless, smell like alcohol, flammable	0.1 ng	Detection of alcohol consumption	[62]
NH_3	Colorless, pungent smell, flammable	10–50 ppb	Cause of liver failure	[63]
CH_4	Extremely flammable	20 ppm	Indication of lactose intolerance	[64]
C_3H_6O	Pungent smell, flammable	10–30 ppb	Diabetes, hyperglycemia	[65]
CO_2	Colorless, odorless	60 ppb	Monitoring of respiration	[65]

Source: Adapted from Zheng and Cheng [66].

gases, and assessment of health conditions. Table 10.2 lists common gas sensors and their applications in health care [66].

Several techniques have been developed to form wide bandgap nanowires, especially metal oxide nanowires, on a flexible substrate such as polymers and textiles, to enable seamless integration of gas sensing device onto tissue or clothes. For instance, semiconductor nanowires can be directly grown on a soft platform where the temperature is typically below the thermal degradation temperature of the substrate. Among several type of flexible substrate, polyimide has attracted significant interest owing to its excellent thermal and mechanical properties along with chemical inertness. Figure 10.13a shows an acetylene gas sensor employing vertical ZnO nanorod array on a polyimide (PI) film [67]. The PI film was initially attached onto a polytetrafluoroethylene (PTFE) for handling, followed by sputtering a thin ZnO template film with a thickness of 120 nm. ZnO nanorods were then grown on the ZnO coated substrate using a hydrothermal method at a low temperature. To enhance the sensitivity, Ag nanoparticles were coated onto the surface of the as-grown ZnO nanorods using radio frequency (RF) magneton sputtering. Finally, a 200 nm thick Au films was deposited on to Ag-coated ZnO nanorods to form a resistive type gas sensor. Due to the flexibility of the PI substrate, the sensor showed excellent mechanical stability with a bending angle of up to 90 °C. No significant degradation was observed when the device was subjected to 50 000 bending and relaxing cycles. The sensors exhibited a good sensitivity (R_{air}/R_{gas}) of 27.2 at 1000 ppm of acetylene gas and short raising and recovery time of 62 and 39 seconds, respectively. Using a similar hydrothermal method, Yi et al. synthesized ZnO nanowires on a soft SU8 substrate for ethanol gas sensing. The sensor consists of a graphene layer transferred onto the vertical ZnO nanowire array via stamping, which functions as the top electrode of a

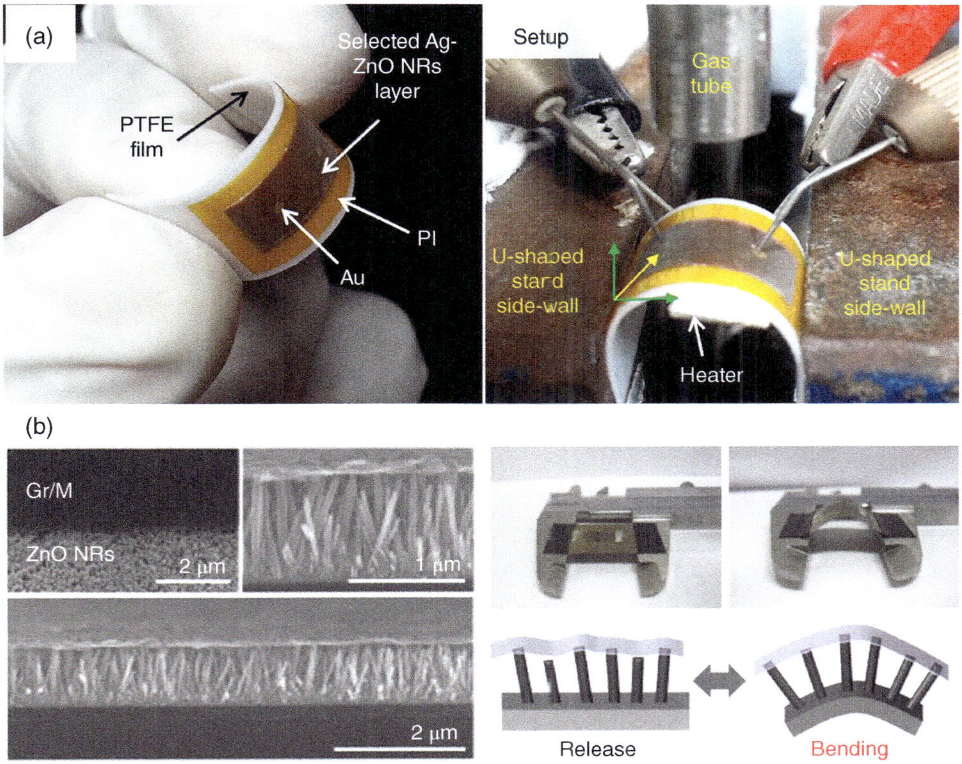

Figure 10.13 Direct synthesis of nanowires on soft substrates. (a) Optical image of a flexible acetylene gas sensor based on PI/PTFE-supported Ag-loaded vertical ZnO nanorods. *Source:* [67], Uddin et. al., (2016) / Elsevier. (b) A hybrid graphene/ZnO nanowire structure for flexible ethanol gas sensing. *Source:* [68], Yi et al. (2011)/Elsevier.

Schottky diode type gas sensor, Figure 10.13b. The graphene/ZnO nanowire hybrid structure shows excellent mechanical deformability, without any electrical failure for bending radius below 0.8 cm (i.e. corresponding to a tensile strain of ~1.3%). The device can detect ethanol with low concentrations down to 10 ppm with a sensitivity (R_{air}/R_{gas}) of 2.1/ppm.

Another approach to form flexible gas sensors is post-growth transfer of metal oxide onto soft substrates [69]. This method is advantage due to its compatibility with standard nanowire synthesis processes on solid substrates that can be taken place at elevated temperature. The growth of nanowire at higher temperatures is also expected to produce a better quality and controllable uniformity. In addition, this method is imperative for some certain types of nanowires where direct deposition on soft polymer at low temperature is not feasible. Typical transfer techniques employ a sacrificial layer in the growth process that facilitates the detachment of nanowires from the templating substrate. For instance, Kumaresan et al. developed flexible indium–gallium–zinc oxide (IGZO) based thin film transistor (TFT) for hydrogen gas sensors, Figure 10.14. The IGZO TFTs were fabricated on the 1.9 μm thick poly(methyl methacrylate) (PMMA) substrate, which was spin-coated on a rigid glass slide. Au/Ti thin films were deposited on the PMMA substrate through a shadow mask to form the bottom gate for the TFTs. The dielectric layer was patterned on the gate electrode by spin coating a thin film of PMMA (0.8 μm thick) followed by the deposition of 50 nm thick IGZO active layer using RF sputtering. The TFT on

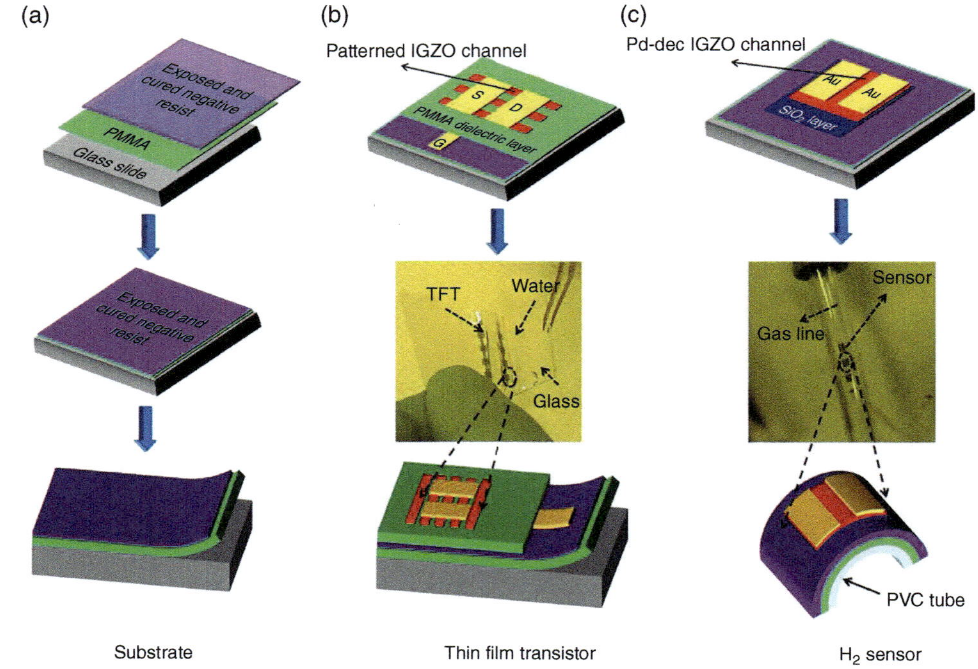

Figure 10.14 Pd decorated IGZO thin film transistor on PMMA for H_2 gas sensor. *Source:* [69], Kumaresan et al. (2018)/Wiley.

PMMA platform was detached from the rigid glass substrate, forming a flexible H_2 sensor. The soft TFT device exhibits an excellent current rectification characteristic, a high mobility of $10.7 cm^2/(V s)$, and an ON/OFF current ratio of 5 \time 10^5. The H_2 sensor show excellent mechanical flexibility that can be wrapped around cylinder tube with diameter of 2 mm. Current measurement showed consistent current over 1000 bending cycles. The use of Pd as a catalyst enabled a high sensitivity of 10^6% at 5% H_2 at room temperature.

References

1 Yamazoe, N. (1991). New approaches for improving semiconductor gas sensors. *Sensors and Actuators B: Chemical* 5 (1–4): 7–19.

2 Fine, G.F., Cavanagh, L.M., Afonja, A., and Binions, R. (2010). Metal oxide semi-conductor gas sensors in environmental monitoring. *Sensors* 10 (6): 5469–5502.

3 Baron, R. and Saffell, J. (2017). Amperometric gas sensors as a low cost emerging technology platform for air quality monitoring applications: a review. *ACS Sensors* 2 (11): 1553–1566.

4 Jaaniso, R. and Tan, O.K. (ed.) (2013). *Semiconductor Gas Sensors*. Elsevier.

5 Tricoli, A., Righettoni, M., and Teleki, A. (2010). Semiconductor gas sensors: dry synthesis and application. *Angewandte Chemie International Edition* 49 (42): 7632–7659.

6 Chen, X., Wong, C.K., Yuan, C.A., and Zhang, G. (2013). Nanowire-based gas sensors. *Sensors and Actuators B: Chemical* 177: 178–195.

7 Paska, Y. and Haick, H. (2012). Interactive effect of hysteresis and surface chemistry on gated silicon nanowire gas sensors. *ACS Applied Materials & Interfaces* 4 (5): 2604–2617.

8 Ramgir, N., Datta, N., Kaur, M. et al. (2013). Metal oxide nanowires for chemiresistive gas sensors: issues, challenges and prospects. *Colloids and Surfaces A: Physicochemical and Engineering Aspects* 439: 101–116.

9 Pearton, S.J., Ren, F., Wang, Y.L. et al. (2010). Recent advances in wide bandgap semiconductor biological and gas sensors. *Progress in Materials Science* 55 (1): 1–59.

10 Kolmakov, A. and Moskovits, M. (2004). Chemical sensing and catalysis by one-dimensional metal-oxide nanostructures. *Annual Review of Materials Research* 34: 151–180.

11 Ramgir, N.S., Yang, Y., and Zacharias, M. (2010). Nanowire-based sensors. *Small* 6: 1705–1722.

12 Andersson, M., Pearce, R., and Spetz, A.L. (2013). New generation SiC based field effect transistor gas sensors. *Sensors and Actuators B: Chemical* 179: 95–106.

13 Feng, P., Shao, F., Shi, Y., and Wan, Q. (2014). Gas sensors based on semiconducting nanowire field-effect transistors. *Sensors* 14 (9): 17406–17429.

14 Zhang, Y., Kolmakov, A., Chretien, S. et al. (2004). Control of catalytic reactions at the surface of a metal oxide nanowire by manipulating electron density inside it. *Nano Letters* 4 (3): 403–407.

15 Luther, B.P., Wolter, S.D., and Mohney, S.E. (1999). High temperature Pt Schottky diode gas sensors on n-type GaN. *Sensors and Actuators B: Chemical* 56 (1, 2): 164–168.

16 Ali, M., Cimalla, V., Lebedev, V. et al. (2006). Pt/GaN Schottky diodes for hydrogen gas sensors. *Sensors and Actuators B: Chemical* 113 (2): 797–804.

17 Yu, J., Ippolito, S.J., Shafiei, M. et al. (2009). Reverse biased Pt/nanostructured MoO_3/SiC Schottky diode based hydrogen gas sensors. *Applied Physics Letters* 94 (1): 013504.

18 Esch, H., Huyberechts, G., Mertens, R. et al. (2000). The stability of Pt heater and temperature sensing elements for silicon integrated tin oxide gas sensors. *Sensors and Actuators B: Chemical* 65 (1–3): 190–192.

19 Hou, M., Suria, A.J., Yalamarthy, A.S. et al. (2016). 2DEG-heated AlGaN/GaN micro-hotplates for high-temperature chemical sensing microsystems. In: *Proceedings of Solid-State Sensors, Actuators and Microsystems Workshop (5–9 June 2016), Hilton Head, United States*, 356–359.

20 Hsueh, T.J., Peng, C.H., and Chen, W.S. (2020). A transparent ZnO nanowire MEMS gas sensor prepared by an ITO micro-heater. *Sensors and Actuators B: Chemical* 304: 127319.

21 Madler, L., Roessler, A., Pratsinis, S.E. et al. (2006). Direct formation of highly porous gas sensing films by in situ thermophoretic deposition of flame-made Pt/SnO_2 nanoparticles. *Sensors and Actuators B: Chemical* 114: 283–295.

22 Zhang, M., Yuan, Z., Song, J., and Zheng, C. (2010). Improvement and mechanism for the fast response of a Pt/TiO_2 gas sensor. *Sensors and Actuators B: Chemical* 148: 87–92.

23 Liu, X., Cheng, S., Liu, H. et al. (2012). A survey on gas sensing technology. *Sensors* 12: 9635–9665.

24 Ng, K.T., Boussaid, F., and Bermak, A. (2011). A CMOS single-chip gas recognition circuit for metal oxide gas sensor arrays. *IEEE Transactions on Circuits and Systems I: Regular Papers* 58 (7): 1569–1580.

25 Prasad, A.K., Kubinskib, D.J., and Gouma, P.I. (2003). Comparison of sol–gel and ion beam deposited MoO_3 thin film gas sensors for selective ammonia detection. *Sensors and Actuators B: Chemical* 93: 25–30.

26 Krutovertsev, S.A., Sorokin, S.I., Zorin, A.V. et al. (1992). Polymer film-based sensors for ammonia detection. *Sensors and Actuators B: Chemical* 7: 492–494.

27 Sett, D. and Basak, D. (2017). Highly enhanced H_2 gas sensing characteristics of Co:ZnO nanorods and its mechanism. *Sensors and Actuators B: Chemical* 243: 475–483.

28 Tan, O.K., Cao, W., Hu, Y., and Zhu, W. (2004). Nano-structured oxide semiconductor materials for gas-sensing applications. *Ceramics International* 30 (7): 1127–1133.

29 Zhang, J., Qin, Z., Zeng, D., and Xie, C. (2017). Metal-oxide-semiconductor based gas sensors: screening, preparation, and integration. *Physical Chemistry Chemical Physics* 19 (9): 6313–6329.

30 Minaee, H., Mousavi, S.H., Haratizadeh, H., and De Oliveira, P.W. (2013). Oxygen sensing properties of zinc oxide nanowires, nanorods, and nanoflowers: the effect of morphology and temperature. *Thin Solid Films* 545: 8–12.

31 Lang, T., Wiemhöfer, H.D., and Göpel, W. (1996). Carbonate based CO_2 sensors with high performance. *Sensors and Actuators B: Chemical* 34 (1–3): 383–387.

32 Star, A., Han, T.R., Joshi, V. et al. (2004). Nanoelectronic carbon dioxide sensors. *Advanced Materials* 16 (22): 2049–2052.

33 Guntner, A.T., Abegg, S., Konigstein, K. et al. (2019). Breath sensors for health monitoring. *ACS Sensors* 4 (2): 268–280.

34 Baik, K.H. and Jang, S. (2020). Carbon dioxide sensing characteristics of AlGaN/GaN high electron mobility transistor with ZnO nanorods. *Sensors and Materials* 32 (8): 2569–2576.

35 Comert, B., Akin, N., Donmez, M. et al. (2016). Titanium dioxide thin films as methane gas sensors. *IEEE Sensors Journal* 16 (24): 8890–8896.

36 Esser, B., Schnorr, J.M., and Swager, T.M. (2012). Selective detection of ethylene gas using carbon nanotube-based devices: utility in determination of fruit ripeness. *Angewandte Chemie International Edition* 51 (23): 5752–5756.

37 Benignus, V.A. (1981). Health effects of toluene: a review. *NeuroToxicology* 2 (3): 567–588.

38 Bunpang, K., Wisitsoraat, A., Tuantranont, A. et al. (2019). Highly selective and sensitive CH_4 gas sensors based on flame-spray-made Cr-doped SnO_2 particulate films. *Sensors and Actuators B: Chemical* 291: 177–191.

39 Das, N., Halder, A.K., Sen, J.M.A., and Maiti, H.S. (2006). Sonochemically prepared tin-dioxide based composition for methane sensor. *Materials Letters* 60: 991–994.

40 Vuong, N.M., Hieu, N.M., Hieu, H.N. et al. (2014). Ni_2O_3-decorated SnO_2 particulate films for methane gas sensors. *Sensors and Actuators B: Chemical* 192: 327–333.

41 Kwak, C.-H., Woo, H.-S., and Lee, J.-H. (2014). Selective trimethylamine sensors using Cr_2O_3-decorated SnO_2 nanowires. *Sensors and Actuators B: Chemical* 204: 231–238.

42 Li, S., Huang, H., and Zhao, D. (2020). GaN nanowires decorated with Pd for methane gas sensor. *IOP Conference Series: Earth and Environmental Science* (4–5 July 2020), Dalian, China, Vol. 558, No. 4, 042037.

43 Hisatomi, T. and Domen, K. (2019). Reaction systems for solar hydrogen production via water splitting with particulate semiconductor photocatalysts. *Nature Catalysis* 2 (5): 387–399.

44 Qi, J., Zhang, W., and Cao, R. (2018). Solar-to-hydrogen energy conversion based on water splitting. *Advanced Energy Materials* 8 (5): 1701620.

45 Mirzaei, A., Kim, S.S., and Kim, H.W. (2018). Resistance-based H_2S gas sensors using metal oxide nanostructures: a review of recent advances. *Journal of Hazardous Materials* 357: 314–331.

46 Dey, A. (2018). Semiconductor metal oxide gas sensors: a review. *Materials Science and Engineering B* 229: 206–217.

47 Phanichphant, S. (2014). Semiconductor metal oxides as hydrogen gas sensors. *Procedia Engineering* 87: 795–802.

48 Samerjai, T., Tamaekong, N., Liewhiran, C. et al. (2011). Selectivity towards H_2 gas by flame-made Pt-loaded WO_3 sensing films. *Sensors and Actuators B: Chemical* 157: 290–297.

49 Liewhiran, C., Tamaekong, N., Wisitsoraat, A. et al. (2013). Ultra-sensitive H_2 sensors based on flame-spray-made Pd-loaded SnO_2 sensing films. *Sensors and Actuators B: Chemical* 176: 893–905.

50 Liewhiran, C., Tamaekong, N., Wisitsoraat, A., and Phanichphant, S. (2009). H_2 sensing response of flame-spray-made Ru/SnO_2 thick films fabricated from spin-coated nanoparticles. *Sensors* 9: 8996–9010.

51 Tamaekong, N., Liewhiran, C., Wisitsoraat, A., and Phanichphant, S. (2009). Sensing characteristics of flame-spray-made Pt/ZnO thick films as H_2 gas sensor. *Sensors* 9: 6652–6669.

52 Ippolito, S.J., Kandasamy, S., Kalantar-zadeh, K.K., and Wlodarski, W. (2005). Hydrogen sensing characteristics of WO_3 thin film conductometric sensors activated by Pt and Au catalysts. *Sensors and Actuators B: Chemical* 108: 154–158.

53 Zhang, C., Boudiba, A., Navio, C. et al. (2011). Highly sensitive hydrogen sensors based on co-sputtered platinum activated tungsten oxide films. *International Journal of Hydrogen Energy* 36: 1107–1114.

54 Samerjai, T., Liewhiran, C., Wisitsoraat, A. et al. (2014). Highly selective hydrogen sensing of Pt-loaded WO_3 synthesized by hydrothermal/impregnation methods. *International Journal of Hydrogen Energy* 39: 6120–6128.

55 Calavia, R., Mozalev, A., Vazquez, R. et al. (2010). Fabrication of WO_3 nanodot-based microsensors highly sensitive to hydrogen. *Sensors and Actuators B: Chemical* 149: 352–361.

56 Zhu, L.F., She, J.C., Luo, J.Y. et al. (2010). Study of physical and chemical processes of H_2 sensing of Pt-coated WO_3 nanowire films. *The Journal of Physical Chemistry C* 114 (36): 15504–15509.

57 Kang, B.S., Ren, F., Gila, B.P. et al. (2004). AlGaN/GaN-based metal–oxide–semiconductor diode-based hydrogen gas sensor. *Applied Physics Letters* 84 (7): 1123–1125.

58 Sanger, A., Jain, P.K., Mishra, Y.K. and Chandra, R. (2017). Palladium decorated silicon carbide nanocauliflowers for hydrogen gas sensing application. *Sensors and Actuators B: Chemical* 242: 694–699.

59 Mridha, S. and Basak, D. (2006). Investigation of a p-CuO/n-ZnO thin film heterojunction for H_2 gas-sensor applications. *Semiconductor Science and Technology* 21: 928–932.

60 Moon, H.G., Jung, Y., Han, S.D. et al. (2016). Chemiresistive electronic nose toward detection of biomarkers in exhaled breath. *ACS Applied Materials & Interfaces* 8: 20969–20976.

61 Vreman, H.J., Stevenson, D.K., Oh, W. et al. (1994). Semiportable electrochemical instrument for determining carbon monoxide in breath. *Clinical Chemistry* 40: 1927–1933.

62 Arena, A., Donato, N., Saitta, G. et al. (2010). Flexible ethanol sensors on glossy paper substrates operating at room temperature. *Sensors and Actuators B: Chemical* 145: 488–494.

63 Gouma, P., Kalyanasundaram, K., Yun, X. et al. (2010). Nanosensor and breath analyzer for ammonia detection in exhaled human breath. *IEEE Sensors Journal* 10: 49–53.

64 Hanf, S., Bögözi, T., Keiner, R. et al. (2014). Fast and highly sensitive fiber-enhanced Raman spectroscopic monitoring of molecular H_2 and CH_4 for point-of-care diagnosis of malabsorption disorders in exhaled human breath. *Analytical Chemistry* 87: 982–988.

65 Righettoni, M., Tricoli, A., and Pratsinis, S.E. (2010). Si:WO_3 sensors for highly selective detection of acetone for easy diagnosis of diabetes by breath analysis. *Analytical Chemistry* 82: 3581–3587.

66 Zheng, X. and Cheng, H. (2019). Flexible and stretchable metal oxide gas sensors for healthcare. *Science China Technological Sciences* 62 (2): 209–223.

67 Uddin, A.I., Yaqoob, U., Phan, D.T., and Chung, G.S. (2016). A novel flexible acetylene gas sensor based on PI/PTFE-supported Ag-loaded vertical ZnO nanorods array. *Sensors and Actuators B: Chemical* 222: 536–543.

68 Yi, J., Lee, J.M. and Park, W.I. (2011). Vertically aligned ZnO nanorods and graphene hybrid architectures for high-sensitive flexible gas sensors. *Sensors and Actuators B: Chemical* 155 (1): 264–269.

69 Kumaresan, Y., Lee, R., Lim, N. et al. (2018). Extremely flexible indium-gallium-zinc oxide (IGZO) based electronic devices placed on an ultrathin poly(methyl methacrylate) (PMMA) substrate. *Advanced Electronic Materials* 4 (7): 1800167.

11

Wide Bandgap Nanoresonators

11.1 Introduction

Micro and nanoresonators are micro/nano electromechanical systems (MEMS/NEMS) applications that can generate and sustain mechanical vibration at a certain frequency and have been widely deployed in several industrial systems [1–3]. A well-known example for the applications of mechanical resonators is the MEMS oscillator that produces highly stable reference frequencies for time measurement [4]. Implementation of multiple oscillators with different frequencies generates clock signals that synchronize most of standard integrated circuit (IC) chips to perform processor timing, disk input/output, and serial input/output [5, 6].

As the resonant behavior of a microdevice can be influenced by an external input by either physical impact or chemical bonding, resonators have attracted great attention to the development of sensing applications [7, 8]. For instance, a small modification in physical parameters such as mass, surrounding pressure, damping, and temperature could alter the resonant frequencies and Q-factors. Measurement of the change in resonant characteristics provides a way to precisely quantify these physical parameters. Compared with microresonators, nanoresonators offer numerous advantages including the large surface to volume ratio, and small effective mass that enable the detection of extremely small mass down to molecules level [9–11]. A combination of chemical functionalization (e.g. surface chemical bonding) with high performance nanoresonators (e.g. with high frequency and high Q-factor) has enabled the development of numerous chemical and biological sensors.

The advantage in the low dimensional geometry, unique mechanical properties (e.g. large Young's modulus), and multiple functions in wide bandgap (WBG) materials (e.g. piezoelectric, piezotronics effects) offer numerous attractive features for nanoresonators [12–14]. The possibility of forming WBG nanoresonators, which are firmly anchored on standard substrate such as silicon have enabled the developed of high-performance resonators for logical circuits and sensing devices [15, 16]. This chapter covers the basic concept of nanoresonators, the operating principles, advantages of WBG nanoresonators, and their sensing applications.

Wide Bandgap Nanowires: Synthesis, Properties, and Applications, First Edition. Tuan Anh Pham, Toan Dinh, Nam-Trung Nguyen, and Hoang-Phuong Phan.

11.2 Principle of Nanoresonators

11.2.1 Resonant Frequency

Figure 11.1 presents a simplified model for nanoresonators with single-side clamped and double-side clamped nanowires. The dynamic of the nanoresonator is described via the displacement as a function of location and time ($u(x,t)$). The equation of motion in nanowires can be represented using the Euler–Bernoulli beam model [17]:

$$\rho A \frac{\delta^2 u(x,t)}{\delta t^2} + EI_y \frac{\delta^4 u(x,t)}{\delta x^4} = 0 \tag{11.1}$$

where ρ is the mass density, A is the cross-sectional area, I is the moment of inertia, and E is Young's modulus of the nanowires. The boundary condition of the Euler–Bernoulli equation depends on the configuration of the beam (single clamped or doubly clamped). For instance, the boundary condition for a doubly clamped nanoresonator is given by [18]

$$\begin{cases} u(0,t) = u(L,t) = 0 \\ \dfrac{du(0,t)}{dt} = \dfrac{du(L,t)}{dt} = 0 \end{cases} \tag{11.2}$$

The solutions of this equation give the normal modes, where the displacement can be separated into a position dependent and a time dependent function:

$$U(x,t) = \sum_{n=1}^{\infty} u_n(x)\cos(\omega t) \tag{11.3}$$

where n is the modal number and ω is the resonant frequency. Figure 11.2 plots schematic sketch of the first four bending mode ($n = \overline{1,4}$) in one-side clamped nanowires (i.e. nano cantilevers).

The resonant frequency at each modal number (n) is given by

$$\omega = \beta_n^{\ 2} \sqrt{\frac{EI_y}{\rho A}} \tag{11.4}$$

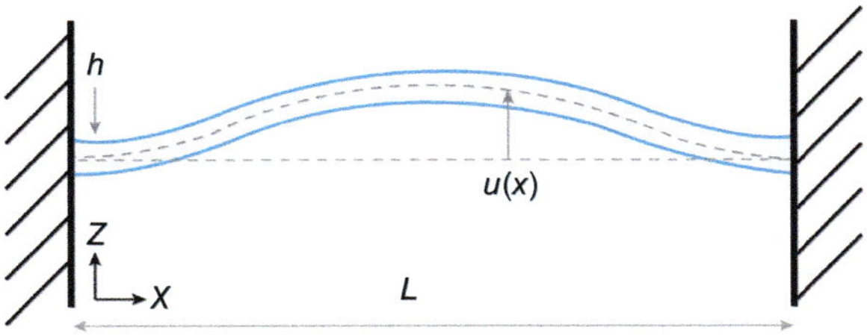

Figure 11.1 A model of doubly clamed nanowire resonator. *Source:* Schmid et al. [17]/Springer Nature.

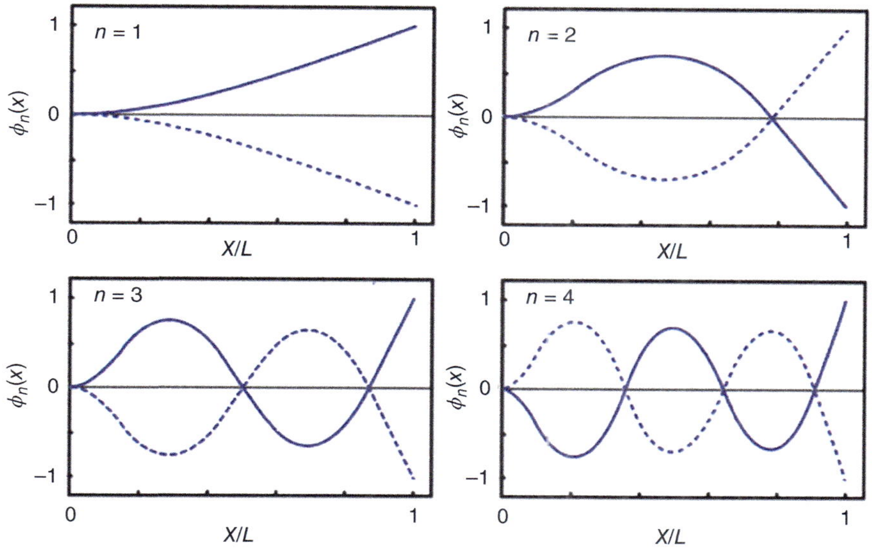

Figure 11.2 Examples of the first four eigen modes in single-side clamped resonator (i.e. cantilever configuration). *Source:* Schmid et al. [17]/Springer Nature.

where β_n is the wavenumber. Equation (11.3) indicates that for nanoresonators free of stress, the resonant frequency is proportional to Young's modulus and reversely proportional to the mass density of the constructing materials. Compared with silicon, several WBG semiconductors such as diamond, SiC, and III-nitride possess much larger Young's moduli, while having a comparable mass density [19]. These properties make WBG materials preferable for the design of high frequency devices (e.g. very high frequency [VHF] and ultrahigh frequency [UHF] resonators). Furthermore, due to its low dimension, nanoresonators have a relatively small total mass (i.e. mass density × volume) [20]. Adding a tiny mass onto the surface of nanoresonators could result in a significant mass change ratio, thereby markedly shifting the resonant frequency of the device. This advantage makes nanoresonators an attractive platform for mass and gas sensing applications.

11.2.2 *Q*-Factor

Quality factor (Q) is a parameter that defines the rate at which a nanoresonator dissipates its energy. In another other word, the quality factor represents the sharpness of a resonant peak, which is quantified as the ratio of the energy stored (W) to the energy lost (ΔW) in one resonant cycle [17, 21]:

$$Q = 2\pi \frac{W}{2\Delta W} \tag{11.5}$$

In most nanoresonator applications, a high Q-factor corresponding to a low loss of energy is desirable. However, due to numerous sources of energy loss such as environmental damping (Q_{damp}), clamping between the resonator and the hosting substrate (Q_{clamp}), and intrinsic mechanical energy dissipation ($Q_{intrinsic}$), maintaining a high Q-factor in mechanical resonators is challenging. Furthermore, due to the losses at the resonator surface due to the increased surface to volume ratio, scaling down the resonators (from microscale to nanowires) could lead to a significant decrease in the Q-factor. Generally, the total quality factor is the sum of these dissipation mechanisms [17]:

$$\frac{1}{Q_t} = \frac{1}{Q_{damp}} + \frac{1}{Q_{clamp}} + \frac{1}{Q_{intrinsic}} + \frac{1}{Q_{other}} \tag{11.6}$$

In some cases, the damping effect can be minimized by providing a vacuum media surrounding the nanoresonators to reduce the damping forces from air and liquid. However, in several applications such as gas sensing, viscosity measurement, and pressure sensing, significant energy lost due to damping is unavoidable. Furthermore, the fabrication process can also affect the final structure of devices such as over etching and misalignment can modify the final structure of a resonator with respect to the initial design [8]. This modification of the anchor area influences the Q-factor due to the change in the clamping loss. For instance, Figure 11.3 presents the mechanical loss in a doubly clamped SiN resonator, in which the anchor area has a clamp-widen design [22]. The widening effect at the supporting areas frequently occurs in MEMS fabrication, where the Si substrate is isotropically etched to release nanoresonators. Two configurations of diagonally oriented and perpendicular string geometries were developed to investigate the relationship between the Q-factor and the radius of widened clamp (R). The finite element analysis (FEA) showed a significant residual stress in both designs, in which the perpendicular resonator exhibited a larger tensile stress at the anchor of up to 500 MPa compared with that of the diagonal resonator (180 MPa), Figure 11.3a,b. The large stress induced at the anchor part estimated from the FEA suggests a possibility of Q-factor enhancement in clamp-widen resonators. However, while the anchor was released from the substrate, the stress concentrated at this part could bend the clamping area, inducing further mechanical loss that could the positive effect of the tensile stress. This phenomenon could lead to a decrease in the Q-factor at a large anchor curvature (R). Figure 11.3c shows the relationship of the Q-factor with the variation in the radius of the free-standing anchor. For diagonal resonator with a window size of 500, the Q-factor increased slightly to 23% when the radius (R) increased from 0 to 25 μm. The Q-factors decreased when further increasing the radius from 25 to 100 μm, which is attributed to the dominance of clamping loss due to mechanical bending at the anchor. A similar trend was also observed in perpendicular resonator with a window size of 800 μm, in which the Q-factor increased with increasing R from 0 to 40 μm, and the decreased when further increasing R up to 100 μm.

11.2.3 Motion Equation with Residual Stress Consideration

Most of WBG nanoresonators are fabricated using a top-down process to ease the integration of the functioning element to the fixed substrate. In this case, nanowires are developed from nanothin film grown on a foreign substrate that allows top-down machining and releasing of nanowires. As the films are deposited on a different substrate, the nanowires typically experience residual stress, a physical property widely employed to enhance the performance of nanowire resonators [23, 24]. The Euler–Bernoulli equation of motion for a doubly clamped nanowire with consideration of tensile stress is given by

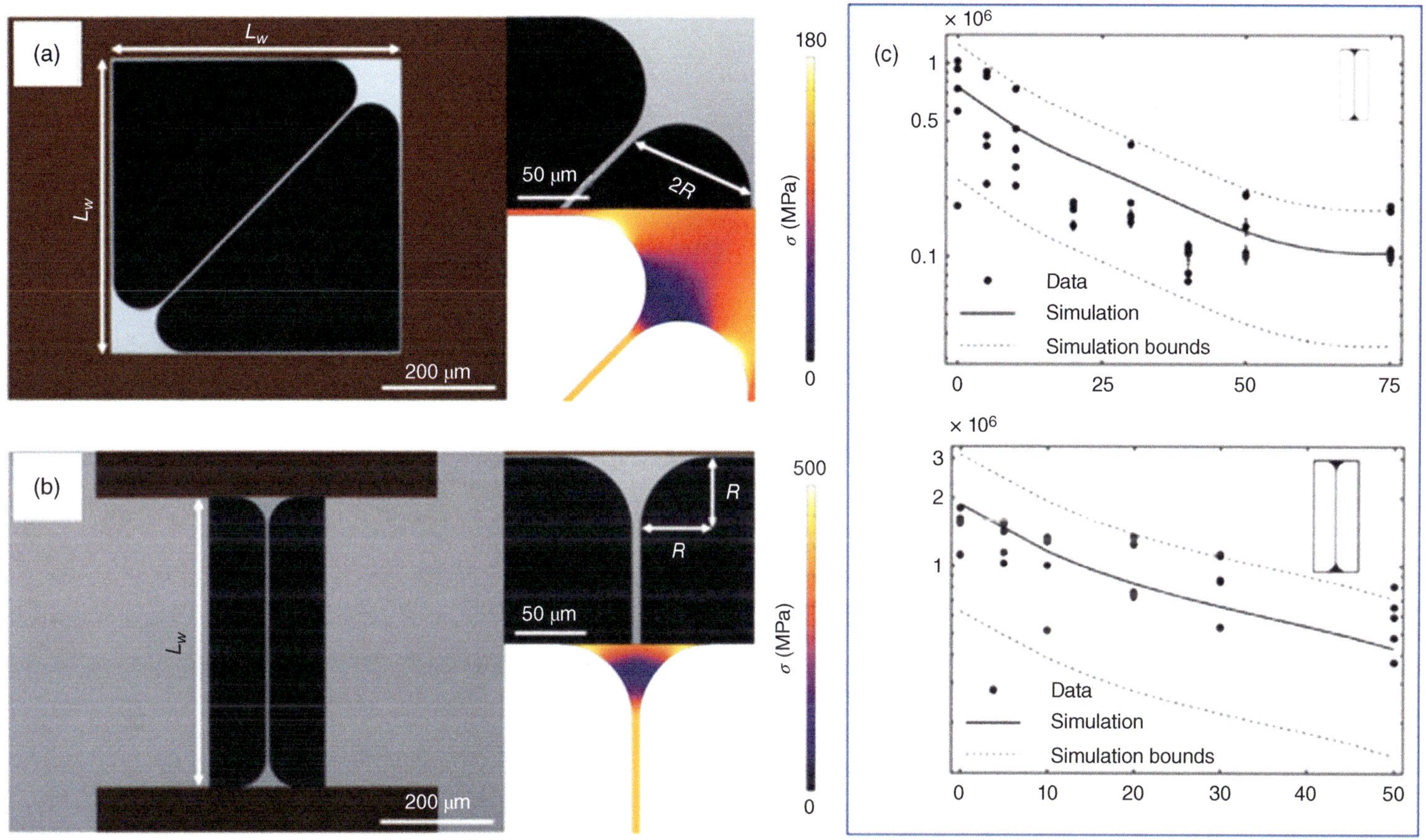

Figure 11.3 Examples of Q-factor reduction due to the clamp-widen design. (a, b) Microscopy image of a 500 **μ**m wide opening window of diagonally oriented and perpendicular SiN resonators. Inset: stress profile at clamping area. (c) Relationship between quality factor and the clamping ratio. *Source:* [22], Sadeghi, et. al., (2019)/AIP Publishing LLC.

$$\rho A\frac{\delta^2 u(x,t)}{\delta t^2}+EI_y\frac{\delta^4 u(x,t)}{\delta x^4}-N\frac{\delta^2 u(x,t)}{\delta x^2}=0 \tag{11.7}$$

where $N = \sigma A$ is the tensile residual force applied on the resonator. The resonant frequency of nanowires under tensile stress is given by

$$f=\beta_n{}^2\sqrt{\frac{EI_y}{\rho A}}\sqrt{1+\frac{\sigma A}{EI_y\beta_n{}^2}} \tag{11.8}$$

Equation (11.8) indicates that utilizing the residual tensile stress in WBG materials post-deposition process, the resonant frequency is enhanced by a factor of $\sqrt{1+\frac{\sigma A}{EI_y\beta_n{}^2}}$.

11.2.4 Why Large $f \times Q$ Matter?

Generally, for sensing and oscillator applications, a large $f \times Q$ value is desirable to obtain high sensitivity as well as to lower measurement noise [25, 26]. Specifically, for mass sensing devices, adding a mass of Δm to a resonator with an original mass of m will change its resonant frequency an amount of approximately: $\Delta f/f = 2\Delta m/m$. Using the Allan variance, the minimum detectable change is proportional to the fractional frequency change and is given by

$$\frac{\delta m}{m}=\sqrt{\frac{k_B T}{mL^2}\frac{1}{Q\omega^3\tau_A}} \tag{11.9}$$

where k_B is the Boltzmann constant, T is the temperature, τ_A is the integration time, and L is the length of the resonator. Evidently, the mass detection resolution of a resonant frequency depends on the value of $f \times Q$, where a smaller $f \times Q$ leads to a smaller detectable range. Therefore, choosing the constructing materials and designing the configuration of nanoresonators are critically important to develop high performance sensing devices.

11.3 Actuation and Measurement Techniques

11.3.1 Electrostatic Actuation

Electrostatic actuation has been employed in numerous MEMS devices, including comb actuators, which are widely used in micro-positioned systems [27]. This is versatile method that can be applied for different types of conductive materials, from metals to semiconductors. The main concept of electrostatic can be represented using a parallel plate model, Figure 11.4. The actuator functions as a capacitance, in which the storage energy (W) under an applied voltage (V) is

$$W=\frac{1}{2}CV^2=\frac{\varepsilon_0\varepsilon_r A}{z}V^2 \tag{11.10}$$

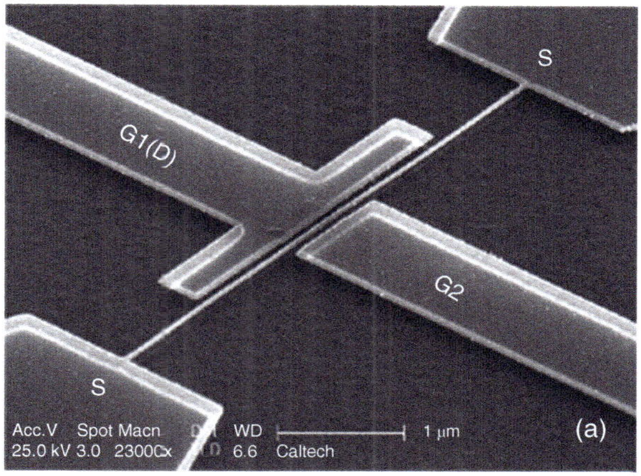

Figure 11.4 An SEM image of an electrostatic nanoresonator. *Source:* [28], Feng, et. al., (2010)/Elsevier.

where C is the capacitance; ε_0 and ε_r are the absolute and relative permittivity of the dielectric materials (e.g. air or liquid); A is the electrodes' surface area; and z is the distance between the two electrodes. The electrostatic force is given by the change of the storage energy (dW) over the displacement (dz) between the two electrodes:

$$F = \frac{dW}{dz} = -\frac{\varepsilon_0 \varepsilon_r A}{z^2} V^2 \tag{11.11}$$

Evidently, the electrostatic force is proportional to the surface area and reversely proportional to the distance between the two electrodes. This force pulls the two electrodes toward each other until it is balanced with the elastic force of the electrodes. While nanowires exhibit relatively small surface areas, they can still be effectively driven using the electrostatic force due to the small distance between the nanowires and electrodes of sub-hundred nanometer. In addition, the surface area can be enhanced by using a long microelectrode parallel with the longitudinal direction of the nanoresonator. Examples for electrostatic actuation in WBG materials include the NEMS switch using silicon carbide nanowire, Figure 11.4 [28]. The nanowire was fabricated from an epitaxial SiC film with a thickness of 50 nm. Using a top-down process, the width of the nanowires was formed as small as 20 nm, and the width between the nanowire and the driving electrode was only 10 nm. The nanowire was employed as a mechanical switch, which is turned on when the nanowire contacts with the drain electrode using an electrostatic force. Another electrode (i.e. gate) was formed on the other side of the nanowire to actively pull the nanowire back to its original position (i.e. the OFF state) if surface forces dominate over the restoring elastic force of the nanowire. The threshold voltage (V_{on}) required to turn on the switch is estimated as: $V_{on} \propto \sqrt{E_Y w^3 g_0^3 / L^2}$. The switching time is estimated as: $t_s = \sqrt{\frac{27}{2}} \left(V_{on} / \omega_0 V_{act} \right)$, where g_0 is the initial coupling gap. Experimental results showed that due to the small gap between the nanowire and the drain electrode, the device can be turned on at voltages as small as 1 V, consistent with the theoretical estimation. Furthermore, the use of SiC with large Young's modulus enables a fast switch time in a sub-microsecond range. These characteristics demonstrate WBG nanowires for low-power loss, and high frequency electronic circuit.

11.3.2 Piezoelectric Actuation

There are numerous WBG semiconductors that exhibit significant piezoelectric effect such as ZnO, GaN, and AlN. The capability of generating mechanical deformation under electrical bias offers a unique pathway to actuate piezoelectric micro and nanoresonators with a relatively simple circuit and device design [29, 30]. Briefly, applying a voltage to a piezoelectric resonator will create a strain and generate a bending moment in the structure. Depending on the orientation of the applied voltage (i.e. polarization), the deformation can occur in out of plane or along the longitudinal direction of the resonator. Applying a continuous AC signal will lead to alternative compressive and tensile strains in each cycle, causing a vibration in the resonator. Generally, the piezoelectric device will be driven into resonance when the frequency of the applied AC voltage matches with the mechanical resonant frequency. Figure 11.5 presents an example of piezoelectric resonators using III-nitride materials [31].

In this design, the piezoelectric layer was a 100 nm-thick GaN film sandwiched between two Mo layers. Prior to sputtering of these GaN and Mo layer, a thin GaN seeding layer with

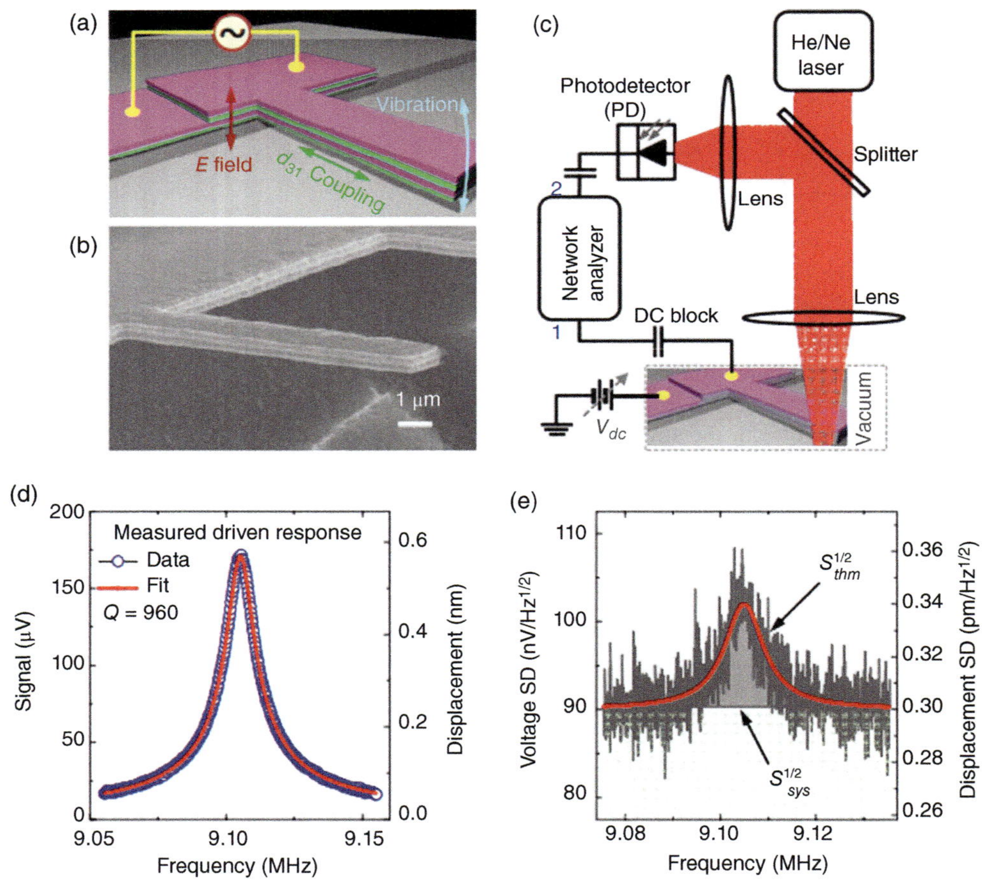

Figure 11.5 Piezoelectric actuators using GaN materials. (a) Schematic sketch of the resonator. (b) SEM image of a top-down fabricated device. (c) Vibration measurement using an optical setup. (d) Measurement of the resonant frequency and Q-factor. (e) Thermomechanical noise spectral density. *Source:* [31], Karabalin, et. al., (2009)/AIP Publishing LLC.

a thickness of 20 nm was deposited on a Si substrate. The resonator was micro-fabricated using standard lithography combined with metal etching and inductive coupling plasma etching of the GaN layer. The resonator was suspended by isotropically etching the Si substrate. The top and bottom Mo layers were employed as the electrode for the piezoelectric resonator, enabling an actuating motion in the d_{31} direction. A strain was generated in the longitudinal direction (S_{xx}) of the resonator and proportional to the piezoelectric coefficient d_{31}: $S_{xx} = d_{31}E_z$, where E_z is the vertical component of the electric field. The 20 nm thick GaN seeding layer (at the bottom) shifts the neutral axis in the active layers (Mo/GaN/Mo). As a consequence, the generation of the longitudinal strain (S_{xx}) leads to an out-of-plane deformation, causing the resonator to vibrate, Figure 11.5a,b. The resonant frequency of the device was measured using an optical readout scheme. Owing to the large Young's modulus in all constructing layers (e.g. 345 GPa in GaN and 329 GPa in Mo), the piezoelectric actuator exhibits a high frequency of 9.11 MHz and a high Q-factor of 960 at a base pressure of ~5 mTorr. The thermal noise density was transduced into a voltage noise density and was found at 48 nV/Hz$^{1/2}$. The measurement system possessed a high mechanical displacement sensitivity of ~0.3 pm/Hz$^{1/2}$.

11.3.3 Magnetomotive Actuation

Electrostatic method can be applied to several electrically conductive materials; however, the requirement to form ultra-small gap between the nanowire and the driving electrode makes the fabrication of electrostatic nanoresonators challenging. The piezoelectric, on the other hand, is only applicable with materials that exhibit significant electrical polarization. The efficiency of electrostatic and piezoelectric methods has been reported to decrease substantially when the resonators are scaled down from micrometer to nanometer regime. Magnetomotive actuation has been proven an effective method to drive nanoresonators with relatively simple design, excellent versatility, and good efficiency [32, 33]. One of the early works on magnetomotive nanoresonators was proposed by the Roukes group on Si nanowires, Figure 11.6A. In this configuration, an AC current $I(\omega)$ is applied between the two terminals of nanowire [34]. When the nanowire is placed under an external magnetic field (e.g. using a permanent magnet), a Lorentz force will be exerted onto the nanowire, that sets it into vibration:

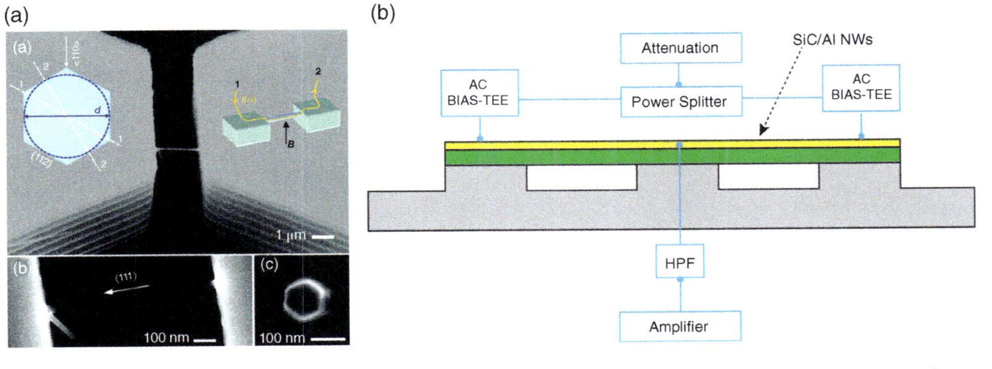

Figure 11.6 Magnetomotive actuation. (a) The first concept was applied on top down Si nanowires. *Source:* [34], Feng, et. al., (2007)/Elsevier. (b) Demonstration on wide bandgap nanostructures (SiC nanowires). *Source:* Jun et al. [35]/IEEE.

$$f(\omega) = LI(\omega)B \tag{11.12}$$

The magnitude of the vibration can be controlled by varying the applied current or by adjusting the external magnetic field (e.g. using a solenoid). Furthermore, the flexural motion of the nanowire under the magnetic field in turn induces an electromotive force V_{EMF}, which is widely known in a standard DC motor:

$$V_{EMF}(\omega) = BLv(\omega)\frac{1}{L}\int_{-\frac{L}{2}}^{\frac{L}{2}} u_0(x)dx = j\omega\eta BLa(\omega) \tag{11.13}$$

Here $v(\omega)$ and $a(\omega)$ are the velocity and displacement at the midpoint of the nanowires. Measurement of the V_{EMF} offers a new route to develop self-actuation and self-sensing nanoresonator using the magnetomotive effect, which is highly suitable for system miniaturization. Jun et al. applied this concept to develop a magnetomotive SiC nanoresonator, Figure 11.6B [35]. The resonator was developed from a SiC film grown on a Si substrate using atmospheric pressure chemical vapor deposition. SiC nanowires with various lengths were fabricated using standard electron beam lithography and reactive ion etching (RIE) to form small widths of 120 nm. A thin Al layer was employed as the capping layer during the RIE process, as well as the current carrying layer for magnetomotive actuation. The radio frequency (RF) driving and readout signals were accomplished with a network analyzer. The resonators were effectively actuated using the magnetomotive effect, while sharp peaks of the resonant frequency and high Q-factors were clearly observed using the electromotive force phenomenon. Specifically, at a similar width, nanowires with lengths ranging from 11 to 15 μm exhibit the first mode resonant frequency varying from 9 to 12 MHz. The experimental results demonstrated that the resonant frequency of the SiC nanowire array can be easily tuned by change the geometrical ratio (length/width). As such the resonant frequency increased when decreasing the length of the resonators. On the other hand, reducing the length of nanowires leads to a decrease in the Q-factor from 4300 to 3100.

11.3.4 Thermal Actuator

Thermal actuation employs the Joule heating effect to induce a thermal stress in nanobeams [36]. The gradient in the thermal stress between different layers in the nanobeam can cause flexural bending and set the device into motion under an AC bias [37, 38]. Under a bias voltage (V), the Joule heating power (P) is given by

$$P_{supply} = \frac{V^2}{R} \tag{11.14}$$

where R is the resistance of the resonator. This supplied power causes a temperature increase in the resonator, which is resulted from the scattering between charge carriers and phonons at high voltages. The temperature of the device increased by a value of ΔT until it reaches the thermal equilibrium, where the supplied power is balanced by the heat loss:

$$P_{supply} = Thermal\ loss = P_{radiation} + P_{conduction} + P_{convection} \tag{11.15}$$

The increase in temperature will cause a thermal expansion in the constructing layers of nanoresonators: $\sigma_T = \varepsilon \Delta T$, where ε is coefficient of thermal expansion (CTE), which is an intrinsic property in each material. The difference in the CTE of each constructing materials along with the thermal gradient caused by the heat conduction effect leads to a non-uniform deformation in the device, making the nanoresonator to vibrate when the voltage is applied in the AC mode. It should be noted that power consumption is one of the main issues in the thermal resonator, since a significantly high temperature is required to generate sufficient bending. However, this can be compensated by reducing the sources of the heat loss (e.g. (i) measurement in high vacuum to minimize the heat convection, (ii) using substrate with low thermal conductivity) or employing materials with large CTE difference to maximize the mechanical deformation.

Figure 11.7 presents an example of thermal resonators using SiC cantilever [38]. The SiC were constructed based on the cantilever shape with two configurations of (i) metal heater and (ii) SiC heater. For the metal heater configuration, a U-shape Pt or NiCr heater was deposited and pattern on a SiC cantilever. On the other hand, the SiC heater configuration uses an open circuit with two parallel Pt or NiCr electrodes. The SiC layer between the two

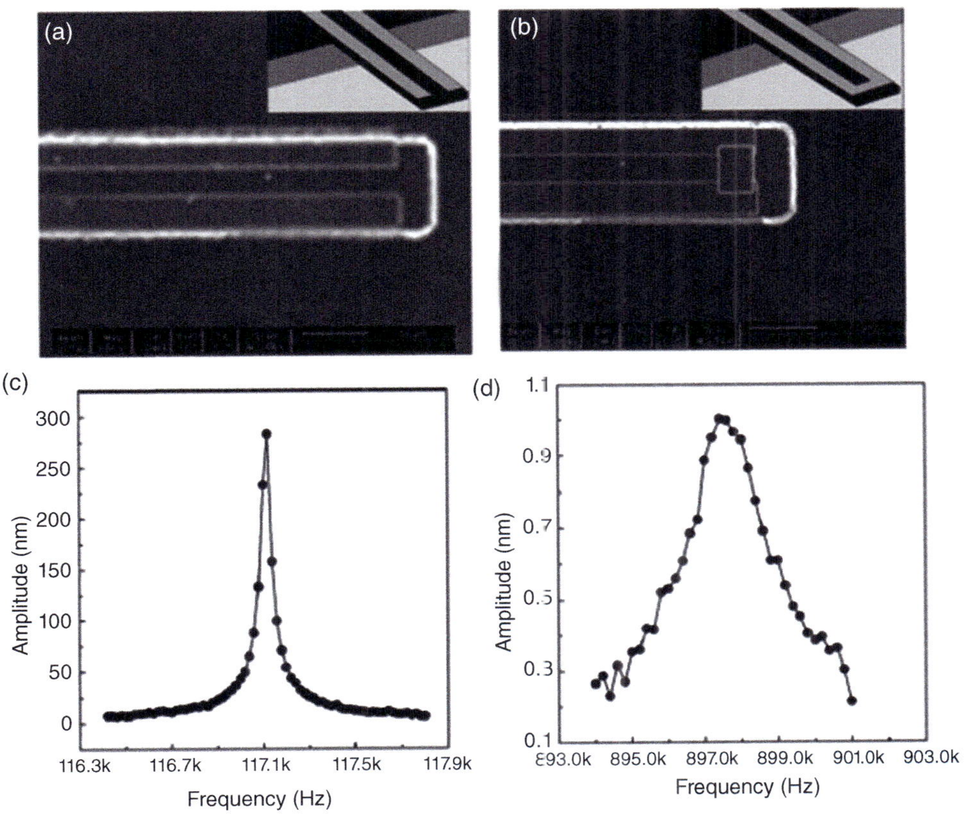

Figure 11.7 SEM images of SiC thermal actuators. (a) Device uses SiC as a heater. (b) Device use metal heater. (c, d) The first mode resonant frequency measured in the Pt/SiC and NiCr/SiC resonators, respectively. *Source:* [38], Jiang, et. al., (2006), Elsevier.

metal pads functions as the heater for the resonator. A mix of AC and DC voltages was applied to actuate the resonator. The supplied power was given by

$$P = \frac{V^2}{R} = \frac{1}{R}\left(2V_{ac}V_{dc}sin\omega_{ac}t + 0.5V_{ac}{}^2\left(1 - \cos 2\omega_{ac}t\right) + V_{dc}{}^2\right) \tag{11.16}$$

where ω_{ac} is the frequency of the applied V_{ac}. The results show that the thermal distribution and heating efficiency significantly depend on the electrical conductivity of metals. In particular, the Pt/SiC resonator exhibits a high temperature profile at the root of the cantilever, while the NiCr/SiC resonator shows a more uniform temperature distribution along the cantilever. As the Pt/SiC cantilever has a non-uniform thermal profile, a higher out of plane bending moment was generated due to difference in thermal expansion. This leads to more effective actuation in Pt/SiC resonator compared with the NiCr/SiC counterpart. As such at a lower actuation voltage (V_{dc} = 0.2V and V_{ac} = 0.3V_{pp}), the Pt/SiC resonator exhibit a sharper resonant peak than that of NiCr/SiC resonator, which was actuated at a higher voltage (V_{dc} = 1V and V_{ac} = 2V_{pp}). The SiC resonators also demonstrated its high resonant frequency of approximately 900 kHz, owing to the excellent mechanical properties in SiC. Furthermore, the resonant frequency slightly decreased with increasing the bias voltage, which was attributed to the change in the mechanical properties of constructing (SiC) and actuating layer (metal). This suggests an interesting route to control the resonant behavior in thermal resonators using bias voltages.

11.4 Engineering the Performance of Nanoresonators Using Wide Bandgap Materials

11.4.1 Residual Stress

Most of the WBG nanoresonators are developed from nanothin films combined with top-down electron beam lithography process. The nanothin films are generally grown on a low cost and worldwide available Si wafer using chemical vapor deposition at high temperatures. The difference in the lattice constant and thermal expansion between WBG thin films and the Si substrate leads to a large residual stress in the as-grown films [24, 39]. This residual stress in some case is undesirable as it may cause wafer bow and crack or can affect the properties of WBG materials. However, in other cases, if properly controlled, the residual stress can be utilized to enhance device performance. Tailoring residual stress to enhance f and Q in nanoresonators has been an active research topic, especially for epitaxial WBG materials.

Information of the residual stress can be assessed using optical measurement method such as Raman spectroscopy or X-ray diffraction. For instance, using Raman spectroscopy Rohmfeld et al. measured the residual strain in 3C-SiC films grown on a Si substrate by the atmospheric pressure chemical vapor deposition (APCVD) method at 1280 °C [40]. The residual strain can be quantified based on the shift in the longitudinal optical (LO) and transverse optical (TO) mode. A calibration experiment was performed by pressurizing a free-standing SiC membrane to correlate the applied pressure (or strain) with the resultant shift in the Raman peaks. The authors found a significant residual strain of 0.1–0.4% in the SiC films, attributed to the thermal stress resulted from the high temperature growth process. Another common technique is the X-ray measurement to directly measure the lattice constant of thin films post deposition. For instance, Harutyunyan et al. investigated the residual strain–stress in GaN films grown on a sapphire substrate using a high-resolution X-ray diffraction measurement [41]. It is reported that at a given thickness of the epilayer and the substrate, the residual stress in GaN films can be engineered by the thickness of the buffer layer, the deposition temperature, the doping concentration, and the compound parameter x of the $Ga_{1-x}N_x$ material. In this work, the authors investigated the relationship between x and the stress in $Ga_{1-x}N_x$ films. The X-ray diffraction method can quantify the lattice constant in different crystallographic planes, based on which the out of plane (ε_c) and in-plan (ε_a) residual strains can be estimated using the following equations:

$$\begin{cases} \varepsilon_c = \dfrac{c_r - c_0}{c_0} \\ \varepsilon_a = \dfrac{a_r - a_0}{a_0} \end{cases} \tag{11.17}$$

where c_0 and c_r are the c-lattice parameters for the strain-free and the actual films, respectively, while a_0 and a_r are the a-lattice parameters for the strain-free and the actual films, respectively. Experimental results showed a higher strain in the a-direction of approximately -2×10^{-3} to -3×10^{-3} in comparison with that in the c-direction (1.35×10^{-4} to 1.16×10^{-3}). The data also showed that the residual strain varied with the content of nitrogen in $Ga_{1-x}N_x$ films. This finding indicates a possibility in manipulating strains in WBG materials via crystal orientations and compound concentrations.

Employing the large residual strain in WBG materials can enhance their f and Q-factors. Figure 11.8 presents an experimental study performed on a GaN doubly clamped resonators, which was released from a Si substrate [42]. Raman spectroscopy shows a red shift at the E_2-high peaks with respect to the standard value of 568 cm^{-1} for the unstrained bulk GaN, indicating a tensile strain in the epitaxial GaN films. Before the resonators were leased from the Si substrate, a significant shift of 2 cm^{-1} was observed in sample A with a thickness of 500 nm in comparison to a thicker film (sample B with a thickness of 3 μm), Figure 11.8a,b. After removing the Si layer underneath of the resonator, the residual stress was partly relaxed, which was reflected in a decrease in the shift of the Raman peak (e.g. from 2 cm^{-1} before releasing to 1.2 cm^{-1} after releasing in sample A). Figure 11.8c shows the schematic of the resonator formed by isotropically etching Si using the HF:HNO_3 (1 : 15) mixed solutions. Figure 11.8d shows the resonant frequency of the resonator A at around 1.2 MHz. The Q-factors of the resonators were extracted from the full width at half maximum (FWHM) analysis FWHMs and the Lorentzian fits, showing a large value of approximately 118×10^3. Compared with sample A, sample B with large thickness and small residual stress exhibit a much lower Q-factor (in a range of 3500–4000). FEA shows that without the residual stress the first mode resonant frequency is 139 kHz, while under a tensile stress of 640 MPa, this value increases to 911 kHz, approximately 6.6 times enhancement. A similar trend was also observed in the simulation of sample B, where a tensile stress of 203 MPa raised the resonant frequency from 509 kHz at the strain-free state to a higher value of 592 kHz. The simulation analysis was consistent with the experimental studies, indicating the role of residual stress in enhancing the performance of WBG resonators.

11.4.2 Mechanical Clamping Enhancement

Mechanical energy dissipation due to clamping configurations is a source of the decrease in Q-factor of nanoresonators [43, 44]. Therefore, to maintain a high Q-factor when scaling down the dimension of resonators, optimizing the clamping structure is of critically importance.

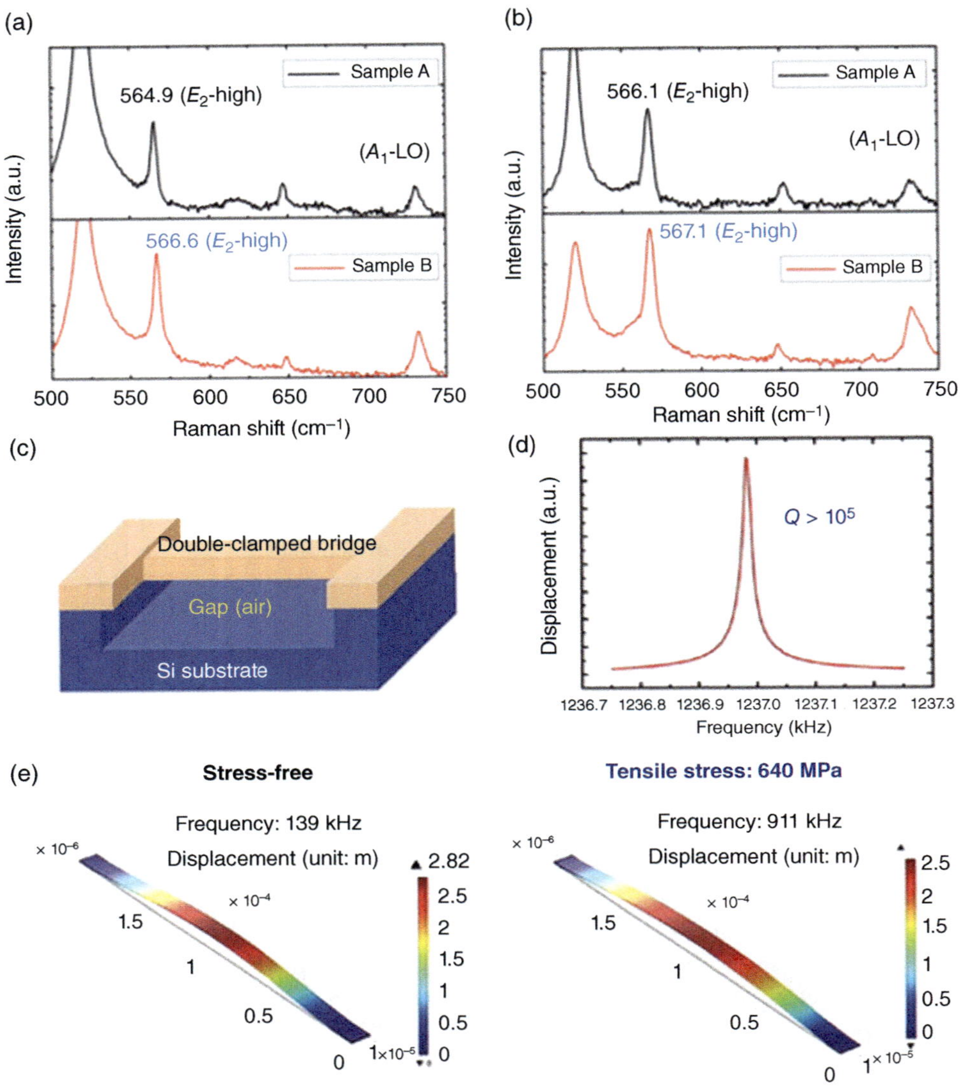

Figure 11.8 Enhancement of the resonant frequency and Q-factors using residual stress. (a, b) Measurement of Raman spectroscopy of GaN films with different thickness (500 nm) and (3 μm) before and after releasing from the substrate. (c) Schematic of the resonator with a doubly clamped configuration after releasing from the Si substrate. (d) Measurement of the resonant frequency at 1.23 MHz with a high Q-factor over 10^5. (e) Simulation of residual stress and its correlation with the performance of the SiC resonators. *Source:* Sang et al. [42]/Taylor & Francis.

Figure 11.9 presents examples of strategies to reduce the clamping loss in resonators [45]. Tao et al. developed single crystal diamond nanoresonators with a thickness of 85 nm and thickness uniformity of better than 20 nm. The clamping structure of the diamond cantilever resonators was enhanced by either bonding the diamond to an oxide film (method 1) or bonding the substrate to a quartz substrate (method 2). In the first method the thermal oxide layer serves as the clamping structure as well as the handing substrate for standard semiconductor-on-insulator fabrication, Figure 11.9a. In the second method, a 5 μm thick diamond ledge was formed at the anchor of the cantilever to reduce the clamping loss,

Figure 11.9b. Figure 11.9h–j presents scanning electron microscopy (SEM) images of the as-fabricated diamond resonators. The resonant frequency and Q-factor were measured using the ring-down method from room temperature down to millikelvin temperatures. The results show an extremely high Q-factor of over one million at room temperature and can be enhanced by 10^6 times when measuring at 0.1 K. The minimized mechanical loss is an important factor contributing to the ultrahigh Q-factor in diamond resonators. The thermal noise of the cantilever was found to be extremely small (~$5\times10^{19}\,N/Hz^{1/2}$) at millikelvin temperatures, demonstrating the potential of WBG nanoresonators for highly sensitive mechanical sensing applications (e.g. mass sensors).

Another example on the enhancement in the Q-factor of WBG material by engineering mechanical clamp is shown in Figure 11.9b [46]. Briefly, in this work, Kermany et al. developed a SiC resonator with a perfect clamp structure (i.e. free of undercut under the anchoring area). The resonators were developed from a (111) 3C-SiC film grown on a Si substrate with a high residual stress of up to 1.5 GPa. The authors showed that using a standard Si isotropic etching process, the anchor part of the resonators was widened, which directly affects the Q-factor of SiC resonators. To avoid the undercut effect, another masking layer was formed to cover the area of the SiC anchor. Specifically, this mask was used to protect the Si at the

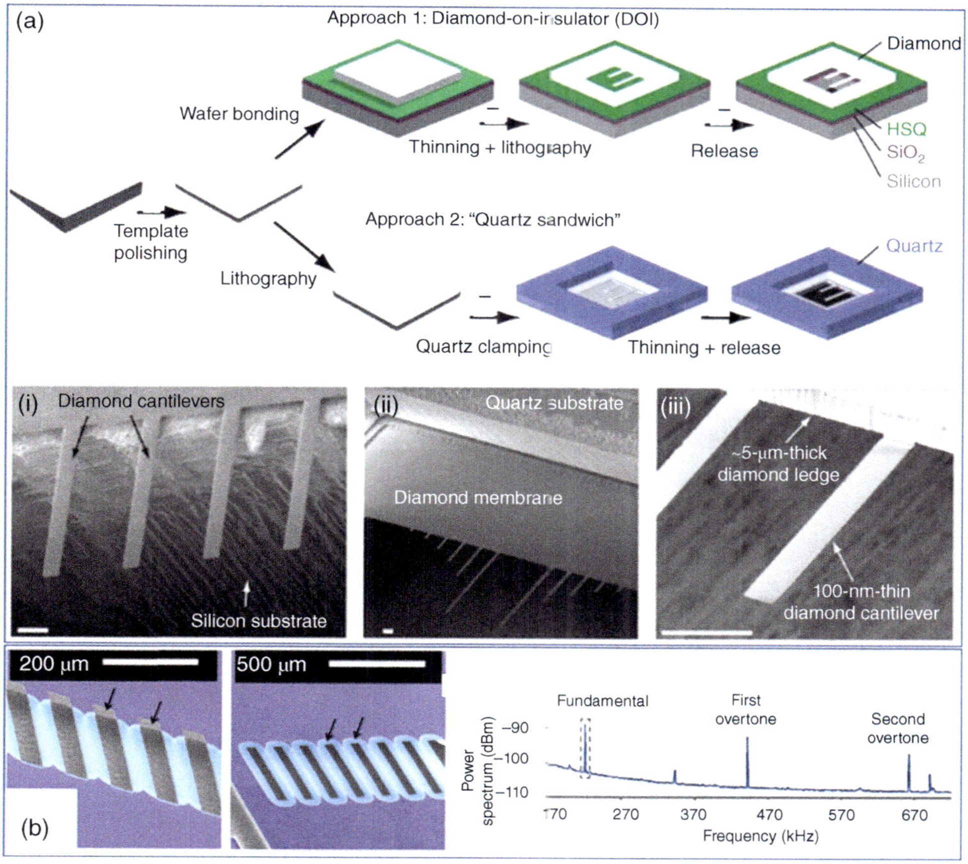

Figure 11.9 Enhancement in Q-factor using optimal clamping designs. (a) Clamping using sandwiched structure. *Source:* [45], Tao, et. al., (2014)/Springer Nature. (b) Avoiding under cut in SiC resonators. *Source:* [46], Kermany, et. al., (2014)/AIP Publishing LLC.

anchor area from being removed due to the isotropic etching behavior of XeF_2 gas. The resonant frequency of the SiC devices were then measured using an optical setup in vacuum conditions. Experimental data shows that increasing length and reducing width leads to an increase in the *Q*-factor, indicating the dominant role of the clamping factor in the performance of SiC resonators. Using the undercut-free design, the authors demonstrated extremely high *Q*-factor over one million, and a high $f \times Q$ of approximately 9.4×10^{11}. This $f \times Q$ is higher than standard state-of-the art SiN-based devices of the same dimensions, demonstrating the advantages of WBG materials for high performance nanoresonators and oscillators.

11.4.3 Tuning Resonant Frequency Using Electrically Driven Forces

As shown in Eq. (11.8), the resonant frequency of nanoresonator depends on applied stresses. This theory indicates that it is possible to tune the output of resonator by driving the external mechanical impacts.

11.4.3.1 Electrothermal Tuning

The resonant frequency and *Q*-factor can be tuned by using thermal expansion that causes a stress in resonator. A simple method to induce thermal stress is the use of the Joule heating effect [47, 48]. The magnitude of the external thermal stress can be controlled by manipulating the DC voltage used in Joule heating. Figure 11.10 shows an example of resonant frequency tuning using the electrothermal approach in SiC resonator [49]. In this work, the resonator was developed from a highly doped 3C-SiC film grown on a Si substrate. This highly electrically conductive SiC resonator functioned as the heater where Al electrodes were formed at the two ends of the resonator. The authors applied DC bias ranging from 0 to 5 V to create a temperature profile in the bridge resonator. Experimental results show that

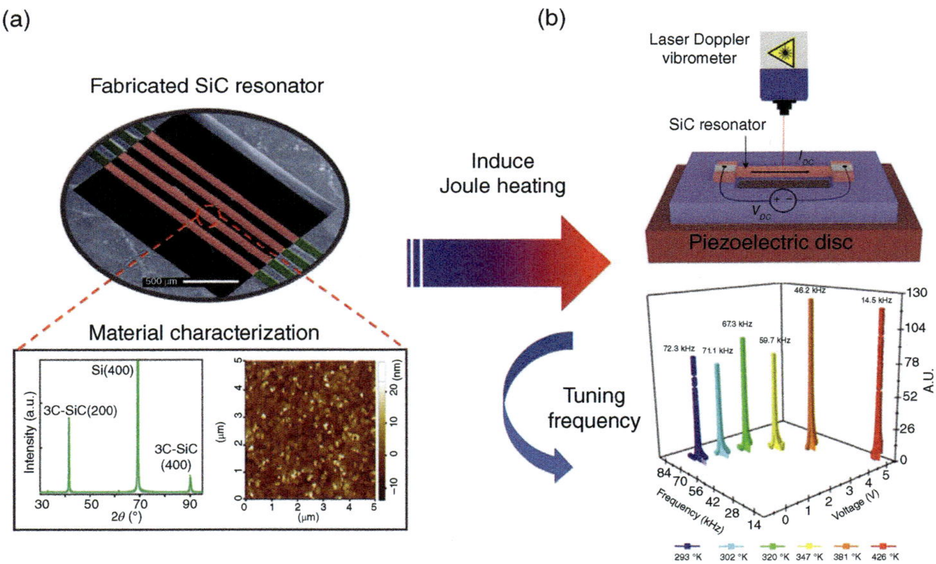

Figure 11.10 Frequency tuning using thermal stress. (a) SEM images of SiC resonators. Inset: X-ray diffraction (XRD) and atomic force microscopy (AFM) measurement of the SiC film. (b) Experimental setup and the change of the resonant frequency under applied bias. *Source:* [49], Guzman, et. al., (2020)/Elsevier.

increasing the applied voltage leads to a significant decrease in the resonant frequency. Specifically, at zero bias, the first mode resonant frequency was measured at 72.3 kHz using a Laser vibrometer. At 5 V, the frequency was reduced down to 14.5 kHz. The reduction in the resonant frequency is attributed to the decrease in the total stress in the SiC device, where the Joule heating induced thermal stress tends to cancel out the intrinsic stress in the film ($\sigma_{total} = \sigma_{intrinsic} - \alpha E_{SiC}\Delta T$, where α and E_{SiC} are the thermal expansion coefficient and Young's modulus of SiC, respectively). The frequency tuning range in this work is larger than other studies that employed metallic heaters. One main reason is that the SiC resonator directly induces compressive stress through its own thermal expansion which is similar to the bulking phenomenon. Furthermore, the use of SiC as both the mechanical component and heater helps avoiding the dependence of the thermal expansion of coefficient (TCE) between the constructing materials. This configuration enables heating the entire nanoresonator and thereby generating more compressive stress for a better frequency tuning performance.

11.4.3.2 Piezoelectric Tuning

The piezoelectric effect can be utilized to actuate the resonator as well as to induce stress to the device to modulate their behavior. The concept of resonant frequency tuning using piezoelectricity is relatively analogous to the Joule heating effect, in which applying an electrical bias generates a mechanical stress along the polarization direction of piezoelectric material, resulting in a change in the resonant frequency [50, 51]. Figure 11.11 shows an example of frequency tuning in a 3C-SiC resonator using the piezoelectric effect in a lead zirconium titanate (PZT) film formed on the surface of the resonator [52]. SiC films were epitaxially grown on a Si substrate, followed by the deposition of a thin SiO_2 layer as the insulator. A stack of Pt/PZT/Pt was then deposited on the SiO_2/SiC to form the piezoelectric element. The SiC resonator with the piezoelectric PZT film was released from the substrate by inductively plasma etching of Si. At zero bias, the SiC device exhibits a high resonant frequency in a range of 1–1.85 MHz. Experimental data showed a right shift in the frequency under a DC bias varying from 0 to 5 V. In particular, the resonant frequency increased by approximately 2500 ppm at 5 V. The magnitude of the frequency shift also depends on the dimension of the length of the electrodes (i.e. Pt on PZT) and resonator dimensions. For instance, a longer electrode offers a larger change. These results demonstrate the capability of piezoelectric stimulation in enhancing the performance of WBG material-based resonators.

11.5 Applications of Nanoresonators

11.5.1 Logic Circuit at High Temperatures

Microprocessors that can tolerate to extreme conditions are imperative for the measurement and control of several industrial systems, including pipelines (with corrosion conditions), and propulsion systems (high temperatures). WBG materials such as SiC, GaN, and diamond like carbon with their excellent mechanical properties and chemical inertness are preferred choices for the development of logic circuits for these microprocessors [20, 53–55]. Specifically, electronics built on WBG materials are expected to have capability to directly operate at high temperatures, thereby reducing the requirement for expensive heat-sink and thermal management. WBG nanoresonators with high resonant frequency have been demonstrated in NEMS switch that can replace existing junction field effect transistors (JFETs) to avoid

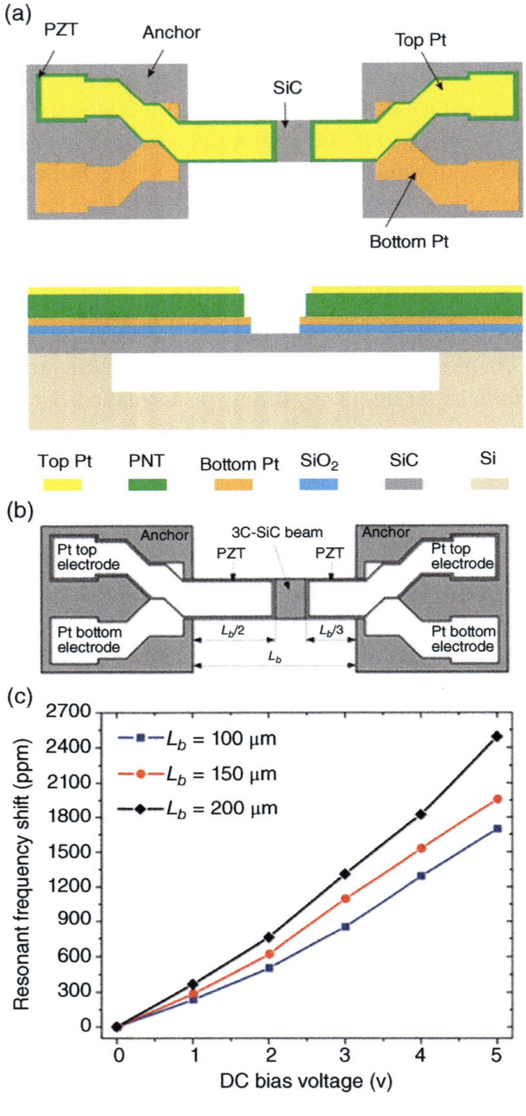

Figure 11.11 Frequency tuning using piezoelectric effect. (a) SEM images of a SiC resonator. (b) Schematic sketch of the resonator with PZT layer. (c) Frequency shift under different bias. *Source:* Sviličić et al. [52]/American Vacuum Society.

problematic issues such as current leakages and low switching speeds. Using SiC nanoswitch, Lee et al. developed a mechanical logic circuit that can work at high temperature up to 500 °C [53–55]. The circuit demonstrated the function of an inverter, which consists of two lateral NEMS switches, Figure 11.12.

Each switch has three terminals of source, drain, and gate, similar to a standard field effect transistor (FET). The gate terminal functions as a mechanical switch to form a conducting path between the drain and source using an electrostatic force. The drain is connected to a positive bias and the source is connected to a negative voltage. Applying a positive voltage at the gate will pull the drain-nanowire (negative bias) toward the output terminal, leading to a low state at the output. On the other hand, when a negative bias is applied to the gate, the electrostatic force between gate and the source (positive bias) will pull the source-nanowire toward the output terminal, resulting in a high state. Figure 11.12b illustrates the operation of the inverter

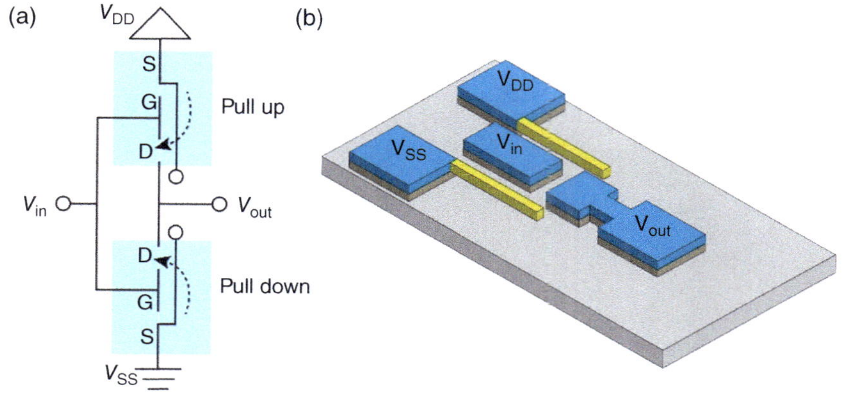

Figure 11.12 Logic circuit at high temperatures. (a) Concept of the logic circuit and SEM images of fabricated devices; (b) output and input waveforms of a SiC inverter. *Source:* Hedayati et al. [53].

at 500 °C, where a high voltage in the input results in a low state in the output and vice versa. As the source and drain electrodes were formed in a shape of nanowires with a small width of approximately 200 nm, the mechanical inverter can switch ON and OFF with a relatively low bias voltage of 3 V or even lower. The large resonant frequency in the SiC nanowires enables the converter to operate at relatively high speed of 500 kHz. The mechanical switch also minimizes the leakage current down to as small as 2 fA, which is mainly caused by the tunneling current. The excellent wear resistance in SiC also allows the device to operate well above two billion cycles at 500 °C without degradation. These results demonstrate an interesting pathway toward high-temperature computation with excellent reliability and energy efficiency.

11.5.2 Mass Sensing Applications

Ultrahigh sensitivity for mass sensing can open intriguing possibilities, including gas detection, and measurement of inertial mass of neutral macromolecules to probe the fundamental properties of materials [20, 56]. VHF nanoresonators using WBG nanowires can offer unprecedented sensitivity for inertial mass sensing. For instance, the Roukes group developed a Zeptogram mass sensor using SiC nanoresonators [57]. The resonator is a SiC nanowire with a dimension of 2.3 μm × 150 nm × 70 nm, fabricated from a SiC on Si wafer, Figure 11.13. The resonant frequency was measured in a reflection scheme using a directional coupler and was found to be 190 MHz with a high *Q*-factor of ~5000. A controlled flux of atoms or molecules was generated onto the surface of the resonator using a glass nozzle with an aperture of 100 μm. Figure 11.13b shows the real-time observation of the frequency shift of the SiC resonator using a VHF phase-locked loop technique. The nozzle shutter was opened for every five seconds, corresponding to an accredit mass of approximately 100 zg for each step, which was clearly captured by the frequency change. Figure 11.13c shows the mass responsivity of two SiC devices with different resonant frequencies of 133 and 190 MHz, demonstrating the capability of detecting different atoms of Xe (for the 133 MHz device), and N_2 (for the 190 MHz device). The highest mass resolution was demonstrated with the 133 MHz device, with an extremely small fractional frequency fluctuation of 5×10^{-8} (i.e. 50 ppb) at 4.2 K. This indicates that the SiC nanoresonators can detect mass resolutions as small as 7 zg, equivalent to 30 Xe atoms. Mass detection using wide band nanoresonators is anticipated to open promising possibilities in the atomic physics and life sciences.

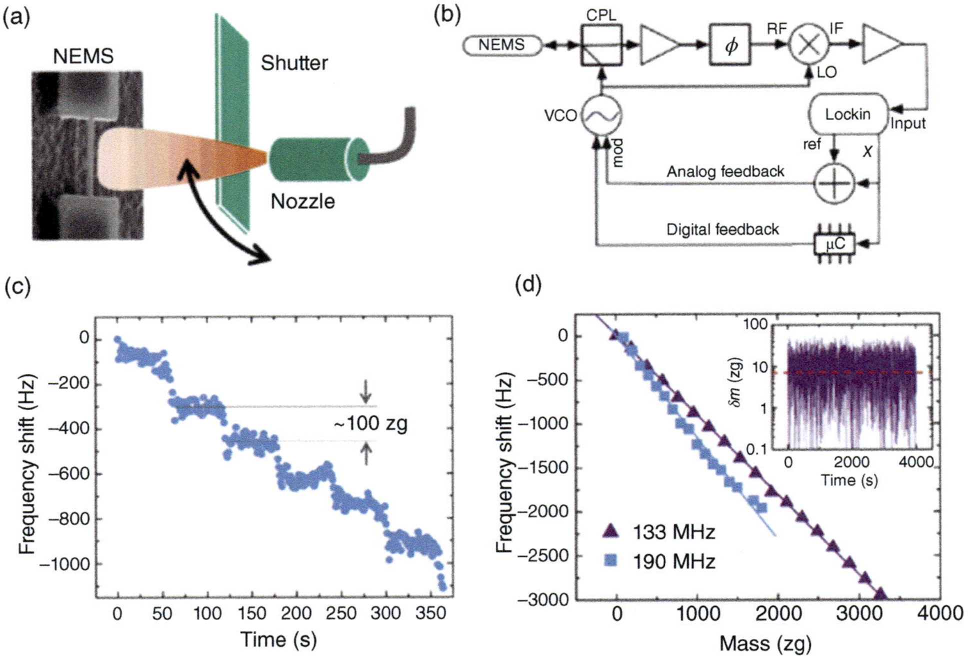

Figure 11.13 Highly sensitive mass sensing using nanoresonators. (a) SEM image of very high frequency (VHF) nanoresonators. (b) Diagram of measurement circuit. (c) Change of frequency for each 100 zg load. (d) The linear relationship between frequency shift and mass. *Source:* [57], Yang, et. al., (2006)/American Chemical Society.

11.5.3 Biosensors

Several WBG materials have proven their biocompatibility, that enables the development of implanted devices or cell culture platforms [58, 59]. WBG resonators with bio-species grown on their surface have been deployed to probe several biological mechanisms and activities. In particular, biospecies attached on the surface of a resonator can induce a mechanical stress or add extra mass that leads to a change in the resonant frequency. A systematic study on the frequency shift can elucidate the biological properties of the targeted species.

Figure 11.14 presents a proof of concept in SiC disk resonators to investigate mechanical properties of breast cancer cells [60]. Cells were found to stick and grow well on the surface of the SiC resonators. The flexural vibrational modes of the SiC device were excited using an amplitude-modulated laser (405 nm). Another He–Ne red laser (633 nm) was utilized to interferometrically measure the resonance of the device. Multiple flexural modes within the high frequency/UHF range (10–200 MHz) were observed with relatively high *Q*-factors of ~2200 in air. When a cell was attached on the surface of the resonator, the first four modes (~60–90 MHz) shifted by 3.79, 2.84, 4, and 3.47 MHz, respectively. The robust multimode responses to single breast cancer cell loading demonstrate the potential of SiC resonators for biosensing applications as well as for single-cell biophysical studies.

11.5.4 Mechanical Sensing

The resonant frequency and *Q*-factors of a resonator are sensitive to external mechanical stress, as shown in Eq. (11.8). This property can be utilized to develop highly sensitive

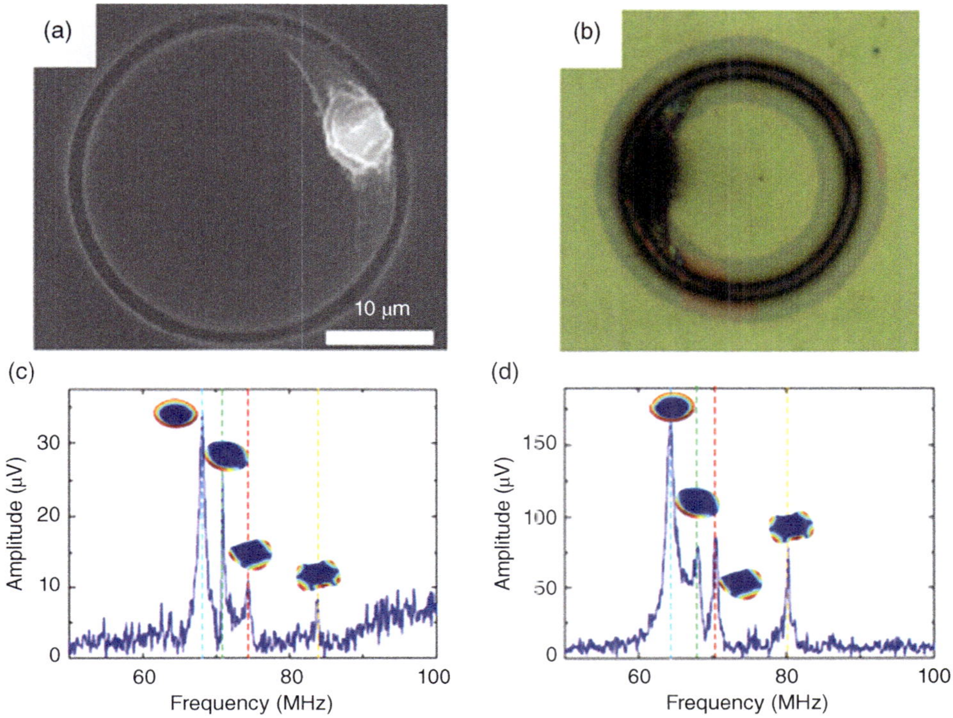

Figure 11.14 Wide bandgap resonators for characterization of bio-species. (a) SEM images showing cell growth and well attach on SiC disk resonators; (b) photograph of SiC resonators with and without cell; frequency response (c) before and (d) after cell attachment.

mechanical sensors such as strain and pressure sensors [61–63]. Figure 11.15 shows a SiC comb-structure resonator for strain sensing with the capability of operating at high temperatures of above 300 °C [64]. The resonator is actuated using the electrostatic force between the two electrodes of the comb resonators. The authors employed a smart design, namely balanced-mass double-ended tuning fork (DETF) that can prevent stiction during a g-shock event by minimizing the deflection. This design enhances the fracture-limited g-shock survival by reducing the peak strain by approximately 25% compared with the standard silicon comb-driven DETF. The sensors were fabricated from a poly SiC film using surface-micromachining techniques. The sensors were shown to survive high accelerations up to 10 000 without any observable damages. It properly functioned when being heated up at 300 °C using an infrared (IR) lamp. To characterize the strain response of the resonator, an on-chip electrostatic actuator was integrated to induce axial forces to the sensors. Figure 11.15 shows a linear relationship between the resonant frequency and the applied strain, with a good sensitivity of 66 Hz/με. The frequency resolution was estimated based on the phase noise density. Accordingly, the strain resolution of the SiC resonator was found to be 0.11 με with a bias voltage of 80 V and a bandwidth of 10–20 kHz, comparable with the state-of-the-art silicon DETF strain sensors. The high sensitivity, excellent shock tolerance, and high-temperature operating capability demonstrate the promise of WBG resonators for mechanical sensing in extreme environments.

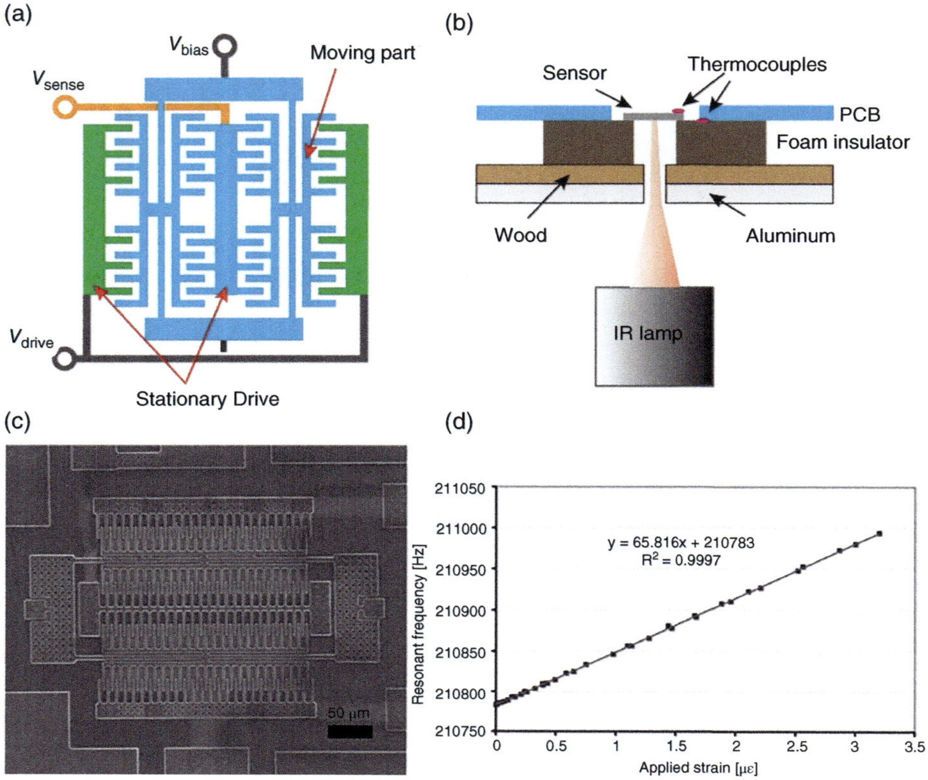

Figure 11.15 Strain sensor using wide bandgap resonator. (a) Concept of a balanced-mass double-ended tuning fork resonator. (b) SEM image of SiC based device. (c) Experiment setup for high temperature. (d) The relationship between resonant frequency and applied strain. *Source:* [64]. Azevedo, et. al., (2007)/IEEE.

11.5.5 Optical Devices

Nanoresonators have attracted great research interest for optical devices such as multiwavelength detectors owing to their high frequency and Q-factor [65, 66]. Multiple wavelength detectors are imperative components in the wavelength division multiplexing (WDM) method that is widely used for near infrared optical fiber based intercontinental communication. [67] The WDM method allows for high-speed and large capacity communication by multiplexing optical signals of several wavelengths into a single optical fiber. To offer high performance WDM, a wavelength detection device, capable of sensing a broad band of frequency, is highly demanded. In this regard, Maeda and Kometani developed a sub-picometer multiple wavelength detector using a highly sensitive nanomechanical resonators [68]. The device includes a narrow-band photonic absorber with an Au photonic crystal structure. This photonic absorber was formed on a diamond-like-carbon-based resonator. Under light illumination, the Au photonic crystal absorbs the incident photon energy with different rates, depending on the wavelength of the light source. This difference in the power absorption leads to a variation in the resultant thermal stress induced into the diamond nanomechanical resonator. By detecting the frequency shift in the resonators, it is possible to measure the wavelength of the incident light, Figure 11.16a–c. The device shows a narrow band or photonic absorption (with FWHM of less than 10 nm), and a high Q-factor of over 20 000 owing to the excellent mechanical property of diamond, Figure 11.16d. A clear resonant

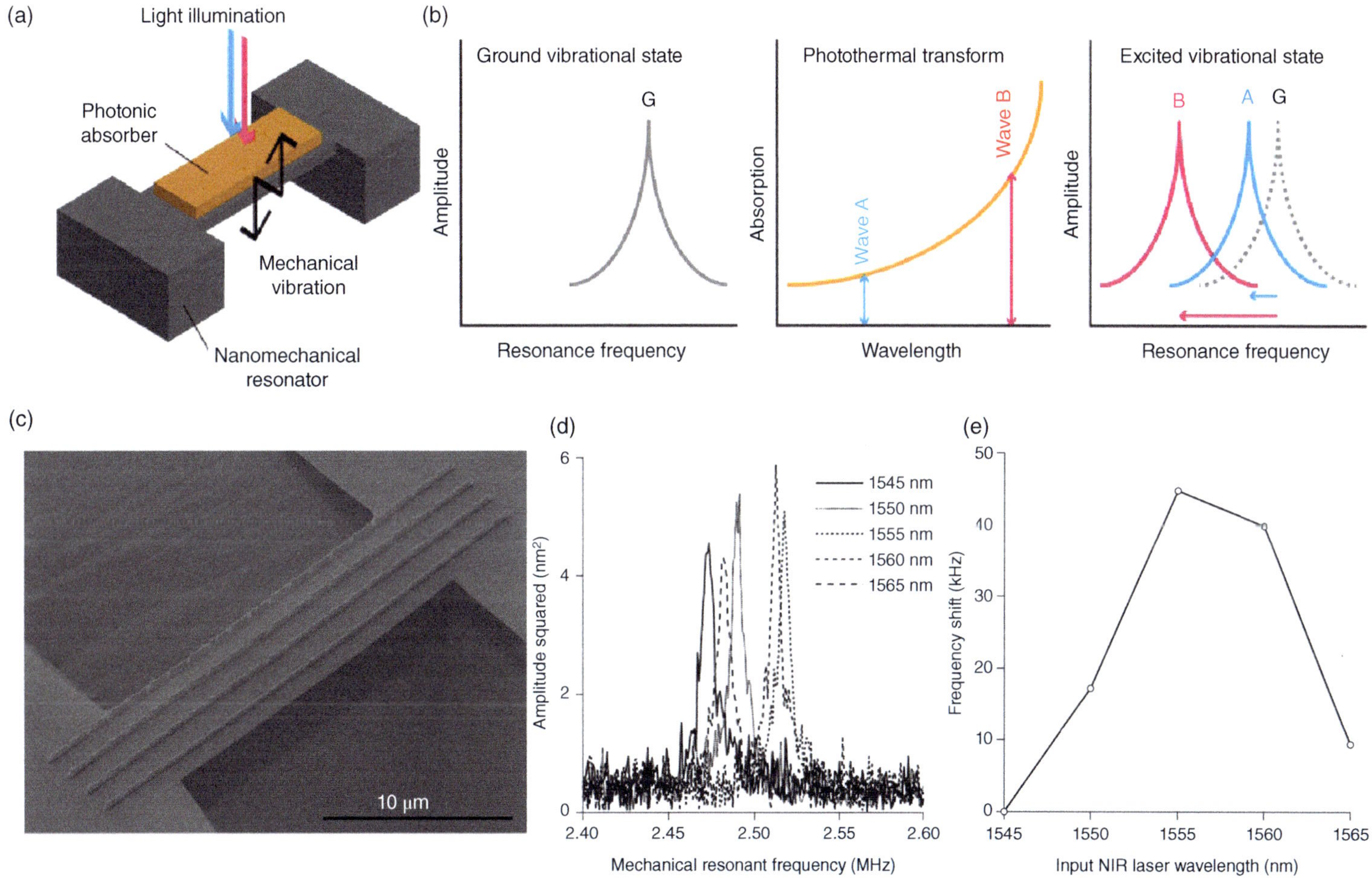

Figure 11.16 Multiwavelength detector. (a) The schematic sketch of the device. (b) The operating principle, where light absorption is detected by the shift in resonant frequency. (c) SEM image of a diamond resonator with Au photonic crystal formed on its surface. (d) Frequency measurement under light illumination using a laser Doppler vibrometer. (e) Frequency shift vs NIR wavelength. *Source:* [68], Maeda, et. al., (2017)/AIP Publishing LLC.

frequency shift was observed under light illumination where a 45 kHz change in the resonant frequency was measured when shifting the incident wavelength from 1545 to 1555 nm (i.e. 10 nm), Figure 11.16d. The wavelength sensitivity was found to be approximately 0.196 pm in the near infrared (NIR) regime with 0.3% coefficient of variation. The high *Q*-factor in the diamond resonator plays an important role in offering a high wavelength sensitivity (i.e. the wavelength sensitivity depends on the frequency resolution). The sub-picometer resolution multi-wavelength detector demonstrates the high potential of WBG nanoresonators for optical applications.

References

1 Zhao, C., Montaseri, M.H., Wood, G.S. et al. (2016). A review on coupled MEMS resonators for sensing applications utilizing mode localization. *Sensors and Actuators A: Physical* 249: 93–111.

2 Wu, G., Xu, J., Ng, E.J., and Chen, W. (2020). MEMS resonators for frequency reference and timing applications. *Journal of Microelectromechanical Systems* 29 (5): 1137–1166.

3 Rhoads, J.F., Shaw, S.W., and Turner, K.L. (2010). Nonlinear dynamics and its applications in micro- and nanoresonators. *Journal of Dynamic Systems, Measurement, and Control* 132 (3): 034001.

4 Van Beek, J.T.M. and Puers, R. (2011). A review of MEMS oscillators for frequency reference and timing applications. *Journal of Micromechanics and Microengineering* 22 (1): 013001.

5 Chen, R., Ng, K.W., Ko, W.S. et al. (2014). Nanophotonic integrated circuits from nanoresonators grown on silicon. *Nature Communications* 5 (1): 1–10.

6 Favero, I. and Karrai, K. (2008). Cavity cooling of a nanomechanical resonator by light scattering. *New Journal of Physics* 10 (9): 095006.

7 Jha, C.M., Bahl, G., Melamud, R. et al. (2007). CMOS-compatible dual-resonator MEMS temperature sensor with milli-degree accuracy. In: *TRANSDUCERS 2007–2007 International Solid-State Sensors, Actuators and Microsystems Conference, 10–14 June 2007, Lyon, France*, 229–232. IEEE.

8 Davis, Z.J., Svendsen, W., and Boisen, A. (2007). Design, fabrication and testing of a novel MEMS resonator for mass sensing applications. *Microelectronic Engineering* 84 (5–8): 1601–1605.

9 Brar, V.W., Jang, M.S., Sherrott, M. et al. (2013). Highly confined tunable mid-infrared plasmonics in graphene nanoresonators. *Nano Letters* 13 (6): 2541–2547.

10 Verre, R., Baranov, D.G., Munkhbat, B. et al. (2019). Transition metal dichalcogenide nanodisks as high-index dielectric Mie nanoresonators. *Nature Nanotechnology* 14 (7): 679–683.

11 Jensen, K., Girit, C., Mickelson, W., and Zettl, A. (2006). Tunable nanoresonators constructed from telescoping nanotubes. *Physical Review Letters* 96 (21): 215503.

12 Zheng, X.Q., Lee, J., Rafique, S. et al. (2017). Ultrawide band gap β-Ga_2O_3 nanomechanical resonators with spatially visualized multimode motion. *ACS Applied Materials & Interfaces* 9 (49): 43090–43097.

13 Sepúlveda, N., Lu, J., Aslam, D.M., and Sullivan, J.P. (2008). High-performance polycrystalline diamond micro- and nanoresonators. *Journal of Microelectromechanical Systems* 17 (2): 473–482.

14 Hofmann, C.E., García de Abajo, F.J., and Atwater, H.A. (2011). Enhancing the radiative rate in III–V semiconductor plasmonic core–shell nanowire resonators. *Nano Letters* 11 (2): 372–376.

15 Placidi, M., Godignon, P., Mestres, N. et al. (2008). Fabrication of monocrystalline 3C–SiC resonators for MHz frequency sensors applications. *Sensors and Actuators B: Chemical* 133 (1): 276–280.

16 Hutchinson, A.B., Truitt, P.A., Schwab, K.C. et al. (2004). Dissipation in nanocrystalline-diamond nanomechanical resonators. *Applied Physics Letters* 84 (6): 972–974.

17 Schmid, S., Villanueva, L.G., and Roukes, M.L. (2016). *Fundamentals of Nanomechanical Resonators*, vol. 49. Berlin: Springer.

18 Bao, M.H. (2000). *Micro Mechanical Transducers: Pressure Sensors, Accelerometers and Gyroscopes*. Elsevier.

19 Nguyen, N.K., Nguyen, T., Nguyen, T.K. et al. (2021). Wide-band-gap semiconductors for biointegrated electronics: recent advances and future directions. *ACS Applied Electronic Materials* 3 (5): 1959–1981.

20 Wang, Q. and Arash, B. (2014). A review on applications of carbon nanotubes and graphenes as nano-resonator sensors. *Computational Materials Science* 82: 350–360.

21 Rieger, J., Faust, T., Seitner, M.J. et al. (2012). Frequency and Q factor control of nanomechanical resonators. *Applied Physics Letters* 101 (10): 103110.

22 Sadeghi, P., Tanzer, M., Christensen, S.L., and Schmid, S. (2019). Influence of clamp-widening on the quality factor of nanomechanical silicon nitride resonators. *Journal of Applied Physics* 126 (16): 165108.

23 Sui, W., Zheng, X.Q., Lin, J.T. et al. (2021). Thermal response and TCf of GaN/AlN heterostructure multimode micro string resonators from −10 °C up to 325 °C. *Journal of Microelectromechanical Systems* 30 (4): 521–529.

24 Phan, H.P., Nguyen, T.K., Dinh, T. et al. (2017). Ultra-high strain in epitaxial silicon carbide nanostructures utilizing residual stress amplification. *Applied Physics Letters* 110 (14): 141906.

25 Uranga, A., Verd, J., and Barniol, N. (2015). CMOS–MEMS resonators: from devices to applications. *Microelectronic Engineering* 132: 58–73.

26 Thakar, V. and Rais-Zadeh, M. (2013). Optimization of tether geometry to achieve low anchor loss in Lamé-mode resonators. In: *2013 Joint European Frequency and Time Forum & International Frequency Control Symposium (EFTF/IFC), 21–25 July 2013, Prague, Czech Republic*, 129–132. IEEE.

27 Dao, D.V., Nakamura, K., Bui, T.T., and Sugiyama, S. (2010). Micro/nano-mechanical sensors and actuators based on SOI-MEMS technology. *Advances in Natural Sciences: Nanoscience and Nanotechnology* 1 (1): 013001.

28 Feng, X.L., Matheny, M.H., Zorman, C.A. et al. (2010). Low voltage nanoelectromechanical switches based on silicon carbide nanowires. *Nano Letters* 10 (8): 2891–2896.

29 Dosch, J.J., Inman, D.J., and Garcia, E. (1992). A self-sensing piezoelectric actuator for collocated control. *Journal of Intelligent Material Systems and Structures* 3 (1): 166–185.

30 Deng, J., Liu, Y., Liu, J. et al. (2018). Development of a planar piezoelectric actuator using bending–bending hybrid transducers. *IEEE Transactions on Industrial Electronics* 66 (8): 6141–6149.

31 Karabalin, R.B., Matheny, M.H., Feng, X.L. et al. (2009). Piezoelectric nanoelectromechanical resonators based on aluminum nitride thin films. *Applied Physics Letters* 95 (10): 103111.

32 Ben-Shimon, Y. and Ya'akobovitz, A. (2021). Magnetic excitation and dissipation of multilayer two-dimensional resonators. *Applied Physics Letters* 118 (6): 063103.

33 Phan, H.P., Dinh, T., Nguyen, T.K. et al. (2017). Self-sensing paper-based actuators employing ferromagnetic nanoparticles and graphite. *Applied Physics Letters* 110 (14): 144101.

34 Feng, X.L., He, R., Yang, P., and Roukes, M.L. (2007). Very high frequency silicon nanowire electromechanical resonators. *Nano Letters* 7 (7): 1953–1959.

35 Jun, S.C., Cho, J.H., Kim, W.K. et al. (2009). Resonance properties of 3C-SiC nanoelectromechanical resonator in room-temperature magnetomotive transduction. *IEEE Electron Device Letters* 30 (10): 1042–1044.

36 Phinney, L.M., Baker, M.S., Serrano, J.R., and Nazul, I. (2012). Thermal microactuators. In: *Microelectromechanical Systems and Devices* (ed. N. Islam), 415–434. InTech.

37 Ouyang, J. and Zhu, Y. (2012). Z-shaped MEMS thermal actuators: piezoresistive self-sensing and preliminary results for feedback control. *Journal of Microelectromechanical Systems* 21 (3): 596–604.

38 Jiang, L., Cheung, R., Hedley, J. et al. (2006). SiC cantilever resonators with electrothermal actuation. *Sensors and Actuators A: Physical* 128 (2): 376–386.

39 Gaydaychuk, A., Zenkin, S., and Linnik, S. (2019). Influence of Al–Si–N interlayer on residual stress of CVD diamond coatings. *Surface and Coatings Technology* 357: 348–352.

40 Rohmfeld, S., Hundhausen, M., Ley, L. et al. (2002). Quantitative evaluation of biaxial strain in epitaxial 3C-SiC layers on Si(100) substrates by Raman spectroscopy. *Journal of Applied Physics* 91 (3): 1113–1117.

41 Harutyunyan, V.S., Aivazyan, A.P., Weber, E.R. et al. (2001). High-resolution X-ray diffraction strain–stress analysis of GaN/sapphire heterostructures. *Journal of Physics D: Applied Physics* 34 (10A): A35.

42 Sang, L., Liao, M., Yang, X. et al. (2020). Strain-enhanced high *Q*-factor GaN micro-electromechanical resonator. *Science and Technology of Advanced Materials* 21 (1): 515–523.

43 Liao, M., Toda, M., Sang, L. et al. (2014). Energy dissipation in micron- and submicron-thick single crystal diamond mechanical resonators. *Applied Physics Letters* 105 (25): 251904.

44 Jandak, M., Neuzil, T., Schneider, M., and Schmid, U. (2016). Investigation on different damping mechanisms on the *Q* factor of MEMS resonators. *Procedia Engineering* 168: 929–932.

45 Tao, Y., Boss, J.M., Moores, B.A., and Degen, C.L. (2014). Single-crystal diamond nanomechanical resonators with quality factors exceeding one million. *Nature Communications* 5 (1): 1–8.

46 Kermany, A.R., Brawley, G., Mishra, N. et al. (2014). Microresonators with *Q*-factors over a million from highly stressed epitaxial silicon carbide on silicon. *Applied Physics Letters* 104 (8): 081901.

47 Jun, S.C., Huang, X.M.H., Manolidis, M. et al. (2006). Electrothermal tuning of Al–SiC nanomechanical resonators. *Nanotechnology* 17 (5): 1506.

48 Svilicic, B., Mastropaolo, E., Flynn, B., and Cheung, R. (2012). Electrothermally actuated and piezoelectrically sensed silicon carbide tunable MEMS resonator. *IEEE Electron Device Letters* 33 (2): 278–280.

49 Guzman, P., Dinh, T., Phan, H.P. et al. (2020). Highly-doped SiC resonator with ultra-large tuning frequency range by Joule heating effect. *Materials & Design* 194: 108922.

50 Ibrahim, S.W. and Ali, W.G. (2012). A review on frequency tuning methods for piezoelectric energy harvesting systems. *Journal of Renewable and Sustainable Energy* 4 (6): 062703.

51 Azizi, S., Ghazavi, M.R., Rezazadeh, G. et al. (2014). Tuning the primary resonances of a micro resonator, using piezoelectric actuation. *Nonlinear Dynamics* 76 (1): 839–852.

52 Sviličić, B., Mastropaolo, E., Chen, T., and Cheung, R. (2012). Piezoelectrically transduced silicon carbide MEMS double-clamped beam resonators. *Journal of Vacuum Science and Technology B* 30 (6): 06FD05.

53 Hedayati, R., Lanni, L., Rodriguez, S. et al. (2014). A monolithic, 500 °C operational amplifier in 4H-SiC bipolar technology. *IEEE Electron Device Letters* 35 (7): 693–695.

54 Elahipanah, H., Kargarrazi, S., Salemi, A. et al. (2017). 500 °C high current 4H-SiC lateral BJTs for high-temperature integrated circuits. *IEEE Electron Device Letters 38* (10): 1429–1432.

55 Lee, T.H., Bhunia, S., and Mehregany, M. (2010). Electromechanical computing at 500 C with silicon carbide. *Science* 329 (5997): 1316–1318.

56 Cimalla, V., Niebelschütz, F., Tonisch, K. et al. (2007). Nanoelectromechanical devices for sensing applications. *Sensors and Actuators B: Chemical* 126 (1): 24–34.

57 Yang, Y.T., Callegari, C., Feng, X.L. et al. (2006). Zeptogram-scale nanomechanical mass sensing. *Nano Letters* 6 (4): 583–586.

58 Phan, H.P., Zhong, Y., Nguyen, T.K. et al. (2019). Long-lived, transferred crystalline silicon carbide nanomembranes for implantable flexible electronics. *ACS Nano* 13 (10): 11572–11581.

59 Pham, T.A., Nguyen, T.K., Vadivelu, R.K. et al. (2020). A versatile sacrificial layer for transfer printing of wide bandgap materials for implantable and stretchable bioelectronics. *Advanced Functional Materials* 30 (43): 2004655.

60 Jia, H., Wu, X., Tang, H. et al. (2015). Culturing and probing physical behavior of individual breast cancer cells on SiC microdisk resonators. In: *2015 28th IEEE International Conference on Micro Electro Mechanical Systems (MEMS)' 18–22 January 2015, Estoril, Portugal*, 698–701. IEEE.

61 Brueckner, K., Niebelschuetz, F., Tonisch, K. et al. (2011). Micro- and nano-electromechanical resonators based on SiC and group III-nitrides for sensor applications. *Physica Status Solidi A Applications and Material Science* 208 (2): 357–376.

62 Sepulveda, N., Aslam, D., and Sullivan, J.P. (2006). Polycrystalline diamond MEMS resonator technology for sensor applications. *Diamond and Related Materials* 15 (2, 3): 398–403.

63 Müller, A., Konstantinidis, G., Giangu, I. et al. (2017). GaN membrane supported SAW pressure sensors with embedded temperature sensing capability. *IEEE Sensors Journal* 17 (22): 7383–7393.

64 Azevedo, R.G., Jones, D.G., Jog, A.V. et al. (2007). A SiC MEMS resonant strain sensor for harsh environment applications. *IEEE Sensors Journal* 7 (4): 568–576.

65 Hofmann, U., Janes, J., and Quenzer, H.J. (2012). High-Q MEMS resonators for laser beam scanning displays. *Micromachines* 3 (2): 509–528.

66 Ren, Z., Chang, Y., Ma, Y. et al. (2020). Leveraging of MEMS technologies for optical metamaterials applications. *Advanced Optical Materials* 8 (3): 1900653.

67 Tao, J.F., Cai, H., Wu, J. et al. (2013). Optical wavelength signal detector via tunable micro-ring resonator for sensor applications. In: *2013 IEEE 26th International Conference on Micro Electro Mechanical Systems (MEMS), 20–24 Jan. 2013, Taipei, Taiwan*, 500–503. IEEE.

68 Maeda, E. and Kometani, R. (2017). Sub-picometer multi-wavelength detector based on highly sensitive nanomechanical resonator. *Applied Physics Letters* 111 (1): 013102.

Index

Wide Bandgap Nanowires: Synthesis, Properties, and Applications, First Edition. Tuan Anh Pham, Toan Dinh, Nam-Trung Nguyen, and Hoang-Phuong Phan.

Printed and bound by CPI Group (UK) Ltd, Croydon, CR0 4YY
16/08/2022
03141870-0001